J. A. H. Oates

Lime and Limestone

WILEY-VCH

Charles Seale-Hayne Library

University of Plymouth

(01752) 588 588

LibraryandITenquiries@plymouth.ac.uk

This b

CH

J. A. H. Oates

Lime and Limestone

Chemistry and Technology,
Production and Uses

 **WILEY-VCH**

Weinheim · New York · Chichester
Brisbane · Singapore · Toronto

J. A. H. Oates
(BSc, CChem, FRSC, FIQ, MIQA)
Limetec Consultancy Services
19, Macclesfield Road
Buxton
Derbyshire, SK17 9AH
England

This book was carefully produced. Nevertheless, author and publisher do not warrant the information contained therein to be free of errors. Readers are advised to keep in mind that statements, data, illustrations, procedural details or other items may inadvertently be inaccurate.

Cover picture: A Beckenbach Annular Shaft Kiln
(courtesy of Beckenbach Wärmestelle GmbH)

Library of Congress Card No. applied for
A catalogue record for this book is available from the British Library

Die Deutsche Bibliothek – CIP-Einheitsaufnahme
Oates, Joseph A. H.:
Lime and limestone : chemistry and technology, production and uses /
J. A. H. Oates. – Weinheim : Wiley-VCH, 1998
ISBN 3-527-29527-5

Composition: Kühn & Weyh Software GmbH, D-79015 Freiburg
Printing: betz-druck GmbH, D-64291 Darmstadt
Bookbinding: Großbuchbinderei J. Schäffer, D-67269 Grünstadt
Printed in the Federal Republic of Germany

Preface

The Lime and Limestone Industries are dynamic industries, with new production methods, new products and new uses continually being developed. They are particularly interesting and challenging, because of the wide variety of products made, and the even wider range of applications.

Both industries are being profoundly affected by the general requirement to improve environmental performance. On the one hand, this presents a challenge, which is leading to heavy capital investment, mergers and even closures. On the other hand, it presents new opportunities to supply environmental control products.

Another major development, which affects both producers and customers, is the widespread reduction of manning levels. While this increases productivity and profitability, it places greater pressures on everyone to cope with less technical assistance than in the past. In consequence, there is a need to develop and broaden individual expertise, and to have improved access to up-to-date information.

One of the aims of producing this book is to present an integrated perspective of the Lime and Limestone Industries, and to indicate how they have been, and still are being shaped by customer-led requirements.

- It describes the many complex interactions, relating to product quality, that exist between suppliers and customers, both within and outside the Industries. This should help production personnel in the Industries to appreciate the impact of their actions and decisions on their customers.

- For users of lime and limestone products, it seeks to give an understanding of the factors which affect product quality and the ways in which the products interact with the processes in which they are used. It also indicates how the Lime and Limestone Industries control product quality, and what actions might be taken to tailor quality for a particular application.

- It may also provide a basis for constructive dialogue between suppliers and customers, thereby facilitating the development of new and improved products.

The formation of the European Union is leading to the preparation of a large number of application-specific Standards, specifying both product quality and test methods. Combining the diverse products and practices used in so many countries, each with its own traditions, is proving to be both demanding and stimulating. While many relevant CEN Standards have already been published, many others are still in preparation at the time of going to press.

The structure of the book and of individual chapters has been designed to present the information in a logical way that gives as coherent an account of the Industries as possible. Some chapters, such as quarrying and the processing of limestone, have much in common with other segments of industry. In such cases, the text has been kept relatively short and reference has been made to more specialist publications. In other chapters, such as the production of lime, and its use in building and construction, the subjects are so broad that it is beyond the scope of this publication to do them justice. In such cases, a brief description has been given, supplemented by a relatively large number of references for further reading. Annex 2 contains a list of journals and reference books of interest to the reader wishing to up-date, or broaden his knowledge of a particular subject.

During the past four decades that the author has been involved with both Industries, there have been many profound changes. No doubt the rate of change will accelerate, bringing new challenges and opportunities. If this work can, in some small way, help those involved to meet those challenges and to exploit the opportunities, it will have achieved its objective.

Buxton, Derbyshire, England   J.A.H. Oates
March, 1998

Acknowledgements

My first and foremost acknowledgement is to my wife Yvonne, without whose encouragement, tolerance and forbearance over the past three years, this book would never have been completed.

I also owe a great debt of gratitude to my former colleagues and friends at Buxton Lime Industries Ltd. (until 1990, the lime business of ICI plc) for their help and advice throughout. They are too numerous to mention by name, but have personally received my heartfelt thanks. I could not have written this book without their support.

I would like to acknowledge the invaluable contributions of the following experts: Mr. D.D. Brumhead, geology and exploration; Mr. D. Rockliff of Tilcon, aggregates; Mr. P.J. Jackson, formerly of Rugby Cement, cement and mortar; Mr. R.H. Llewelyn and his colleagues at the Agricultural Development and Advisory Service, agriculture; Messrs. E. Perry and P. Richards of Redland Aggregates, dolomitic lime; Messrs. A.D. Russel and W. Harrison, formerly of the Building Research Establishment, mortar; Mr. J. Saunders, formerly of Celcon and Ryarsh Brick, aircrete and sandlime bricks; Mr N.H. Groocock, formerly of Severn Trent Water, water treatment, and Mr. R. Collins of National Power, flue gas desulfurisation.

Messrs. A.S. Anthony and F. Leitch of the British Lime Association and Dr. B. Oppermann and Mr. N. Peschen of the Bundesverband der Deutschen Kalkindustrie provided invaluable assistance with general information and contacts. Messrs A.R. Mears and B. Feldmann of the British Standards Institution helped to ensure that I was fully informed of recent developments in British and European Standards.

I particularly welcomed the contributions, help and encouragement from scores of people throughout the world (in the Lime and Limestone Industries, as well as suppliers to and customers of those Industries), who generously shared their knowledge and expertise, and provided diagrams and photographs. I hope that the finished product does justice to their inputs.

With the help of the above-mentioned people, and others who have assisted with producing diagrams, typing and proof-reading, I have endeavoured to produce a text that is as comprehensive, accurate and up-to-date as possible. However, the responsibility for any ommissions or errors is mine alone. Finally, I would like to thank the publishers for having the courage to commission me, and faith in my ability to deliver the required product.

Tony Oates

Contents

Part 3   Production of Quicklime

Part 4 Production of Slaked Lime

Annexes

1 Introduction

1.1 General

- *Limestone* is a naturally occurring mineral that consists principally of calcium carbonate. Part of the calcium carbonate may have been converted to dolomite by replacement with magnesium carbonate as a secondary component (up to 46 % by weight). Many limestones are remarkably pure, with less than 5 % of non-carbonate impurities. Limestone is found in many forms and is classified in terms of its origin, chemical composition, structure, and geological formation. It occurs widely throughout the world, and is an essential raw material for many industries.

- *Quicklime* is produced by the thermal dissociation of limestone. Its principal component is calcium oxide. Its quality depends on many factors including physical properties, reactivity to water and chemical composition. As the most readily available and cost-effective alkali, quicklime plays an essential part in a wide range of industrial processes.

- *Slaked lime* is produced by reacting, or "slaking" quicklime with water, and consists mainly of calcium hydroxide. The term includes *hydrated lime* (dry calcium hydroxide powder), *milk of lime* and *lime putty* (dispersions of calcium hydroxide particles in water). Slaked lime is widely used in aqueous systems as a low-cost alkali.

The generic term, *lime* includes quicklime and slaked lime, and is synonymous with the term *lime products*. "Lime" is, however, sometimes used incorrectly to describe limestone products (e.g. agricultural lime): *this is a frequent cause of confusion*.

Because the quarrying of limestone and the production of quick- and slaked lime are long-established industries, they have generated many traditional terms. These are explained in the Glossary of Terms (Annex 1).

1.2 Importance of Lime and Limestone

1.2.1 Limestone

Because limestone deposits are widely distributed throughout the world, a high proportion of humanity has ready access to the material. No reliable figures appear to have been published for the world-wide use of limestone, but the author estimates that it is about 4,500 million tonnes per annum (tpa).

In most countries, the major uses of limestone are as an aggregate in construction and building and as the primary raw material for the production of cement.

The *amount* of limestone used in construction and building varies widely from one locality to another and depends on its availability and cost relative to other aggregates, such as gravel and crushed hard rocks. In the USA, for example, limestone sales amounted to about 800 million tonnes in 1994 — about 72 % of the crushed rock sales.

The *proportion* of limestone quarried that is used in construction and building is also affected by availability and cost. In many countries, the level is around 40 to 50 %, whereas in the USA and the UK, where limestone is widely available and relatively inexpensive, the level is over 70 %.

While limestone is not an essential raw material for the production of cement, it is generally the cheapest source of calcium oxide. On the basis of the global production of cement, the limestone used in its production probably amounts to about 1,500 million tpa, or one third of the total extracted.

Some limestones contain over 95 % $CaCO_3$. Such "chemical grade" materials are particularly suitable for lime production, flue gas desulfurisation and a range of other processes. The quantities involved, however, amount at most to a few percent of the total extracted.

Very finely divided limestones (whiting) and precipitated calcium carbonate are used as fillers. While the tonnages involved are minute when compared with the total, they are very high added-value products that play important roles in a wide range of industries.

The main market outlets for limestone products are outlined in chapter 7 and are described in more detail in chapters 8 to 12.

1.2.2   Lime

As mentioned in section 1.1, lime is the least expensive and most widely-used alkali. The global production of lime products is believed to be over 200 million tpa. This amount includes an estimated production in China of about 20 million tpa (although much higher rates for that country have been quoted).

Lime is one of the most heavily used chemicals. In the USA, for example, about 15 million tpa of lime are produced, making it the fifth largest-selling chemical on a tonnage basis (the four chemicals with larger sales are sulphuric acid at 40 million tpa, nitrogen at 36 million tpa, with oxygen and ethylene both at 17 million tpa).

In most industrialised countries, the major uses of lime products are in steelmaking, and the construction and building industry. In the European Union, for example, some 38 and 36 %, respectively is used in those industries.

The remaining lime is used in a large number of industries. The main market outlets are outlined in chapter 25 and described in more detail in chapters 26 to 32.

1.3 History [1.1–1.3]

1.3.1 Limestone

Limestone has undoubtedly been used since the Stone Age, although primitive man probably found uses for it before that time. The first records relate to the Egyptian Second Dynasty (some 5,800 years ago), when it was employed in the construction of the Giza Pyramids. Marble, a highly crystalline form of limestone, was used by the Greeks shortly after this period for statues and the decoration of buildings. Limestone was widely used by the Romans for building roads.

Over the centuries, limestone has been used extensively as an aggregate in building and construction. It has been used as aggregate in lime-based concrete since Roman times, and, more recently, in cement-based concrete. Although limestone cannot readily be dressed, it has been used extensively in building in both the rough-hewn form and as cut dimension stone.

The benefits of "liming" soils with marls and soft chalks was known to the Romans in the first century A.D. Pliny reported that the Ubians, north of Mainz, used "white earth" (a calcarious marl) to fertilise their fields.

The high purity of some limestones has been exploited for many centuries by the lime-burning, glass-production, and metals-refining industries. The development of Portland cement in the 19th century caused a major expansion in the demand for limestone, both as a raw material and as an aggregate. This expansion permitted the exploitation of some of the softer and/or less pure deposits such as chalk and marl.

1.3.2 Lime

Some of the earliest evidence for the use of lime dates back some 10,000 years. Excavations in Cajenu in Eastern Turkey, uncovered a Terrazzo floor, which had been laid with lime mortar. That site dated from 7,000 to 14,000 years ago. In some cases, the lime had been used in conjunction with gypsum ($CaSO_4 \cdot 2H_2O$), which raises the question as to whether the lime had used as a binder in its own right, or had arisen as a result of contamination of the gypsum with calcium carbonate.

Nevertheless, there is firm evidence of the use of lime in the Near East, dating from about 8,000 years ago, and in Lepenski Vir, in the former Yugoslavia, a floor, dated 6,000 B.C., was excavated in the 1960s. That consisted of a type of mortar made from lime, sand, clay and water.

Lime stabilisation of clay was used in Tibet, over 5 000 years ago, in the construction of the pyramids of Shersi. It was also used in conjunction with limestone by the Egyptians in the construction of the pyramids and by the Chinese when building the Great Wall.

By about 1,000 B.C., there is evidence of the wide-spread use of quick- and hydrated lime for building by many civilisations, including the Greeks, Egyptians, Romans, Incas, Mayas, Chinese, and Mogul Indians.

Perhaps the earliest excavated lime kiln was at Khafaje in Mesopotamia which was dated at about 2450 B.C. A battery of six lime kilns, excavated at a legionary site at Iversheim, Germany, showed that the Romans produced lime in quantity on military sites. The production of lime in kilns was mentioned by Cato in 184 B.C. Pliny the Elder (ca. 17 A.D.), in his "Chapters on Chemical Subjects" described the production, slaking and uses of lime, and stressed the importance of chemical purity.

The Romans employed hydraulic lime and lime-pozzolan mixtures in many construction projects, including the Appian Way. They developed the technology of lime burning and the use of mortar, cement and concrete, using lime as the binder. They built the first "lime factories", which were operated by legionaries and managed by a "Magister Calcariarum".

Lime was also well known to the Romans as a chemical reagent. In 350 B.C. Xenophon referred to the use of lime for bleaching linen. Almost all of the Mediterranean peoples were familiar with lime as a paint. Lime was used for tanning leather, and was mixed with organic substances to produce putty and glue. The Assyrians described the importance of lime in their recipes for glass. Lime was also used in glazes for pottery. A medical use of lime was recorded by Dioscorides in 75 A.D.

Little is known regarding the condition of the lime industry in medieval times, but a knowledge of its properties and its use for building purposes is reflected in the writings of the day. For example Trevisa (1398) wrote "Whyle lyme is colde in handlyng it conteyneth prevely wythin fyre and grete hete." "Lyme Kilns" and "Lymbrenners" are also mentioned in many ancient church and municipal records.

Quicklime was used in the Middle Ages for offensive purposes in war — there are records that the English hurled it in their enemies' faces at a naval battle in 1217. It was also used by alchemists for "causticising" the alkali metal carbonates in wood ashes and for other purposes, but it was so familiar a material that it was seldom thought to be worth recording. During the 1400s, the use of lime in building spread throughout Europe.

In the 1700s Joseph Black gave the first sound technical explanation of the calcination of limestone including the evolution of carbon dioxide. Lavoisier confirmed and developed Black's explanation. In 1766 De Ramecourt published a detailed account of "the art of the lime burner", which described the design, operation and economic aspects of limestone quarrying and lime burning.

Debray, in 1867, carried out the first measurements of the dissociation pressure of calcium carbonate. He heated Iceland Spar in a tube to the temperature of boiling mercury, sulphur, cadmium and zinc (357, 445, 767, 907 °C respectively). He found no decomposition at the first two, but measurable pressures at the boiling points of cadmium and zinc. The first exact measurements of the dissociation pressure were made by Le Chatelier in 1886.

In 1935 Searle [1.2] described some 40 designs of lime kiln. Since then, a large number of designs have been developed. A great variety of designs are still operated, but only a limited number continue to be commercially viable. The more important of these are described in chapter 16.

1.4   References

[1.1]   R.S. Boynton, "Chemistry and Technology of Lime and Limestone", John Wiley & Sons, 1980.
[1.2]   A.B. Searle, "Limestone and its Products", Ernest Benn, 1935.
[1.3]   N.V.S. Knibbs, "Lime and Magnesia", Ernest Benn, 1924.

Part 1 Production of Limestone

2 Formation, Classification and Occurrence of Limestone

2.1 Formation of Limestone

2.1.1 Origins of Calcium Carbonate

The chemical components of calcium carbonate — dissolved calcium ions and carbon dioxide — are widely distributed. Calcium is the fifth most common element in the earth's crust (after oxygen, silicon, aluminium and iron). It was extracted from early igneous rocks by the combined effects of erosion by the weather and corrosion by acidic gases (oxides of sulfur, oxides of nitrogen and carbon dioxide dissolved in rain water). Carbon dioxide makes up about 0.03 % by volume of the earth's atmosphere and is dissolved in both fresh and sea water. Combination of dissolved calcium ions and carbon dioxide resulted in the sedimentary deposition of calcium carbonate, which was subsequently converted into limestone rock. Early limestones (Precambrian — Table 2.1) are believed to have been deposited as precipitates of $CaCO_3$, and/or as a result of the biochemical activity of very simple organisms, such as bacteria.

Table 2.1. Table of geological periods

Era	Period	Maximum age 10^6 years
Cenozoic	Quaternary	1
	Tertiary	75
Mesozoic	Cretaceous	135
	Jurassic	180
	Triassic	225
Palaeozoic	Permian	270
	Carboniferous	350
	Devonian	400
	Silurian	440
	Ordovician	500
	Cambrian	600
Precambrian	—	> 600

2.1.2 Carbonate Sedimentation

The sedimentation of calcium carbonate occurs by two mechanisms – *organic* and *inorganic*. The *organic* route involves a wide variety of organisms, which build shells, skeletons or secrete carbonate. The *inorganic* route involves the direct pre-cipitation (or crystallisation) of carbonate.

Most commercially viable deposits of carbonate were formed by the *organic route*. Carbonate-secreting organisms (e.g., bivalves, gastropods, brachiopods, corals, sponges, bryozoans, echinoderms, ostracods, foraminifera and various algae) have existed in all of the world's seas. The factors which controlled the rate of carbonate production (calcium, magnesium and carbon dioxide concentrations, temperature, salinity, water depth and turbidity) resulted in most of the thick deposits being produced in shallow seas (i.e., in the photic zone) between 30° north and south of the equator. Such deposits may now be outside that band as a result of continental drift.

At least eight mechanisms can cause the surface layers of the sea to become super-saturated with respect to aragonite, calcite and dolomite [2.1]. The rate of formation of dolomite, however, is very much slower than those of calcite and aragonite. As a result, while some organic species produce aragonite structures and others make calcite, none produce dolomite directly. The aragonite structures are generally very low in magnesium (typically less than 0.5 % $MgCO_3$). Depend-ing on the organism and on the chemistry of the water (principally the ratio of calcium to magnesium), calcite structures are generally either low in magnesium with less than 4 % $MgCO_3$, or high with, typically, 11 to 19 % $MgCO_3$.

The above process, coupled with the fact that most carbonate-secreting organ-isms only thrive in clear waters — remote from rivers carrying significant amounts of solids washed from the land — accounts for the remarkably high pur-ities of many carbonate deposits, which often exceed 98 % of calcium plus magnesium carbonates.

Carbonate sediments are also produced in a similar way by organisms in inland waters, but the resulting deposits are generally not as extensive, nor as commer-cially important as those produced in the marine environment.

Inorganic precipitation of calcium carbonate occurs from both sea and inland waters (as used by geologists, "precipitation" refers to the relatively slow process of crystal growth on surfaces). This route has resulted in some commercially significant deposits, the most common of which are oolitic limestone and traver-tine (see section 2.2.1). Some minor dolomite sediments have been formed by direct precipitation from sea and lake waters.

2.1.3   Sedimentary Environments

Most carbonate sediments were formed in situ, in shallow water, accumulating where the grains were formed, or were subjected to limited transport, for example down a gently-sloping sub-tidal shelf. Descriptions of depositional environments can be found in the literature [2.1, 2.2]. The major environments are illustrated in Figure 2.1.

The variety of environments (which includes beaches, tidal and sub-tidal flats, lagoons, reefs, shelves, slopes and deep basins) gave rise to many types of deposit, whose characteristics are related to the particular environment in which they were formed [2.1–2.4].

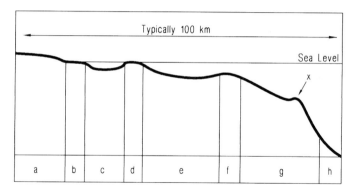

Figure 2.1. Composite diagram of major carbonate depositional environments
(a) sub-aerial karst; (b) tidal flat; (c) lagoon; (d) barrier reef; (e) shelf; (f) shelf-margin reef;
(g) slope with reef mound (x); (h) deep basin

2.1.4 Diagenesis

Diagenesis is the conversion of sediments into rock by organic, physical and
chemical processes. Six main processes have been identified for limestone [2.3] —
microbial micritization, cementation, neomorphism, dissolution, compaction and
dolomitization.

- *Microbial micritization.* Many organisms bore into carbonate deposits. The
  most important one is cyanobacteria, but others include cliona sponges, bivalves, polychaetes and fungi. The bore-holes become filled with a calcium
  carbonate structure called micrite, which typically forms an envelope around
  the skeletal grains.

- *Cementation* results from the passage of water, super-saturated with respect to
  calcite, through porous limestone deposits, leading to the growth of calcite
  crystals in the pores, thereby binding together the components of the deposit.
  The most common cement in medium- to coarse-grained limestones is *sparite,*
  or calcite spar (which fills interstitial spaces in fine-grained limestones). Silica,
  in the form of quartz crystals, also acts as a cement in some limestones.

- *Neomorphism* involves recrystallisation. As aragonite has a higher solubility in
  water than calcite, it progressively recrystallises over time to produce a very
  low-magnesium calcite. Calcite recrystallises into larger crystallites — in doing
  so, under many conditions, the magnesium in high-magnesium calcites slowly
  dissolves, leaving low-magnesium deposits (but see dolomitisation below).

- *Dissolution* generally occurs when unsaturated ground waters flow through
  deposits. On the surface it causes typical karst scenery. At greater depths, it
  produces caves, as well as secondary porosity. The latter increases the capacity
  of a deposit to act as a reservoir for oil, water, or gas.

- *Compaction* occurs during the burial process and is a combination of physical effects, such as closer packing and crushing of particles, and dissolution/recrystallisation under high pressure.

- *Dolomitisation* results in the formation of the double carbonate $CaCO_3 \cdot MgCO_3$. It may be produced near the surface, soon after the deposit has formed, or in the burial stage at a much later date. Most ancient limestones, formed in the Precambrian era are predominantly dolomitic. Dolomites typically contain up to 90 % of the double carbonate, with the remainder being mainly calcite.

The mechanisms of dolomitisation are not well understood [2.3, 2.4], but undoubtedly involve passage of sea water through the pores of the limestone over long periods. The dissolved magnesium is able to replace alternate calcium ions in the crystal lattice, because dolomite is a little more stable than calcite. Moreover, as the crystal density of dolomite is about 10 % higher than that of calcite, dolomitization is accompanied by an increase of porosity. This facilitates further dolomitisation. It also raises the capacity of a deposit to act as a reservoir for oil, water, or gas.

De-dolomitization, can occur when ground water, low in magnesium, passes through dolomite for prolonged periods.

Most of the above diagenetic processes can occur in any of three environments. In the *marine* environment, the deposit is in contact with sea-water, which contains dissolved magnesium and may be either un-saturated, or super-saturated with respect to calcium carbonate species. In the *ground-water* environment, it is in contact with water, which is low in dissolved magnesium and is generally unsaturated with calcium carbonate. In the *burial* environment it is subject to high pressures and possibly elevated temperatures (or hydrothermal conditions) and is in contact with water, which may have a widely varying composition.

The interactions between the processes and the environment can be particularly complex and have a marked influence on the characteristics of the limestone [2.3, 2.4]. Some limestones (e.g soft chalk) have not been fully compacted, are highly porous, micro-crystalline and contain well-defined fossils. Others (e.g. marble) have been completely metamorphosed, have very low porosities, are highly crystalline and contain no discernable fossils.

2.1.5   Geological Structures

Bedding planes (Fig. 2.2) are the most common structures in sedimentary limestones. They usually represent changes in, or interruptions to, the conditions of sedimentation. Thin clay seams are often associated with bedding planes. (For example, in the Carboniferous limestone of the English Peak District, an important sequence of clay seams along bedding planes represents successive falls of volcanic ash which interrupted the deposition of carbonate sediments). The original horizontal bedding planes may be disturbed by tectonic activity, producing structures such as anticlines, synclines, overthrust folds and faults. Although

Figure 2.2. Bedding planes in a limestone outcrop (by courtesy of D.D. Brumhead)

Figure 2.3. A diagonal fault in a quarry face (by courtesy of Buxton Lime Industries Ltd.)

bedding planes were generally formed as depositional surfaces, they can also be produced after burial by dissolution of limestone.

Faults (Fig. 2.3), caused by vertical displacement of blocks of limestone relative to each other, can disturb the continuity of bedding planes by tens of metres or more.

Current structures, such as dune bedding, ripples and cross-bedding, are sometimes found in limestone. They were produced by the fossilisation of features produced in sediments which had been subjected to limited transport by water currents.

Fossil karstic surfaces are often found at bedding plane discontinuities, and are particularly well represented in Carboniferous limestones. They mark a period during which the sediment surface was above water, when dissolution of the limestone by surface water produced a pot-holed surface, or limestone pavement. Under such circumstances, an over-lying clay seam may represent a former thin soil cover. Karstic dissolution of the limestone can also occur tens or hundreds of metres below the ground surface, producing cavity structures varying in size from vugs (small cavities) to large pot-holes (or shake-holes) and caves. Cavities near the surface are often filled with clay or over-burden washed down from the surface.

2.2   Classification of Limestones

2.2.1   Types of Limestone

Limestone takes many forms. The following eleven descriptions include the great majority of significant limestone deposits.

- *Biosparites* are the most widespread type of massive, well bedded limestones. They consist of fragments of calcareous skeletons and small shells in a matrix of recrystallised calcite. They are typical of many Carboniferous limestones.

- *Micrites* are limestones originating from carbonate mud or silt.

- *Biomicrites* are limestones consisting of skeletons or fragments of organic debris in a micrite matrix.

- *Reef* limestones are mounds and units of organic debris, often consisting of complete fossils. They are highly fossiliferous and lack a bedding structure, having been formed as thick accumulations topographically higher than their surroundings. In some cases, such reef limestones may occur as fringing (or apron) reefs, which may extend for many kilometres.

- *Algal* limestones are biosparites, or biomicrites, resulting from the activities of algae.

- *Oolitic* limestones (or ooliths) are made up of tiny spherical grains (known as ooids) of 1 mm or less in diameter, precipitated by algal action in turbulent waters and cemented in calcite. Under a hand lens, they have the appearance of fish roe.

- *Dolomite* is used to describe both the mineral and the rock. Dolomite deposits often occur as distinct beds within limestone, but can also occur in thick-bedded units.

- *Chalks* of the Cretaceous era are found in north-west Europe and parts of North America. They were deposited on comparatively deep (50 to 400 m) sea-beds. They are white, relatively soft and contain very little terrigenous material (silt originating from the land). Under the microscope, they are seen to be largely composed of plates of algae (coccoliths) and foraminifera and often contain large fossils of echinoderms, bivalves and brachiopods.

- *Marble* is metamorphosed limestone (i.e. limestone which has been fully recrystallised and hardened under hydrothermal conditions). Marbles derived from pure limestones consist simply of white calcite. When impurities such as dolomite, iron and clay are present, the marble acquires its characteristic mottled or veined appearance. (It should be noted that stone-masons often apply the term marble to any rock which can be easily polished, even to some granites.)

- *Travertine* is composed of calcite deposited by chemical precipitation from natural hot-water springs. It has a characteristic banded appearance and is used as a decorative building stone.

- *Tufa* is also produced by chemical precipitation from natural springs. It is typically deposited over rocks and is softer and more porous than travertine.

2.2.2 Other Classifications

Most classifications have been developed by and for geologists. However, when selecting a limestone deposit for quarrying, the developer is concerned with its physical and chemical properties and their influence on the suitability of the products for the intended end-uses. Similarly, a lime producer would be interested in the characteristics of the limestone fed to the kilns, how it responds to heating, and the physical and chemical properties of the resulting lime products. These aspects are considered in greater detail in later chapters.

Many ways of classifying limestone have been developed to describe the nature of the deposit. Six of these are described briefly below:

a) is based on the average grain size [2.4, 2.5]:

micro-grained (or calcilutite)	—	less than $4\,\mu m$
fine-grained (or calcilutite)	—	4 to $60\,\mu m$
medium-grained (or calcarenite)	—	60 to $200\,\mu m$
coarse-grained (or calcirudite)	—	over $200\,\mu m$ (to about $1000\,\mu m$)

b) is based on the micro-structure [2.3, 2.4] and recognises three components, namely: allochems (grains), matrix (chiefly micrite) and cement (chiefly sparite)

c) is based on the texture [2.3]:
mudstone (calcitic) for cemented calcitic mud (or micrite) with few coarse grains,
wakestone where over 10 % of coarse grains are distributed in a calcitic mud cement,
packstone which is mainly coarse grains in close contact with calcitic mud cement,
grainstone, which is mainly grains without mud cement,
boundstone, where the components were bound together during deposition (e.g. in a reef)

d) is also texture-based [2.6] and uses terms such as:
compact — crystalline — earthy — saccharoidal — cherty — pisolitic — conglomeratic — unconsolidated

e) is based on the principal impurities [2.4]:
carbonaceous (with carbon)
arenaceous (or sandy)
ferruginous (with iron)
argillaceous (or clayey)
siliceous (with silica)
phosphatic (with phosphorous)
(N.B. see section 3.3 for information about the nature and level of impurities)

f) is based on the carbonate content [2.6]:
ultra-high-calcium (more than 97 % $CaCO_3$)
high-calcium, or chemical-grade (more than 95 % $CaCO_3$)
high purity carbonate (more than 95 % $CaCO_3 + MgCO_3$)
calcitic (less than 5 % $MgCO_3$)
magnesian (5 to 20 % $MgCO_3$)*
dolomitic (20 to 40 % $MgCO_3$)*
high magnesium dolomite (40 to 46 % $MgCO_3$)
(* N.B., these ranges may not be widely recognised)

2.2.3   Terminology

Many other terms are used in describing limestone. The more common ones, including those already mentioned, are listed in the Glossary of Terms (Annex 1).

2.3   Occurrence of Limestones

Limestone deposits cover about 10 % of the earth's land surface [2.4, 2.7] and are found in the majority of countries. Three major inter-related factors have had a major influence on the occurrence of limestones around the world, namely climate, geotechnics and sea level.

a) *Climate* has directly influenced the rate of carbonate deposition, by both the organic and inorganic routes. As already mentioned, most limestones were deposited in shallow seas in the band 30 °S to 30 °N. Low rainfall favours carbonate production by reducing the amount of terrigenous material washed from the land, lowering the level of turbidity and increasing the depth of the photic zone, thereby increasing the area over which carbonate deposits are produced. Conversely, turbidity and deposition inhibits the growth of most carbonate-producing organisms.

b) *Geotechnics* directly influenced the occurrence of limestone in three ways — continental drift, regional subsidence and regional up-lift.
Because the original ancient continents formed in relatively high southerly latitudes, those parts of the continental plates which have remained south of 30 °S generally have little limestone. Conversely, those continents which

drifted into and, in some cases, across the band 30 °S to 30 °N, are relatively rich in limestone deposits. As an example, the massive Carboniferous limestone deposits in England were formed in equatorial and sub-tropical latitudes, while the more recent Jurassic limestones and the Cretaceous chalk deposits were formed in the warm temperate zone at progressively higher latitudes. Limestones are less common on the old continental cores (or shields), although, in places where shallow seas were formed, some carbonate was deposited on the ancient rocks and was subsequently converted into limestone.

Regional subsidence, coupled with rising sea levels, had two effects on carbonate deposition. It resulted in the formation of shallow seas, which, given the requisite combination of temperature, salinity and low turbidity, were favourable for carbonate production. Where there was progressive subsidence of the sea bed over prolonged periods, deposits many hundreds of metres deep were able to accumulate (e.g., the thickness of the Triassic deposits in the western Dolomites is more than 2 000 m).

Conversely, regional up-lift, coupled with lowering sea levels, raised the deposits above sea level and into the positions in which they are now found.

An indirect effect of geotechnics was to change the volume of the mid-ocean ridges, which in turn influenced sea level.

c) The *sea level* has varied as a result of the effects mentioned above and has had a marked effect on the area of shallow seas around the globe. Historically, sea levels were generally higher than at present (possibly over 100 m higher in the Cretaceous period) and the area of shallow seas capable of sustaining carbonate producing organisms was considerably greater than at present.

As a result of the above factors, every deposit of limestone has a unique history in terms of the carbonate-producing organisms/mechanisms, the sedimentary environment, the diagenetic mechanism, and geotechnics involved.

2.4 References

[2.1] J.L. Wilson, "Carbonate Facies in Geologic History", Springer-Verlag, 1975, ISBN 3-540-07236-5.
[2.2] M.E. Tucker, V.P. Wright, "Carbonate Sedimentology", Blackwell Scientific Publications, London, 1990, ISBN 0-632-01471-7 (0-632-01472-5 pbk).
[2.3] M.E. Tucker, "Sedimentary Petrology, An Introduction to the Origin of Sedimentary Rocks", Blackwell Scientific Publications, London, 1991, ISBN 0-632-02959-5 (0-632-02961-7 pbk).
[2.4] T.P. Scoffin, "An Introduction to Carbonate Sediments and Rocks", Blackie, 1987, ISBN 0-216-91789-1.
[2.5] R.S. Boynton, "Chemistry and Technology of Lime and Limestone", John Wiley & Sons, 1980.
[2.6] M. Wingate, "Small-scale Lime-burning", Intermediate Technology Publications, 1985.
[2.7] D.J. Wiersma, "The Geology of Carbonate Rocks", Proc. International Lime Congress, Rome, 13–14 Sep. 1990.

3 Physical and Chemical Properties of Limestone

The physical and chemical properties of limestones vary widely depending on the route by which they were formed, the sedimentary environment and the changes brought about by diagenesis (see section 2.1). For this reason, many of the properties given below should be regarded as typical.

The Chemical Abstracts Service (CAS) registry numbers for limestone, $CaCO_3$ and $MgCO_3$ are 1317-65-3, 471-34-1 and 546-93-0 respectively. The EINECS (European Inventory of Existing Commercial Substances) reference for limestone is 207-439-9.

3.1 Physical Properties

- *Molecular weight*. The values for $CaCO_3$ and $MgCO_3$ are 100.09 and 84.32 respectively.

- *Colour*. The colour of limestone often reflects the levels and nature of the impurities present. White deposits are generally of high purity. Various shades of grey and dark hues are usually caused by carbonaceous material and/or iron sulfide. Yellow, cream and red hues are indicative of iron and manganese. Impurities in marble often produce a variety of colours and patterns.

- *Odour*. Limestone often has a musty or earthy odour, which is caused by its content of carbonaceous matter.

- *Texture*. The texture of limestones varies widely. Four of the methods of classifying limestone (see section 2.2) are based on texture and structure. All limestones are crystalline, with average grain sizes ranging from less than 4 μm to about 1000 μm. The distribution of grain sizes affects the texture, and ranges from mudstone (mainly calcitic mud with few coarse grains) to grainstone (mainly coarse grains with little mud cement).

- *Crystal structure*. Calcite and dolomite have rhombohedral structures, while aragonite is rhombic [3.1].

- *Specific gravity*. The specific gravities of the crystalline forms of calcium carbonate and dolomite at 20 °C are: calcite 2.72 g/cm³, aragonite 2.94 g/cm³ and dolomite 2.86 g/cm³ [3.1].

- *Porosity.* The porosity of limestone varies considerably depending on the degree of compaction and the structure. Typical values (by volume) are 0.1 to 2 % for marble, 0.1 to 30 % for limestones, 15 to 40 % for chalks, and up to 50 % for marls. The porosity of dolomites is generally in the range 1 to 10 % [3.2].

- *Water absorption* depends on the porosity, the distribution of pore sizes and the level of carbonaceous matter. Thus a dense limestone may contain 0.4 % of water by weight, while a chalk may contain 20 % of water.

- *Apparent density.* This is a function of the porosity, the crystal density and the amount of water in the pores. For limestones dried at 110 °C, typical values are 1.5 to 2.3 g/cm^3 for chalk, up to 2.7 g/cm^3 for dense high calcium limestones and 2.7 to 2.9 g/cm^3 for dolomite.

- *Bulk density.* The bulk density depends largely on the apparent density of the limestone, the particle size distribution and on the particle shape. Thus crushed and screened limestone with an apparent density of 2.7 g/cm^3 and a ratio of top to bottom size of 2:1 typically has a bulk density of 1400 to 1450 kg/m^3. A slabby particle shape tends to increase bulk density, while a cubical shape reduces it. A high fines content can increase the bulk density by up to 25 %. The presence of water within the pores can significantly raise the bulk density of porous limestones. Surface water between particles can encourage close packing of products containing a high fines content, thereby increasing the bulk density.

- *Angle of repose.* The external angle of repose for screened limestone is generally in the range 35 to 45°, but is affected by factors such as the size distribution, cleanliness and moisture content. The angle can approach 90° for moist products containing significant levels of fines. Such materials are prone to compaction and to "rat-holing" in bunkers. It should be noted that the internal angle of repose can differ significantly from the commonly measured external angle and this can affect the live capacity of bunkers and stockpiles.

- *Strength.* The compressive strength of limestones varies from 10 MN/m^2 for some marls and chalks to 200 MN/m^2 for some marbles. A typical crushing strength for a dense high calcium limestone is 180 MN/m^2. Methods for the measurement of the strength of aggregates are described in section 6.4.3.

- *Hardness.* The hardness of limestones generally lies in the range 2 to 4 Mohs.

- *Abrasion resistance.* The resistance of aggregates to abrasion is measured in terms of the aggregate abrasion value and the polished stone value (see section 6.4.3). An alternative measure is the Bond Work Index [3.3], which ranges from 4 kW·hr/t for soft limestones, to about 10 for typical dense limestones.

- *Specific heat.* The specific heat of calcite is given in Table 3.1 [3.4]. Its average value over various temperature ranges may be determined from Figure 16.8. Information for dolomite appears to be more sparse. A value of 0.22 cal/g · °C has been reported for the temperature range of approximately 20 to 100 °C [3.5, 3.9].

Table 3.1. Specific heat of calcite

Temperature °C	Specific heat cal/g · °C
0	0.191
200	0.239
400	0.270
600	0.296
800	0.322

- *Coefficient of thermal expansion.* Reported values for limestone range from 4 to $9 \times 10^{-6}/°C$, while those for marble are 3 to $15 \times 10^{-6}/°C$ [3.4].

- *Thermal conductivity.* This depends on the porosity and structure. A range has been quoted for marbles at 30 °C of 0.0050 to 0.0077 cal · cm/cm^2 · sec · °C. Values for high calcium limestone, dolomitic limestone and chalk (at 130 °C) are 0.0039, 0.0034 and 0.0022 cal · cm/cm^2 · sec · °C, respectively [3.6]. The conductivities decrease with increasing temperatures (e.g.,from 0.00614 cal/cm^3 · sec · °C at 50 to 100 °C, to 0.00415 cal · cm/cm^2 · sec · °C at 150 to 200 °C [3.9]).

- *Velocity of sound.* A value of 3810 m/sec has been quoted for marble [3.1].

- *Brightness* is an important parameter for whiting and precipitated calcium carbonate, when used as fillers and pigments in paints and paper coatings. Whiting generally has a brightness value in the range 75 to 95, while a typical value for PCC is over 95 [3.7, 3.8] (see section 6.4.2 for test methods).

3.2   Chemical Properties

- *Stability.* Aragonite is pseudo-stable with respect to calcite under ambient conditions, and consequently has a higher solubility in water. In the presence of water, aragonite slowly re-crystallises into calcite (section 2.1.4). It also re-crystallises into calcite at 400 to 500 °C in the absence of water [3.1, 3.9].
  Calcite is metastable with respect to dolomite in the presence of sea-water (which contains dissolved magnesium). The process of dolomitisation is slow, even in geological terms. It is, however, slowly reversed in the presence of fresh water.

The thermal decomposition of calcium and magnesium carbonates is described below.

- *Solubility in carbon dioxide-free water.* The solubility of calcite in distilled water free of carbon dioxide is 14 mg/l at 25 °C, rising to 18 mg/l at 75 °C. That of aragonite increases from 15.3 mg/l at 25 °C to 19.0 mg/l at 75 °C [3.1]. However, these values are only of academic interest, as natural water contains dissolved carbon dioxide.

- *Reaction with carbon dioxide.* The increase in "solubility" of limestones in the presence of carbon dioxide is due to reversible chemical reaction (3.1) and (3.2), which form calcium and magnesium bicarbonates. For example, at 20 °C approximately 30 mg/l of calcite will dissolve in distilled water at equilibrium with the atmospheric carbon dioxide [3.9].

$$CaCO_3 + H_2O + CO_2 \rightleftarrows Ca(HCO_3)_2 \tag{3.1}$$

$$MgCO_3 + H_2O + CO_2 \rightleftarrows Mg(HCO_3)_2 \tag{3.2}$$

On heating a solution of calcium/magnesium bicarbonates, carbon dioxide is evolved and calcium/magnesium carbonates precipitate (this mechanism accounts for the formation of scale in kettles and boilers in hard water areas).

- *pH in water.* Calcitic limestones give pH values of between 8.0 and 9.0. Dolomite gives values of 9.0 to 9.2 [3.9].

- *Reaction with aqueous acids.* In general, limestones react readily with acids and are used for acid neutralisation. High calcium limestones react readily with dilute hydrochloric and nitric acids at ambient temperatures, whereas dolomite and dolomitic limestones only react readily when the dilute acid is heated. The reaction of limestone with sulfurous acid (formed by the dissolution of sulfur dioxide in water) is the basis of a flue gas desulfurisation process (see section 12.5.2). The reactions with acids which form insoluble or sparingly soluble calcium salts (e.g., sulfurous, sulfuric, oxalic, hydrofluoric and phosphoric acids) are inhibited by the reaction product.

- *Heat of reaction with acids.* The heat of reaction of calcite with hydrochloric acid has been reported to be 4495 cal/mole, or 18.8 kJ/mole [3.10].

- *Reaction with acidic gases.* Limestones react readily with gaseous hydrogen chloride and hydrogen fluoride, forming calcium chloride and fluoride respectively. Dry sulfur dioxide reacts with limestone at 95 °C and above to produce calcium sulfite. Sulfur trioxide also reacts with limestone to produce the sulfate.

- *Thermal decomposition.* Calcium carbonate decomposes into calcium oxide and carbon dioxide when heated. The heat of dissociation of $CaCO_3$ is 1781 kJ/

kg relative to 25 °C, and 1686 kJ/kg relative to 900 °C (426 kcal/kg and 403 kcal/kg respectively). The corresponding values for magnesium carbonate are 1439 kJ/kg $MgCO_3$ relative to 25 °C and 1362 kJ/kg relative to 700 °C.

The carbon dioxide dissociation pressure for calcite is shown in Table 3.2. At approximately 900 °C, the pressure reaches 1 atmosphere [3.11] — this is generally referred to as the decomposition temperature of calcium carbonate. The decompositions of dolomite and dolomitic limestones are complex and are described more fully in section 15.2.

Table 3.2. Pressure of carbon dioxide above calcite

Temperature °C	CO_2 pressure (atmospheres)
600	0.003
700	0.026
750	0.079
800	0.24
850	0.50
900	1.00

3.3 Impurities

Because the levels of impurities and trace elements have such a great influence on the suitability of limestones for many applications, this section deals with the topic in some detail with reference to the influence of geological factors.

Impurities could be introduced at any of the stages of deposition and subsequent diagenesis. However, in connection with making limestone products, it is useful to recognise three main processes:

a) The inclusion of non-carbonate matter into the sediment at the time of deposition. Examples include carbonaceous matter from decaying organisms, and particles of clay, silt and sand.
b) Short-term deposition of terrigenous sediment, which was sufficiently substantial to interrupt the production of carbonate, and which is manifested as discrete beds within the limestone.
c) The transfer of water-borne suspended materials (mainly clay and silt) and dissolved elements (such as magnesium, silicon, fluorine, lead, iron and other heavy metals) into faults. Dissolved elements may then have migrated from the fault into the deposit through cracks and pores in the limestone.

Impurities introduced by process (a) are effectively homogeneous and cannot be avoided by selective quarrying, or removed by washing. Those introduced by (b) are heterogeneous and are generally reduced by washing and/or screening. Depending on the nature of the deposit, it may be possible to reduce them further by selective quarrying. Many of the impurities introduced by (c) are concentrated in faulted areas and may largely be avoided by selective quarrying.

As explained in section 2.1.2, *magnesium carbonate* levels in calcite deposits are generally either less than 4 % or between 11 and 19 %. The level can be raised by dolomitisation (see section 2.1.4), which proceeds via process (c). For most uses of limestone, magnesium is not regarded as an undesirable impurity. There are, however, some applications which require less than 4 % $MgCO_3$ (e.g., cement production and the production of lime for the ammonia soda and the autoclaved aerated concrete processes). Other processes can only use hydrated lime containing more than 5 % $MgCO_3$, if the magnesium component is fully converted into $Mg(OH)_2$ by pressure hydration (e.g., the production of mortar, plaster and sand-lime bricks).

Migration of dissolved substances from a fault into the body of the limestone can result in mineralisation of the rock in the immediate vicinity of the fault. For routine limestone production, such rock is generally tipped. However, commercially important metal ore deposits are found in certain types of limestone. They include *galena* (PbS), *sphalerite* (ZnS), *barite* ($BaCO_3$), *haematite* (Fe_2O_3) and *fluorite* (CaF_2). A few deposits contain commercially viable concentrations of trace metals, the more important of which are *silver* and *mercury*.

Silica and *alumina*, in the form of clay, silt and sand are commonly found as heterogeneous impurities in features such as faults and bedding planes, and also occur as homogenous impurities. When limestones containing 5 to 10 % of clayey matter are calcined, they produce feebly hydraulic limes: those containing 15 to 30 % produce highly hydraulic limes [3.12] (see also section 26.9.2).

Chert (a form of silica) forms as a result of silicification. In this process, dissolved SiO_2, which may be of organic or inorganic origin, migrates into and/or within the limestone, either as a replacement mineral, or as a primary deposit. Chert often concentrates at particular horizons, forming layers. *Flint* is a variety of chert, which is only found in chalk.

Iron is found homogeneously as iron carbonate and heterogeneously as pyrite (FeS_2) and limonite (FeO(OH)). *Sulfur* from sulfates, and *carbon* from organic residues are mainly found as homogeneous impurities.

Typical levels of impurities and trace elements are summarised in Table 3.3. They are based on information viewed by the author and on published spectrographic analyses of 25 high-calcium American limestones [3.13]. Further information about the skew distributions of the concentration of trace elements is given in section 28.1.8.

Table 3.3. Typical ranges of impurities/trace elements in commercial limestones

Impurity or trace element	Typical range		
	low	high	units
Silica (as SiO_2)	0.1	2.0	%
Alumina (as Al_2O_3)	0.04	1.5	%
Iron (as Fe_2O_3)	0.02	0.6	%
Sulfur (as $CaSO_4$)	0.01	0.5	%
Carbonaceous matter	0.01	0.5	%
Manganese (as MnO_2)	20	1000	mg/kg
Antimony	0.1	3	mg/kg
Arsenic	0.1	15	mg/kg
Boron	1	20	mg/kg
Cadmium	0.1	1.5	mg/kg
Chromium	3	15	mg/kg
Copper	1	30	mg/kg
Fluoride	5	3000	mg/kg
Lead	0.5	30	mg/kg
Mercury	0.02	0.1	mg/kg
Molybdenum	0.1	4	mg/kg
Nickel	0.5	15	mg/kg
Selenium	0.02	3	mg/kg
Silver	0.2	4	mg/kg
Tin	0.1	15	mg/kg
Vanadium	1	20	mg/kg
Zinc	3	500	mg/kg

3.4   References

[3.1]   "CRC Handbook of Chemistry and Physics", 77th ed., 1996–97.
[3.2]   F. Schwarzkopf, "Lime Burning Technology — a Manual for Lime Plant Operators", 3rd ed., Svedala Industries, Kennedy Van Saun, 1994.
[3.3]   "FGD Chemistry and Analytical Methods Handbook", Vol. 2, EPRI report CS 3612, July 1984.
[3.4]   J. Murray, "Specific Heat Data for Evaluation of Lime Kiln Performance", Rock Prod., Aug. 1947, 148.
[3.5]   "Tables of physical and chemical constants", 16th ed., Harlow Longman, 1995.
[3.6]   F. Birch, J. Schairer, H. Spicer, "Handbook of Physical Constants", 2nd ed., Geol. Soc. Am., Special Paper 36, 1950.
[3.7]   D. Stoye et al., "Paints and Coatings", Ullmann's Encyclopedia of Industrial Chemistry, 5th ed. on CD-ROM, 1997.
[3.8]   R. Patt et al., "Paper and Pulp", Ullmann's Encyclopedia of Industrial Chemistry, 5th ed. on CD-ROM, 1997.
[3.9]   R.S. Boynton, "Chemistry and Technology of Lime and Limestone", John Wiley & Sons, 1980, ISBN 0-471-02771-5.

[3.10] H. Backstrom, J. Am. Chem. Soc. **47**, 1925, 2432–2443.

[3.11] A.B. Searle, "Limestone and its Products", Ernest Benn, 1935.

[3.12] K.A. Gutschick, "Lime and Limestone", Kirk Othmer Encyclopedia of Chemical Technology, 4th ed., 1995.

[3.13] J.A. Murray et. al., J. Am. Ceram. Soc. **37**, No 7, 1954, 323–328.

4   Prospecting and Quarrying

4.1   Introduction

The techniques used in prospecting for and extracting limestone are similar to those for many other hard rocks and have been fully described elsewhere [4.1–4.4]. This chapter, therefore, only seeks to give an appreciation of the processes involved.

Limestone is extracted by both surface quarrying and under-ground mining operations. However, economic and safety considerations result in almost all limestone being extracted by quarrying, which also recovers a higher proportion of the rock. Some examples of successful limestone mines are given in section 4.3.7.

A small tonnage of limestone is used in the production of dimension stone. The process is outlined in section 4.7.

4.2   Prospecting

4.2.1   Planning

Before looking for a suitable deposit in the field, consideration needs to be given to:

a) establishing the minimum acceptable size and quality of the deposit,
b) collecting the geological and other related information that is already available,
c) obtaining details of government regulations and restrictions on the extraction of minerals, and
d) which locations might be particularly favourable from the viewpoints of geography and infrastructure.

- *Size and quality.* The minimum acceptable size of the deposit is related to the expected maximum extraction rate and the minimum life of the site required to give an acceptable return on the investment. When estimating the maximum extraction rate, allowance should be made for losses in quarrying, stone processing and (where appropriate) lime production. Where investment is low, or where a mobile plant is used, a relatively short life might be acceptable. Where investment is high and includes fixed plant and perhaps lime kilns, a life of over 40 years is usually sought. The quality of limestone required depends on the uses to which it will be put (see chapters 7 to 12 and 14).

- *Geological.* The governments of most countries have a Geological Survey, or Mines Department, which is responsible for collecting information about the natural resources of the country. It is likely to employ geologists, who can give general information about limestone deposits. Detailed information can usually be obtained about the physical and chemical characteristics of specific deposits. Geological maps, aerial photographs and satellite images may also be available.

- *Planning permission.* Government regulations and restrictions relating to mineral developments should be assessed at an early stage as they may rule out some deposits and present severe problems with others. Conversely, grants may be available in development areas.

- *Geography and infrastructure.* To be commercially attractive, deposits should be close to the main markets or have a convenient low-cost means of transport. In general, with lime and limestone products, there are fairly well defined limits on the distance to which they may be transported economically. In many parts of the UK, for example, the economic limit for hauling limestone by road for construction projects is only 40 to 80 km. That distance can be extended to over 150 km with rail transport direct from the quarry to the customer and to even greater distances if ships or barges can be used.

Where the proposed development includes cement/lime kilns, it is generally more economic to locate the kilns at the quarry, and transport the products to a remote market, than to site the kilns close to the market, and transport the greater tonnage of limestone to the kilns.

4.2.2 Field Surveys

Because of the many potential complexities of the geology and of the variability of limestone deposits, it is necessary to employ the services of a geological consultant to carry out and interpret the field surveys [4.1].

The amount of field work required depends on a number of factors, including:

a) the information already available about the deposit(s),
b) the minimum acceptable size of the deposit,
c) the proposed capital investment associated with the development, and
d) the likely geological complexity of the deposit(s).

The objective of the field survey is to reduce the geological uncertainties to an acceptable level. Investment in this stage should, therefore, be proportional to the size of the project and the degree of uncertainty about the suitability of the deposit(s).

For the purposes of the initial assessment, the geologist needs the information necessary to assess the extent of the deposit (including the thickness and nature of the over-burden), together with an understanding of the original depositional environment and the subsequent diagenetic history. If the deposit does not contain adequate natural, or man-made exposures, some exploratory core drilling may be required to provide the necessary information. Where drilling is not feasible, digging shallow trenches perpendicular to the strike of the beds may provide useful information and samples, although the surface rock may be weathered and, therefore, unrepresentative of the under-lying deposits.

Having established that the deposit is sufficiently large and of a suitable quality, evidence should generally be obtained to establish the structure of the deposit by the use of core drilling. This should include:

a) the dip and strike of the bedding,
b) the thickness of the beds,
c) the physical and chemical qualities of the beds,
d) the extent of mineralisation,
e) the type and intensity of any folding,
f) the extent of jointing and faulting,
g) the extent of solution and weathering effects,
h) the effective level of the water table.

Recommended procedures for core drilling [4.5, 4.6] should be followed and adequate samples should be taken for assessment by standard testing procedures (chapter 6).

If the interpretation of the initial results is favourable, the geologist may recommend that additional cores be drilled to clarify the nature and extent of particular geological features (such as major faults) and/or to determine the hydrogeology of the area.

4.2.3   Interpretation of Field and Test Results

The prospecting report should give an estimate of the workable volume of the deposit, making due allowance for:

a) overburden removal,
b) the profiles of the perimeter faces,
c) sterilisation of deposits by access ramps and any permanent plant,
d) tipping of sub-standard rock, including fault material,
e) environmental, land-use and planning constraints.

The report should also recommend where to start quarrying and how to develop the quarry. It should highlight the major geological features and how they might affect the development. It should assess the physical and chemical qualities of the various beds in the deposit in the context of the proposed end-uses and indicate whether selective quarrying might be necessary. In addition, it should address the question of reinstatement when the quarry is worked out, which is likely to be a condition of planning consent.

4.3   Quarrying

4.3.1   Introduction

The quarrying process can be divided into five operations – overburden removal, drilling, blasting, loading and hauling to the processing plant.

Historically, these operations were very labour intensive. However, mechanisation, which started in the late 1940's in Europe, and is currently being applied in most of the developing countries of the world, has dramatically reduced labour requirements. This section, therefore, concentrates on mechanised quarrying practices.

4.3.2   Overburden Removal

The thickness of overburden can vary from less than 1 m to tens of metres. Indeed, if the thickness is considerable, the costs of overburden removal can make the development of an open-cast quarry uneconomic and may force the developer to consider mining (see section 4.3.7). The overburden generally consists of top-soil and sub-soil, but may also include rock overlying the limestone.

Disposal of the overburden can be a significant operation. Top-soil should be handled and stored in such a way as to preserve its fertility and permit its later use in landscaping schemes [4.7, 4.8]. Sub-soils and over-lying rock may be tipped, in which case the tips must be designed to be stable and have adequate drainage [4.9, 4.10]. In some situations, it may be possible to sell part of the overburden as in-fill.

Overburden may be removed as an on-going operation, or in campaigns. In many countries, soil is best removed in summer, when it is drier, able to bear the weight of earth-moving equipment and is in a better condition for handling and storage.

4.3.3   Drilling

The capital, operating and maintenance costs of primary fragmentation of the rock are low in relation to those of the subsequent operations to produce saleable limestone products. It is essential, therefore, to design the drilling and blasting operations to produce optimum fragmentation of the rock and a rock pile profile suitable for the loading equipment [4.11].

The pattern of drilling depends on the properties of the rock, on the geological structures (e.g. bedding planes), on the diameter of the drill holes and on the required rock pile profile. Holes may be inclined at up to 20° to the vertical (Fig. 4.1) and positioned to give controlled spacing and burden (Fig. 4.2). Inclined holes generally:

a) produce a safer, more stable face, allowing higher benches to be blasted safely,
b) reduce the incidence of "toes" (stumps of unbroken rock at the foot of the face) and
c) result in a better rock pile profile for loading [4.1].

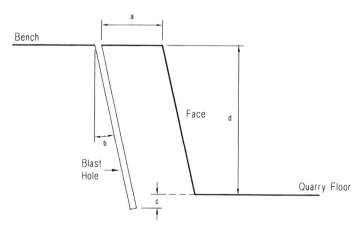

Figure 4.1. Cross-section of a blast hole
(a) burden; (b) angle of inclination; (c) sub-grade drilling; (d) height of face

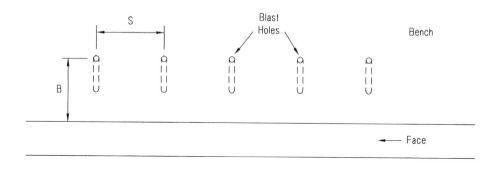

Figure 4.2. Plan of blast holes. B = burden; S = spacing

The optimum height of the quarry faces depends on a number of site-specific factors. In many countries, the regulatory authorities recommend a maximum height of 15 m for safe working.

The burden and spacing should be equal for optimum fragmentation [4.11]. Some quarries find that increasing the spacing by up to 25 % above this level gives adequate fragmentation and minimises overall quarrying costs. The burden is generally related to the diameter of the drill holes, which are typically between 100 and 150 mm. In hard rock, a burden to diameter ratio of 30 is often appropriate. Thus, with a 150 mm drill, the burden would be about 4.5 m. In softer rock, ratios of up to 40 are used [4.11].

Sub-grade drilling (i.e. drilling to below the quarry floor level), by 20 to 30 % of the burden, is general practice. It helps to ensure that the rock is adequately fractured at that level, enabling a flat quarry floor to be produced. It also helps to avoid leaving toes, which interfere with loading and have to be broken by second-ary blasting or hydraulic hammers.

During drilling, any abnormalities, such as cavities or unexpected layers of clay, should be recorded. When the holes have been completed, each should be sur-veyed to check whether it has been placed as intended, or whether there has been any deviation in angle or direction.

Most limestone quarries use "down-the-hole" rotary hammer drills, in which the hammer and the rotary drive are operated pneumatically (Fig. 4.3). Such drills give high levels of accuracy coupled with a relatively fast drilling rate. Some ope-rators use top-hammer rotary percussive drills, which are lower in capital cost than the down-the hole types, but tend to be less accurate, slower and noisier. The "drifter" type of drill, which can be set to drill at any angle, is sometimes used in smaller quarries for both primary and secondary drilling, because of its greater flexibility. In larger quarries, such drills are used for secondary drilling to remove toes and for development of access roads etc.

A recent development involves the replacement of pneumatic drives with hydraulics. This increases drilling rates and reduces energy consumption.

Figure 4.3. A drill rig (by courtesy of Halco Drilling International Ltd.)

4.3.4 Blasting

The explosives used in quarrying may be divided into two types:

a) high explosives (including TNT-based dynamites and gels), which are used to
   initiate the blast, and

b) blasting agents (such as ANFO — a mixture of ammonium nitrate and fuel oil — and slurries/emulsions — mixtures of ammonium nitrate and hydrocarbons), which provide the main explosive effort.

- *High explosives* are relatively easy to initiate and are available in water-resistant packages. Detonators are used to initiate the blast. They consist of a small amount of a sensitive explosive compound that can be fired easily and reliably by heat (which, increasingly, is produced by non-electrical initiation systems). For safety reasons, detonators are usually used in pairs. The detonators initiate a small primer charge, which in turn initiates the blasting agent. In addition, at the bottom of the drill hole, a base charge of high explosive is generally used to provide the extra energy required to ensure efficient fragmentation of the rock at and below the level of the quarry floor (Fig. 4.4).

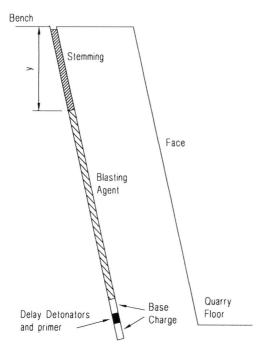

Figure 4.4. Cross-section of a charged blast hole.
y = length of stemming

- *Blasting agents* are considerably cheaper than high explosives. Their use is, therefore, maximised as far as practicable. In wet holes, however, it is necessary to replace part of the blasting agent by high explosive.

- *Delay detonators* have built-in time delays of, for example 25, 42 etc. milliseconds. Such time delays are used to initiate the blast safely, and to sequence the initiation of each hole, thereby reducing ground vibration (section 33.9.2) and permitting multi-row blasting. They can also be used to sequence the detonation within a hole in connection with "decking" (intermediate stemming within the height of a hole). This technique can be used to alter fragmentation.

The force of the explosion is contained within the hole by "stemming" the top of the hole using fine stone (6 to 4 mm has been found to give optimum containment with minimal emission of dust, although the grit/dust produced by the drilling operation has also been used). The length of the stemming is usually between 70 and 100 % of the thickness of the burden. By containing the force of the explosion, the stemming also reduces the air-overpressure (or noise level) produced by the blast (section 33.9.1).

The design of the blast includes the types, quantity and distribution of explosive placed in each drill hole. Before the design of the blast is finalised, an accurate survey of the profile of the quarry face and detailed logs of the drill holes are required to identify how the blast deviates from the target pattern so that corrections to the standard design can be made.

4.3.5 Secondary Breaking

Secondary breaking is required when boulders are produced which are too large to load or to feed into the crushing plant, and when toes are left by primary blasting. Over-sized boulders are broken by three techniques — hydraulic hammers, drop-balls and secondary blasting.

There are two types of secondary blasting. In "pop" blasting, a hole is drilled into the boulder or toe, high explosive is placed in the hole, and stemming is rammed into the drill hole to contain the force of the explosive. In "plaster" blasting, the high explosive is placed on the surface of the rock and fired. While pop blasting breaks the rock more effectively than plaster blasting and creates less noise, it does produce more "fly rock".

4.3.6 Ripping

Some limestones, notably the softer chalks, can be extracted by ripping. While ripping is not widely used, it is considerably cheaper than drilling and blasting. Various techniques are used, depending on the nature of the deposit. Some soft rocks can be extracted by hydraulically operated shovels and backhoes. Somewhat harder limestones can be broken by a ripper tooth fitted to the rear of a tractor-dozer.

4.3.7 Mining

Underground mining of limestone uses the "room and pillar" technique. As mining costs per tonne of saleable product are generally much higher than those of quarrying, there have to be special circumstances to justify operating in this way. For example:

a) the largest limestone mine in the USA produces over 3 million tonnes of a magnesian limestone [4.12], which is particularly suitable for the "MagLime" flue gas desulfurisation process (see section 29.2.5), and
b) in an Italian quarry, increasing overburden thickness necessitated a switch from quarrying to mining to avoid writing off capital equipment [4.13].

In addition, environmental considerations are understood to have been a determining factor for the development of a small number of mines rather than quarries in the USA.

While it is technically feasible to mine limestone, for the vast majority of limestone deposits there are major concerns over the risk of the mine roof collapsing as a result of weaknesses caused by faulting, jointing and bedding of the limestone.

4.4   Loading

The equipment used for excavating and loading the limestone at the quarry face needs to be selected in the context of the hauling equipment and the primary crusher, as well as site conditions.

Rope operated shovels have been used extensively for many years, but are being displaced by hydraulic loaders.

Hydraulic shovels have more digging power (higher break-out force) and faster cycle times than rope shovels. Other benefits include higher bucket filling efficiencies and lower maintenance costs (Fig. 4.5). Hydraulic backhoes can be used for excavating overburden as well as loading rock. This flexibility can be an advantage in smaller quarries.

Figure 4.5. A tracked hydraulic face shovel loading a dump truck (by courtesy of Brøyt Rockloader Ltd.)

Rubber-tyred front end loaders are widely used for loading at the quarry face (Fig. 4.6). They require well fragmented rock and a level, firm and dry quarry floor. They are highly manoeuvrable, have fast cycle times and can be used for load-and-carry operations (see section 4.6). Being highly mobile, they can switch rapidly from one duty to another (e.g. from loading dump trucks at the quarry face to loading road vehicles from stockyards). Tyre wear can be a problem when operating on abrasive rock, but this can be contained by fitting chain meshes over the tyres.

Figure 4.6. A rubber-tyred front-end loader and a dump truck (by courtesy of Volvo Construction Equipment GB Ltd.)

4.5   Hauling

Haulage is a major variable cost in producing limestone products. The most widely used vehicle for hauling rock from the quarry face to the primary crusher is the rigid-bodied dump truck, which is available in sizes ranging from 15 to over 150 tonnes capacity. Articulated dump trucks are more manoeuvrable and are often favoured in smaller quarries.

4.6   Current Trends in Quarrying

A relatively recent trend is to install in-pit crushers (Fig. 4.7). These may be fed directly from the quarry face by rubber-tyred shovels in a load-and-carry operation. The crushed rock is transported to the processing plant by conveyers, which generally have lower capital and operating costs than dump trucks.

Figure 4.7. A track-mounted in-pit crusher (by courtesy of Nordberg (UK) Ltd.)

4.7   Dimension Stone

Marble and other limestones are used in relatively small tonnages as dimension stone for ornamental purposes such as facing buildings and tombstones.

The production of dimension stone is a specialised operation. It is cut from shallow benches using either a channelling machine or a wire saw. A vertical cut, some 3 m deep and 15 to 30 m long, is made about 1 m behind the face of the bench. The base of the block is then cut, using wedges to support the cut section. The block is then sawn at convenient intervals, typically 1 m, to produce smaller blocks with dimensions of about 1 × 1 × 3 m. The smaller blocks are transported to a workshop for further cutting, or shaping, and finishing.

4.8   References

[4.1]   M.R. Smith, L. Collis, "Aggregates", The Geological Society, Bath, 1993, ISBN 0-903317-89-3.
[4.2]   B. Kennedy, "Surface mining", Seely Mudd series, AIME (Society of Mining Engineers), New York, 1990.
[4.3]   H. Hartmann, "Introductory Mining Engineering", Wiley, New York, 1987.
[4.4]   IMM, "Surface Mining", Proceedings of International Symposium, IMM, London, 1983.
[4.5]   Geological Society, "Procedures for core drilling", 1970.

[4.6]   BS 5930: "Code of practice for site investigations", 1981.
[4.7]   S. Corker, "Bulk soil handling techniques", M & Q Environment, **1**, 1987, 15–17.
[4.8]   P. Samuel, "Land restoration to agriculture", Quarry Management, Feb., 1990, 25–33.
[4.9]   Walton Practice, "Handbook on the design of tips and related structures", Department of the Environment, HMSO, London, 1990.
[4.10]   G. Garrard, G. Walton, "Guidance in the design and inspection of tips and related structures", Trans. Inst. Mining and Metallurgy, 1990, A115–A124.
[4.11]   M. Ball, "A review of blast design considerations in quarrying and opencast mining", Quarry Management, June, 1988, 35–43; July, 1988, 23–27.
[4.12]   B. Weaver, "Digging Deep for Lime", Pit and Quarry, April, 1990.
[4.13]   L. Chissale, "Transfer from Open-pit Quarrying to Underground Mining" (in Italian), Proceedings of the European Lime Association's Technical Conference, 1992, Cologne.

5   Processing and Dispatch of Limestone

5.1   Introduction

Processing and dispatch of limestone products can be divided into five main operations: crushing/grinding, sizing, beneficiation, storage/loading-out and transport. All of these operations are common to the production of crushed aggregates from hard rock and have been fully described elsewhere e.g., [5.1, 5.2]. This chapter, therefore, only seeks to give an appreciation of the processes involved.

When designing a stone processing plant, careful consideration needs to be given to the complex inter-actions between:

a) the characteristics of run-of-quarry rock,
b) the crushing, sizing and beneficiation equipment,
c) the conveying and storage equipment, and
d) the yield and characteristics of the products.

A mass balance for the plant should be determined to estimate the quantities of each size of stone that will be produced and how each size will pass through the plant. This enables estimates to be made of the quantity and size distribution of each of the process streams and helps to ensure that individual items of plant are correctly specified, particularly in terms of the throughput rate and the input/output sizes.

Consideration should also be given to providing adequate surge capacity between the quarrying operation and the primary crusher (e.g., by installing an apron feeder) and between the various stages of the processing plant. This helps to divorce the operation of the various stages, so that a stoppage in one does not immediately cause the preceding and succeeding stages to stop. Similarly, consideration should be given to establishing stocks of run-of-quarry and partially processed limestone, which can be re-introduced to the plant via feed hoppers. Such measures enable the processing plant to operate more steadily, thereby increasing its sales capacity.

Balancing the production of the various size grades of stone with the demand for those grades is an important factor in determining the economics of a quarrying and stone processing operation. Operations, supplying a closely graded stone for lime burning, may convert less than 60 % of the rock into kiln feedstone, whereas up to 98 % of the quarried limestone can be converted into saleable products in fully integrated operations. A typical quarry producing aggregates would expect to make about 70 % of saleable products. The losses arise from:

a) the quarrying operation, where stone from faulted areas is often unsuitable for processing,

b) "scalpings" (typically clay-rich material removed by "grizzly" screens prior to the primary crusher), and
c) surplus/non-saleable grades of limestone (typically fine fractions).

Because transport costs often exceed the production cost, it is important for the quarry to have good road, rail and/or water transport links and for the most appropriate and cost-effective form of transportation to be used.

A typical flowsheet for a plant producing crushed limestone is shown in Figure 5.1.

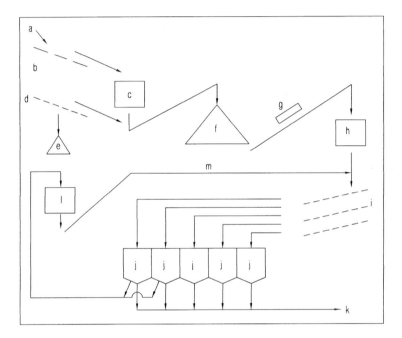

Figure 5.1. Flowsheet of a basic limestone processing plant
(a) feed from dump truck; (b) grizzly; (c) primary crusher; (d) screen; (e) rejects stockpile; (f) stockpile; (g) magnet; (h) secondary crusher; (i) multiple screens; (j) bunkers for "single size" stone; (k) loading out; (l) tertiary crusher; (m) recycle of re-crushed coarse fractions

5.2 Crushing

5.2.1 Introduction

Virtually all of the size reduction in limestone processing is done using crushers. Crushers may be divided into two categories – those using impact to break the rock and those using compression. Impact crushers produce more cubical particles than compression crushers. They are generally lower in capital cost, but require more electrical energy. They also produce more fines. Large amounts of fines can be a significant disadvantage when the principal customer is a lime-burning operation, requiring relatively large feedstone. When selecting crushing equipment, the factors which need to be considered include:

a) the properties of run-of-quarry rock,
b) the required reduction ratio,
c) the particle size distribution of the run-of-crusher product.

It is usual to carry out crushing in two or three stages, because of the limitations of the various designs of crusher, such as:

a) the reduction ratio that can be achieved and
b) the inter-actions between that ratio and the shape and particle size distribution of the product.

• *Primary crushers* [5.3] need to be capable of handling run-of-quarry rock, which may have lump sizes of up to 1 or even 2 m. The fine fraction produced by the primary crusher is generally removed by screening as it falls within the saleable size range.

• *Secondary crushers* are generally used to reduce the coarse fraction from the primary crusher into the saleable size range.

• *Tertiary crushers* are used to help balance the demand between sizes. By selecting primary and secondary crushers that produce more coarse material than is required on average, the amount of fines produced is kept to a minimum. Tertiary crushers are generally used to crush surplus quantities of larger products into the smaller sizes.

In most operations, crushers are selected with care to produce the required amounts of aggregate-sized particles, without over-production of fines (below 3 mm and including dust). Where large amounts of fines are required, impact crushers are often used. When impactors cannot produce the required fineness, it becomes necessary to grind the limestone.

5.2.2   Jaw Crushers

Jaw crushers consist of a fixed and a moving jaw (Fig. 5.2). As the moving jaw is driven towards the fixed jaw, it crushes those particles which are in contact with both, and when it is driven away, the particles move downwards. The size of the feed stone is limited by the "gape" between the upper edges of the jaws, while the size of the crushed product is determined by the "open side setting" between the lower edges of the jaws. Reduction ratios of between 4:1 and 6:1 can generally be obtained. The jaws are faced with replaceable liner plates of manganese steel.

Two mechanisms are used to drive the moving jaw. The simplest uses a "single toggle", in which an eccentric drive causes the jaw to pivot about the upper edge of the jaw. This results in movement with a small vertical component relative to the fixed jaw, which causes abrasion, but also increases throughput. The "double toggle" mechanism produces higher compressive forces in the upper part of the crushing chamber and virtually eliminates the relative vertical movement. It is

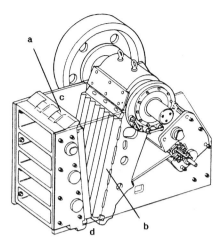

Figure 5.2. Cross-section of a jaw crusher
(by courtesy of Nordberg, UK, Ltd.)
(a) fixed jaw; (b) moving jaw; (c) gape;
(d) open-side setting

particularly suitable for crushing strong, abrasive rocks. For most limestones the single toggle mechanism gives adequate performance at a lower cost.

Jaw crushers tend to produce slabby particles, particularly when operating at high reduction ratios. Where slabbiness is an issue, it can be reduced to a certain extent by using profiled jaws and by operating at lower reduction ratios.

5.2.3 Gyratory Crushers

The gyratory crusher consists of an inverted conical crushing chamber, within which is suspended a conical "mantle" (Fig. 5.3). Both are lined with manganese steel plates. The mantle gyrates, or more strictly oscillates with a precessing motion, within the chamber. As the mantle moves towards one wall of the chamber, it crushes lumps of rock in front of it and allows lumps behind it to move downwards until trapped once again between the mantle and chamber wall. The gyratory crusher has similar characteristics to those of the single toggle jaw crusher (e.g. a reduction ratio of 4:1 to 8:1), but is more suitable for high throughputs.

5.2.4 Impactor Crushers

Impactor crushers consist of a lined crushing chamber inside which a rotor revolves on a horizontal axis. Attached to the rotor are either swing hammers or fixed blow-bars (Fig. 5.4). Stone enters the crushing chamber down the feed chute and is broken by impact with the hammers/blow-bar and with the breaking plates. Reduction ratios of up to 20:1 can be obtained. More complex crushers with double rotors can give ratios of 40:1.

Impactor crushers have many advantages over jaw and gyratory crushers — in addition to high reduction ratios, they produce a cubical particle shape, have lower capital costs, require less head-room and are more easily installed. Their

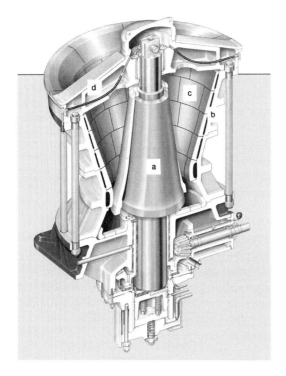

Figure 5.3. A gyratory crusher
(by courtesy of Nordberg, UK, Ltd.)
(a) mantle; (b) shell; (c) wear plates;
(d) spider

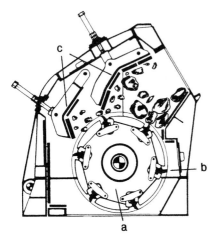

Figure 5.4. Cross-section of an impact crusher
(by courtesy of Hausherr UK Ltd. and Hazemag &
EPR GmbH)
(a) rotor; (b) blow breaker; (c) impact aprons

main disadvantages are their higher energy costs per tonne of product and the
production of more fines. Wear rates are not a problem with most limestones.

Fixed blow-bar crushers are used as primary crushers and can accept lumps up
to 1 m in size. Swing hammer mills are generally used for producing smaller sizes
and are effective down to 3 mm. When crushing hard rocks, impactor crushers are
generally operated in open circuit (i.e., without recycling of oversized lumps).

5.2.5 Cone Crushers

Cone crushers can be regarded as a development of the gyratory crusher. The mantle is similar in shape but the crushing chamber consists of an inverted cone with a steeper angle than the mantle. The mantle rotates more rapidly than in gyratory crushers with the result that breakage occurs by both compression and impact. This increases the energy-efficiency of the crushing process.

Cone crushers are increasingly being used as secondary and tertiary crushers and are frequently operated in "closed circuit" (i.e. with over-sized particles being screened out and recycled to the crusher) to produce a well-defined top size, with minimum fines. They have similar reduction ratios to gyratory crushers (4:1 up to 6:1) and produce a more slabby product than impactors.

5.2.6 Rolls Crushers

Rolls crushers consist of two counter-rotating rolls, one of which is fixed and the other is spring-loaded. They draw feed material into a "nip point", where it is crushed (Fig. 5.5). The rolls may be smooth, or profiled, which helps to draw the rock through the crusher and to reduce the slabbiness of the crushed product. Reduction ratios of only 4:1 are achievable, but the production of fines is low. The head-room requirement is lower than for impactors, but the capital cost is higher.

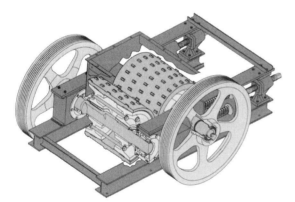

Figure 5.5. Diagram of a rolls crusher (by courtesy of Svedala, Ltd.)

5.2.7 Other Crushers

The *rotary pick "mineral sizer"* has been used as a secondary crusher for soft to moderately hard limestone. The construction is similar to that of the rolls crusher, but the rotors are fitted with alloy steel teeth, which break the rock by a combination of compression, tension and shear.

In *vertical shaft impactors* the stone is fed on to a rapidly rotating horizontal rotor, which flings the particles outwards on to hardened alloy anvils around the crushing chamber. A variant includes a shelf which increases the retention of particles within the chamber, increasing the amount of breakage caused by impact of one particle on another. These designs are particularly suitable for producing cubical particles of abrasive rock, but produce relatively large amounts of material less than 300 µm.

5.3   Pulverising and Grinding

Although most quarrying operations seek to limit the amount of fine limestone produced, there is a demand for pulverised and ground limestone.

Pulverised limestone may be produced economically in impact crushers, such as hammer and beater mills. For use in agriculture, the product should be at least 95 % less than 3.35 mm and substantially less than 600 µm (see chapter 10).

Ground limestone, or "whiting", includes a large number of products which are widely used as fillers and extenders. Particle sizes range from relatively coarse grades with 90 % less than 50 µm to fine grades with a "specific surface diameter" of less than 5 µm. Whiting was originally produced by wet milling, but is now mainly made by dry grinding and air classification. Suitable grinding equipment includes tube mills, vertical roller mills and roller presses, all of which are more efficient if operated in closed circuit with an air classifier.

5.4   Production of Precipitated Calcium Carbonate

The production of precipitated calcium carbonate (PCC) is categorised as a use of slaked lime and is described in section 31.2. However, PCC is clearly a "limestone" in terms of its chemical analysis and competes with whiting in a number of applications (the applications of both whiting and PCC are outlined in section 12.9).

5.5   Sizing

5.5.1   Introduction

Limestone may be sized for three reasons:

a) to produce accurately-sized aggregates or kiln feed,
b) to segregate coarser from finer particles as part of the crushing process, and
c) to reduce impurities, which often concentrate in finer fractions.

Two techniques are used — screening and classification. *Dry screening* is rout-inely practised at sizes down to about 3 mm and can be extended to as low as 300 μm with dry products. Screening at smaller apertures causes blinding of the screen. *Wet screening* is widely used for sizes as low as about 0.5 mm. *Classification* of suspensions in water or air is used to divide at smaller particle sizes.

5.5.2 Screening

5.5.2.1 General

Screening involves passage of particles over apertures in a screen "deck" (or "mat"), which allow particles smaller than the aperture size to pass through the deck, thereby dividing the product into undersize and oversize fractions [5.4–5.6]. The apertures may be square, rectangular, or circular.

For a particle to pass through a screen mesh deck, the two smaller orthogonal dimensions of the particle (b and t cm – see Fig. 5.6) must be less than the width/length of the aperture (x × y cm). The maximum dimension of the particle (l cm) can, however, be greater than the aperture size: in practice, it can often approach double the maximum dimension of the aperture. Particles which are much smaller than the aperture size have a high probability of passing through the screen. As the particle size approaches that of the aperture, the probability of passing decreases. The efficiency of a screen deck increases with its area [5.7]. In practice, its is generally acceptable for 95 % of the true undersize to pass through a deck.

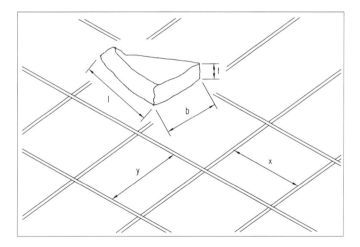

Figure 5.6. Relationship between particle dimensions and screen aperture
(l) length; (b) breadth; (t) thickness; (x) and (y) screen aperture

The process of screening can be adversely affected by "pegging", "blinding" and the appearance of "holes". *Pegging* (Fig. 5.7) occurs when a particle, with a size close to that of the aperture, becomes wedged in an aperture and reduces the effective area of the deck. *Blinding* generally arises from moisture in the product causing particles to adhere to the screen surface and to each other. Eventually the build-up closes the apertures either partially, or completely, thereby reducing screen efficiency. *Holes* arise from wear and from damage and allow some over-sized particles to pass into the fine fraction.

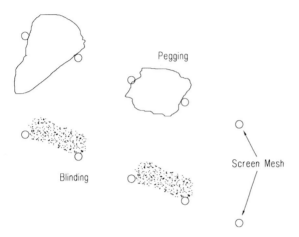

Figure 5.7. Illustration of pegging and blinding of a screen deck

5.5.2.2   Screen Deck Materials

The screen deck may be made from *woven or welded steel wire mesh* with square apertures. Such decks have a high proportion of open area and are relatively inexpensive. However, they have a low resistance to wear and corrosion and produce high noise levels when vibrated. Woven wire decks with rectangular apertures are sometimes used to reduce pegging.

Punched steel plates are very robust, but have a low proportion of open area. They are often used for screening larger sizes of stone and in trommel screens (see section 5.5.2.3), where robustness is important. Increasingly, *moulded decks* made from rubber or polymers (e.g. polyurethane) are used because of their resistance to wear and corrosion, and their surface properties which reduce blinding and noise emission. Their disadvantages are higher initial cost and a lower proportion of open area.

Other screening surfaces include *"rod decks"*, which reduce pegging and blinding, and *"piano-wire"* screens, which are less prone to blind when screening fine, moist products. An alternative approach for moist materials, which are prone to blinding, is to use *heated decks*. These use stainless steel (which has a higher electrical resistance than mild steel), woven into a wire mesh, through which an electrical current is passed.

5.5.2.3 Types of Screen

The most widely used design is the *inclined vibrating* screen. These use decks at some 15° to 20° to the horizontal mounted on a vibrating frame to transport materials across and through the screen. Such screens can be used in series to produce a number of sized fractions. Alternatively, *multi-deck* screens can be used, in which two, three, or even four decks are mounted on one frame, with the largest aperture mat being at the top and the smallest at the bottom of the frame (Fig. 5.8). With all designs, the material should be fed uniformly across the width of the deck at a steady rate to maximise the effectiveness of the screen.

Figure 5.8. A triple-deck screen with dust-proof enclosure (separating into four sizes) (by courtesy of Tema Isenmann Ltd.)

Several mechanisms are used to vibrate inclined screens, including rotation of eccentric shafts and unbalanced weights and the use of electromagnetic vibrators. The vibration may be circular, elliptical or linear and its direction can be selected either to throw the material forward, to increase throughput rate, or reversed to throw it backwards, thereby reducing its velocity and increasing the screening efficiency.

Horizontal vibrating screens require less headroom than the inclined type, but, because they do not utilise gravity, require higher levels of vibration and need to be more robustly built. The mats also need to be more resistant to abrasion. They are generally actuated by linear vibration inclined at 45° towards the discharge of the screen to help transport the material.

Vibrating screens are particularly prone to blinding when moisture is present. With stone below 10 mm, one solution is to use *wet screening*, in which water is sprayed on to the mat so that the fine fraction is washed through as a slurry.

Trommel screens are rotating, inclined cylinders, the walls of which are generally made of punched plate. Stone is fed to the upper end of the cylinder and is transported by the rotation of the cylinder over the apertures. Plates with progressively larger apertures may be fitted to produce a number of size fractions. Over-sized particles are discharged from the lower end of the screen. Trommel screens are robust and relatively inexpensive. They are not prone to pegging or blinding and encourage dispersion of agglomerates. Their disadvantages, relative to other designs of screen, are that they are larger and require more headroom, have a lower screening efficiency, and a higher energy consumption. Increasingly, however, they are used as washers to remove clay and surface contamination (see section 5.6.3).

Grizzly screens are robust devices for separating fines from coarse stone. Two designs are used. One consists of fixed bars inclined at an angle of about 50°. The other, described as the "live roll" grizzly, uses rotating rollers, which may be shaped to create apertures, or may be elliptical, to convey the stone, while removing fines. Grizzlies are often placed before the primary crusher to remove the smaller lumps of stone and clay from run-of-quarry stone, which do not need to be crushed (see Fig. 5.1).

Ball-deck screens are used where pegging is a problem. Elastomer balls are trapped below the upper deck by a coarser lower deck which causes the balls to impinge on the lower surface of the upper deck, thereby displacing pegged particles.

Probability screens are used to overcome blinding. The type designed by Mogensen uses steeply-inclined vibrating screens with rectangular mesh mats, which may be heated. The angle of inclination results in the cut size being smaller than the greater dimension of the apertures. This enables damp material to be screened at (say) 3 mm, using mats with apertures of 3 mm × 6 mm. The "Ropro" design uses stainless steel spokes mounted radially on a rotating vertical shaft. The feed material falls through the spokes. Coarse particles are driven outwards to the oversize chute, while fines fall through the spokes into a second chute.

5.5.3 Classification

Suspensions of particles in water (typically less than 2 mm), produced in limestone scrubbing plants, can be classified to produce saleable products such as coarse sand, fine sand and a clay-rich suspension of fines. Classifiers may be divided into two groups — one uses gravity to separate coarse from fine particles, while the other uses centripetal forces.

There are many designs of *gravity classifier*, each with its own characteristics [5.1] — for example, one withdraws sand from an inclined trough using a screw, which ensures a degree of de-watering, another includes a "teeter zone", which ensures a sharp cut size. In general, gravity classifiers are not suitable for separation at sizes below 75 µm.

Centrifugal classifiers, of which the best known design is the hydrocyclone [5.8], use a pump to provide the motive force. They are capable of sizing particles at as low as 5 µm. *Air classifiers* are widely used in dry grinding plants to control the degree of fineness of the product. Particles which are larger than the cut size are returned to the mill, while finer particles are removed in cyclones and/or bag filters. In many classifiers, the cut size may be adjusted by changing the degree of swirl within the classifier.

5.6   Benefication

5.6.1   Introduction

Beneficiation refers to improving the physical and/or chemical qualities of lime-stone products. In practice, three processes are used:

a) scalping and screening,
b) washing and scrubbing and
c) sorting.

5.6.2   Scalping and Screening

The use of grizzly screens to produce "scalpings" rich in unwanted clay has already been mentioned (section 5.5.2.3). The decision whether to wash scalpings depends on site specific factors, such as the quantity of scalpings, the availability of water, whether the washings can be processed in slurry ponds, and the demand for washed and un-washed scalpings.

Other impurities, such as lead and other heavy metals, are generally found to concentrate in the finer fractions and can be reduced by screening.

5.6.3   Washing and Scrubbing

Most limestone quarries do not find it necessary to wash their stone. Some which produce limestone aggregates may wash the smaller grades below 10 mm to pro-duce limestone sand. A significant proportion of those producing chemical qual-ity limestone for lime kilns and other critical uses, however, find that washing coarser grades is necessary to achieve the required purity. Various devices are used, including rotary barrel washers, trommel screens and screens fitted with high pressure sprays.

5.6.4   Sorting

Many uses of aggregates require "cubical" particle shapes (see sections 8.3.3 and 8.5.4). Products which are excessively flaky may be brought within specification by the use of de-flaking screens. These may have transverse, elongated or slotted apertures, and may incorporate weirs or baffles to improve the presentation of slabby particles. In some cases, such screens simply remove a size fraction which has been found to be particularly flaky.

Mineral sorting devices scan individual pieces using a variety of sensors. The inspection system collects information which is analyzed with respect to the selec-ted parameter and actuates a rejection device, such as an air jet or a mechanical deflector [5.2]. Such devices may be used, for example, to remove flints from chalk, but, because of the need to present pieces individually, they are more economical for larger lumps.

5.7   Storage and Loading Out

Storage of run-of quarry and partially processed limestone is necessary to provide surge capacity between the various processing stages. Storage of fully processed limestone is also required to service despatches, to provide a strategic reserve and to store surplus grades.

Limestone is stored in elevated silos (bunkers, or bins) and also in stockpiles. Short-term storage, typically sufficient to service one day's production, is often provided in silos, which are filled directly from the processing plant, and are used to load road and rail vehicles. Reserve stocks and surplus quantities of positively screened grades are transported to uncovered stockpiles.

As screened limestone is an inert, free-flowing material, there are few restrictions on the materials of construction used for silos and hoppers. A limited quantity of such storage is often provided within the stone processing plant to cope with day-to-day requirements and to reduce the amount of double handling, which is costly and causes breakage. Screened limestone handles well through bunkers with valley angles of not less than 60°. When the product contains significant quantities of fine material, and particularly if it is wet, its angle of repose can increase dramatically and suitable designs of bunkers need to be used.

However, the greatest tonnage of limestone is stored on the ground in uncovered stockpiles. Most stockpiles are fed by an elevated conveyer, resulting in a conical shape. Radial elevated conveyers can service three or four stockpiles, each consisting of a different grade of stone. Alternatively, stone can be transported to the stockpile by dump truck or shovel.

Feeding limestone to a stockpile can cause two problems – breakage of stone as it falls on to the pile and wind-whip of dust from the falling stream of stone. Various devices, including conveyers that can be elevated as the height of the pile increases, "trippers" which move along a gantry, forming a prismatic pile, and anti-breakage chutes, have been developed to minimise the free-fall distance and wind-whip. The stone may also be conditioned with water (with or without a wetting agent) to reduce dust emission.

Limestone may be reclaimed from stockpiles by a variety of mechanisms, including conveyers in tunnels below the stockpile, scraper reclaimers and mobile shovels.

Increasingly, the choice of which grades can be stockpiled is influenced by the need to control dust emission arising from wind-whip. "Best practice", as defined in [5.9] requires that products containing significant amounts of material less than 3 mm are kept in covered storage for normal day-to-day use, with certain exceptions such as sand, scalpings and specified materials which have been conditioned with water before deposition. It also requires that all reception hoppers are partially enclosed within a structure consisting of "at least three walls and a roof". Fuller details on environmental protection measures are given in chapter 33.

As many aggregate specifications require specific particle size distributions, processing plants generally screen the stone into "single sized" fractions with a size ratio of approximately 1.4 (e.g. 20 to 28 mm, 14 to 20 mm and 10 to 14 mm) and store them separately. The fractions are then blended, using sophisticated weighing-out equipment, to meet the size distributions specified by the customers.

5.8 Transport

As most limestone products are relatively inexpensive, transport costs can often amount to over 50 % of the delivered price. It is, therefore, essential to select the most cost-effective transport option to keep the delivered price as competitive as possible.

In many countries, well over half of the tonnage of screened grades of limestone is delivered from the quarry to local users by tipper truck. This option has the advantage of being relatively cheap and very flexible. Control of dust emission during transportation is achieved by sheeting the load.

Despatches by rail can be cheaper per tonne of product and more acceptable from an environmental viewpoint. They require suitable rail reception facilities and are particularly suitable for supplying distribution depots and large-offtake, long-term customers. In general, rail becomes more competitive relative to road transport as the distance from the quarry to the customer increases.

Those quarries fortunate enough to have good access to a navigable river, or to the sea, can barge or ship their products over considerable distances, at a low cost per tonne, to major customers and depots with suitable reception facilities.

Powdered limestone products are generally transported by air pressure discharge vehicles. Whiting and precipitated calcium carbonate are also dispatched in bags.

5.9 Specifications

Details of specific requirements relating to particular applications are given in chapters 7 to 12 and 14. In general terms, limestone products should have the required particle size distribution and, for most applications, be free of excessive surface contamination with clay or other fine particles. For use as an aggregate, it is essential that the strength and durability are adequate. Particle shape may also be a factor. The chemical analysis is important for certain applications.

5.10   References

[5.1]   M.R. Smith, L. Collis, "Aggregates", The Geological Society, Bath, 1993.
[5.2]   H.E. Cohen, "Mineral Sorting", Ullmann's Encyclopedia of Industrial Chemistry, 5th ed. on CD-ROM, Wiley-VCH, 1997.
[5.3]   F.W. McQuiston, R.S. Shoemaker, "Primary crushing plant design", Society of Mining Engineers (AIME), New York, 1978.
[5.4]   A.C. Partridge, "Principles of screening", Mine and Quarry, Dec., 1977, 33–38.
[5.5]   J. Freebury, "Practice of screening", Mine and Quarry, June, 1978, 56–62.
[5.6]   G.J. Brown, "Principles and practice of crushing and screening", Quarry Managers' Journal, March–September, 1963 (7 articles).
[5.7]   A. Jowett, "Assessment of screening efficiency formulae", Colliery Guardian, Oct., 1963, 423–427.
[5.8]   L. Svarovsky, "Solid-Liquid Separation", Butterworths, London, 1981.
[5.9]   "Secretary of State's Guidance — Quarry processes", PG3/8(96), May, 1996, HMSO, ISBN 0-11-753279-7.

6 Sampling and Testing of Limestone

6.1 Introduction

Sampling of limestone may be done for a variety of reasons, ranging from the assessment of a deposit and the evaluation of a product, to quality control and monitoring compliance with a specification. Testing may be of the mechanical, physical, thermal, weathering, chemical, or general characteristics of the limestone.

Because most limestone is used locally, many countries have their own standards for sampling, sample preparation and testing. This chapter refers mainly to European (CEN) and British Standards, to illustrate the principles and procedures involved. Where appropriate, the reader should adopt the requirements of the relevant national or international standards.

A more comprehensive treatment of sampling and testing of aggregates is given in [6.1].

6.2 Sampling

6.2.1 General

It is important to understand the principles of sampling and be aware of the precautions that are necessary to obtain a representative laboratory sample. Definitions of the terms increment, sample, spot sample, composite sample and laboratory sample are given in the Glossary. Figure 6.1 illustrates how they are related.

BS 5309 [6.2], describes the general principles that should be observed when designing sampling procedures. They include:

a) adopting a proven procedure for taking increments,
b) ensuring that each increment is of an appropriate quantity, and
c) taking an appropriate number of increments.

Close attention to each of these points is important, as inadequate sampling practices are a frequent cause of invalid results and discrepancies between laboratories. The sampling procedures adopted will also depend on the required accuracy and precision of the result.

For the average practitioner in the limestone industry, the above reference will be more theoretical and complicated than is required. However, guidance is given in a number of other standards, which is based on practical experience and which

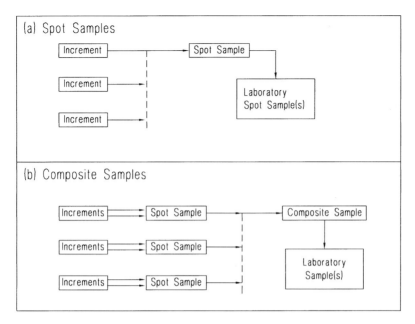

Figure 6.1. Relationships between increments and the various types of sample

has been found to give the accuracy required by the limestone industry and its customers.

Before designing a sampling procedure, the objective should be clearly defined. For example, different procedures would be required for obtaining:

a) a spot sample for the purposes of process control,
b) a spot sample to establish whether the contents of a train-load of product are within specification, and
c) a composite sample to establish the average quality of a product over a period of a week.

The guidance on sampling, specified in the various standards cited, should be adequate for a wide range of situations. In a specific situation, adequate accuracy and precision may be obtained by adopting simplified procedures. Such procedures should, however, be proved by comparison with the standard methods before being adopted.

6.2.2   BS 812

BS 812 [6.3], Part 101 gives guidance on the sampling of aggregates coarser than 75 μm and on the applicability of the standard as a whole. Material finer than 75 μm should be sampled in accordance with EN 196-7 [6.9].

BS 812, Part 102 describes how to take samples suitable for testing in accordance with the other parts of BS 812. Table 6.1 summarises the specified minimum number and size of increments, and the minimum mass of a spot sample representing a batch of material (see also EN 932).

Table 6.1: BS 812 — Minimum size and number of increments

Nominal size of aggregate (mm)	Minimum number of increments	Minimum volume of increment (l)	Approx. minimum mass of sample (kg)
> 28	20	2	50
5 to 28	10	2	25
< 5	10	1	10

A procedure for sampling from stockpiles is described in BS 812 [6.3]. Sampling from stockpiles should, however, only be adopted as a last resort, as experience shows that the precision of the results so obtained is generally much lower than those of alternative procedures.

6.2.3   BS 6463

BS 6463 [6.5], Part 101, applies to natural calcium carbonate (and also to quicklime and hydrated lime). It follows similar principles to BS 812, but is of more general applicability. It includes guidance on sampling from conveyers, bulk containers, packages, and silos. It also refers to the use of mechanical samplers.

The minimum size and number of increments of granular limestone, required to produce representative spot and composite samples, are specified in BS 6463 and are reproduced in Table 6.2.

The procedure adopted by BS 6463 for sampling powdered limestone is that specified in BS EN 459-2 [6.6]. It requires a sufficient number of increments to be taken to produce a spot sample of 20 ± 5 kg.

Table 6.2. BS 6463 — Minimum number and size of increments

Maximum particle size (mm)	Number of increments	Minimum volume of increment (l)
63	10	5
50	10	4.5
37.5	10	4
28	10	3.5
20	10	3
14	10	2.5
10	10	2
6.3	10	1.5
3.35	10	1

6.3   Sample Preparation

BS 812 [6.3], Part 103, and 6463 [6.5], Part 101, specify procedures for preparing laboratory samples, involving the combination of increments and spot samples, drying, blending, reducing, dividing, crushing and grinding. They incorporate a number of precautions which have been found to be necessary to obtain valid samples.

6.4   Testing

Test methods for limestone aggregates and natural calcium carbonates are specified in the relevant parts of BS 812 [6.3] and BS 6463 [6.5]. The former is being replaced by CEN Standards that are in preparation (see section 6.5). BS 6463, Part 102 (also in preparation) refers to CEN test methods, where appropriate (see below).

The following tests methods are widely used to characterise aggregates and calcium carbonates.

6.4.1   Physical Testing — Aggregates

– Particle size distribution: BS 812 [6.3], Part 103
– Particle shape: BS 812 [6.3], Part 105
– Density: BS 812 [6.3], Part 2
– Moisture content: BS 812 [6.3], Part 109
– Compactability: BS 5835 [6.7], Part 1

6.4.2   Physical Testing (Whiting and PCC)

– Particle size distriribution: BS 3406 [6.4], Part 2
– Brightness: DIN 5033 [6.11], Part 9; DIN 53163 [6.12]

6.4.3   Mechanical Testing — Aggregates

• Strength
– Aggregate Crushing Value (ACV): EN 1097-2 [6.13]
– Ten Percent Fines Value (TFV): BS 812 [6.3], Part 111
– Aggregate Impact Value (AIV): BS 812 [6.3], Part 112
– Los Angeles Test: ASTM C131 [6.8]

• Durability
– Aggregate Abrasion Value (AAV): BS 812 [6.3], Part 113
– Polished Stone Value: EN 1097-2 [6.13]

− Drying Shrinkage: EN 1367-4 [6.14]
− Soundness: EN 1367-2 [6.14]
− Frost-heave: BS 812 [6.3], Part 124

6.4.4   Chemical Tests — Aggregates

− Chloride content: EN 1744-1 [6.15]
− Sulfate content: EN 1744-1 [6.15]
− Acid-soluble material: EN 1744-1 [6.15]
− Alkali-silica reactivity: EN 1744-2 [6.15]

6.4.5   Chemical Tests — Natural Calcium Carbonates

− Loss on ignition: EN 196 [6.9], Part 2
− Silicon, iron, aluminium (traditional methods) : EN 196 [6.9], Part 2
− Aluminium, iron, magnesium, manganese, silicon (by flame atomic absorption spectrometry) EN 12485 [6.10]
− Calcium, magnesium, carbonate: EN 12485 [6.10]
− Cadmium, chromium, lead, nickel: EN 12485 [6.10]
− Antimony, arsenic, selenium, mercury: EN 12485 [6.10]
− Fluoride: BS 6463 [6.5], Part 102
− Neutralising value: BS 6463 [6.5], Part 2
− Total sulfur: BS 6463 [6.5], Part 2
− Copper, silver, tin and zinc: BS 6463 [6.5], Part 102

6.5   CEN Standards for Test Methods

• EN 932: Tests for General Properties of Aggregates (in preparation)
   Part 1   Methods for sampling
   Part 2   Methods for reducing laboratory samples
   Part 3   Procedure and terminology for simplified petrographic description
   Part 4   Methods for description and petrography – Quantitative and qualitative procedures
   Part 5   Common equipment and calibration
   Part 6   Definitions of repeatability and reproducibility
   Part 7   Conformity criteria for test results

• EN 993: Tests for Geometrical Properties of Aggregates (in preparation)
   Part 1   Determination of particle size — Sieving method
   Part 2   Determination of particle size — Test sieves, nominal size of apertures
   Part 3   Determination of particle shape of aggregates — Flakiness index
   Part 4   Determination of particle shape of aggregates — Shape index

Part 5 Assessment of surface characteristics – Percentage of crushed or broken surfaces in coarse aggregates

Part 6 Determination of shape/texture — Flow coefficient for coarse aggregates

Part 7 Determination of shell content — Percentage of shells for coarse
aggregates

Part 8 Assessment of fines — Sand equivalent method

Part 9 Assessment of fines — Methylene blue test

Part 10 Determination of fines — Grading of fillers (air jet sieving)

- EN 1097: Tests for Mechanical and Physical Properties of Aggregates (in prep-
  aration)
  Part 1 Determination of the resistance to wear (micro-Deval)
  Part 2 Methods for the determination of resistance to fragmentation
  Part 3 Determination of loose bulk density and voids
  Part 4 Determination of the voids of dry compacted filler
  Part 5 Determination of water content by drying in a ventilated oven
  Part 6 Determination of particle density and water absorption
  Part 7 Determination of the particle density of filler — Pycnometer method
  Part 8 Determination of the polished stone value
  Part 9 Determination of the resistance to wear by abrasion from studded
          tyres: Nordic test
  Part 10 Water suction height

- EN 1367: Tests for Thermal and Weathering Properties of Aggregates
  (in preparation)
  Part 1 Determination of resistance to freezing and thawing
  Part 2 Magnesium sulfate test
  Part 3 Determination of volume stability (Sonnenbrand)
  Part 4 Determination of drying shrinkage
  Part 5 Determination of resistance to heat

- EN 1744: Tests for Chemical Properties of Aggregates (in preparation)
  Part 1 Chemical analysis:
          (a) chloride,
          (b) acid soluble sulfate,
          (c) impurities affecting setting and hardening of cement,
          (d) impurities that affect surface finish,
          (e) water solubility,
          (f) loss on ignition,
          (g) slag unsoundness,
          (h) free lime,
          (i) fulvo acid test,
          (j) sodium hydroxide test
  Part 2 Determination of resistance to alkali reaction
  Part 3 Water susceptibility of fillers

- prEN XXX: Tests for (Bituminous Bound Fillers) Filler Aggregate (in preparation)
  Part 1   Delta ring and ball test
  Part 2   Bitumen number

6.6   References

[6.1]   M.R. Smith, L. Collis, "Aggregates", The Geological Society, Bath, 1993, ISBN 0-903317-89-3.

[6.2]   BS 5309: "Methods for sampling chemical products", Part 1: "Introduction and general principles", 1976; Part 4: "Sampling of solids", 1976.

[6.3]   BS 812: "Testing aggregates".
Part 2: "Methods for determination of density", 1995.
Part 100: "General requirements for apparatus and calibration", 1990.
Part 101: "Guide to sampling and testing aggregates", 1984.
Part 103: "Methods for determination of particle size distribution", 1985/89.
Part 105: "Methods for determination of particle size distribution", 1989/90.
Part 111: "Methods for determination of ten per cent fines value (TFV)", 1990.
Part 112: "Method for determination of aggregate impact value (AIV)", 1990.
Part 113: "Method for determination of aggregate abrasion value (AAV)", 1990.
Part 124: "Method for determination of frost-heave", 1989.

[6.4]   BS 3406: "Methods for determination of particle size distribution".
Part 2: "Recommendations for gravitational liquid sedimentation methods for powders and suspensions", 1984.

[6.5]   BS 6463: "Quicklime, hydrated lime and natural calcium carbonate".
Part 101: "Methods for preparing samples for testing", 1996.
Part 102: "Methods for chemical analysis" (in preparation).
Part 103: "Methods for physical testing" (in preparation).

[6.6]   BS EN 459-2: "Building lime — Test methods", 1995.

[6.7]   BS 5835, Part 1: "Compactability test for graded aggregates", 1980.

[6.8]   ASTM C131: "Test for resistance to abrasion of small size coarse aggregate by use of the Los Angeles machine", 1989.

[6.9]   BS EN 196: "Methods of testing cement".
Part 2: "Chemical analysis of cement", 1993.
Part 7: "Methods of taking and preparing samples of cement", 1992.

[6.10]   EN 12485: "Chemicals used for treatment of water intended for human consumption — calcium carbonate, high-calcium lime and half-burnt dolomite — Test methods" (in preparation).

[6.11]   DIN 5033, Part 9: "Colorimetry: reflectance standard for colorimetry and photometry", 1982.

[6.12]   DIN 53163: "Testing of pigments and extenders: Determination of lightness of extenders and white pigments in powder form", 1988.

[6.13]   EN 1097: "Tests for mechanical and physical properties of aggregates".

[6.14]   EN 1367: "Tests for thermal and weathering properties of aggregates".

[6.15]   EN 1744: "Tests for chemical properties of aggregates".

Part 2 Uses and Specifications of Limestone

7 Overview and Economic Aspects of the Limestone Market

7.1 General

The size and relative importance of the various market segments for limestone (including products based on dolomite) vary widely from one country to another. They depend on many factors, including the degree of industrialisation in general, on the specific industries which have become established, on the quality/availability of limestone, and on traditional building methods.

The economic factors which shape the limestone market are common to most countries. However, specific factors such as its availability in relation to competitive hard rocks, labour costs, availability of capital, the size of the market in both geographical and tonnage terms and the quality of the infrastructure (transport and communication) can markedly affect the structure of the market.

This chapter aims to describe the main features to be considered when assessing a particular market. It uses information from individual countries to illustrate some of these features. It also provides a context for chapters 7 to 12 and 14 on the uses of limestone. Table 7.1 lists the applications mentioned in those chapters.

Table 7.1. Uses of limestone

Section	Application	Section	Application
8.3	Aggregate – concrete	11.3	Smelting of ores copper, lead, zinc, antimony
8.5	– roads unbound		
8.6	– asphalt	11.4	Alumina extraction from bauxite
8.7.2	– rail track	12.2	Glass
8.7.1	– drainage/filtration	12.3	Ceramics
8.4	Sand for mortars	12.4	Mineral wool
8.7.3	Armourstone	12.5	Flue gas desulfurisation
9.2	Cement – ordinary Portland		Hydrogen fluoride absorption
9.3	– composite	12.6	Pulp (sulfite process)
9.4	– masonry	12.7	Organic chemicals
9.5	– calcium aluminate	12.8	Rock dust for mines
10.2	Arable land and pasture	12.9	Whiting/fillers
10.3	Fertilisers	12.10.1	Water treatment
10.4	Animal feedstuffs	12.10.2	Sewage filtration
10.5	Poultry grits	12.10.3	Effluent neutralisation
10.6	Neutralising acid rainfall	12.11	Sodium dichromate
11.1	Iron	12.12	Calcium zirconate
11.2	Steel	14.2	Limeburning

7.2   Market Overview

Limestone is a major mineral product. Estimating the level of world-wide use presents problems, as the statistics for some major countries are not published and as different bases are used in various countries. The author estimates that it is over 4 500 million tonnes per year (tpa). Of that total, The United States uses about 870 million tpa, Japan about 208 million tpa and the United Kingdom about 120 million tpa. Table 7.2 summarises available information for various countries [7.1–7.4].

Table 7.3 summarises the proportions of limestone used in the main market segments in Germany, Japan and the United Kingdom [7.5, 7.6].

- *Construction and Building* (chapter 8). In most countries, the major use of limestone is as a crushed rock for use in construction, mainly as an aggregate and filler in concrete and roadstone. Its physical properties are important; the products should be clean, have a good crushing strength and a cubical, rather than slabby, particle shape. Limestone competes with other hard rock aggregates in this segment. Some limestone sand is also used in concrete and mortar. Limestone can be cut and used as dimension stone, but other rocks are generally preferred. Un-dressed limestone is used to a very limited extent in building.

- *Cement* (chapter 9). Generally the second largest use of low-magnesium limestone is in the manufacture of cement. In this segment, a wide range of both physical and chemical properties can be accepted, providing the magnesium carbonate does not exceed 5 % of the calcium carbonate content (higher levels of magnesium cause expansion after setting and reduced strengths). To keep raw material costs as low as possible, the limestone quarry should be close to suitable deposits of the other major raw materials.

- *Agriculture* (chapter 10). Ground limestone is widely used to raise the pH of acid soils. In most countries, it has largely replaced waste quicklime for this purpose. It is also used in fertilisers and animal foodstuffs.

- *Metal Refining* (chapter 11). Considerable quantities of limestone are used in blast furnaces to remove some of the impurities as a molten slag. A typical usage rate is 200 kg/t pig iron. Limestone can be used to form slags in steel-making (e.g., in the obsolescent Open Hearth process and to a limited extent in modern processes), but the heat absorbed in decomposing it to lime generally limits its use. It is widely used as a fluxing agent in the smelting of metals, and in the extraction of bauxite in the production of aluminium.

- *Quicklime Production* (section 14.2). This is a significant use of limestone in industrialised countries.

- *Glass* (section 12.2). Limestone is one of the main components of glass.

Table 7.2. Estimated extraction of limestone in various countries (1994)

Country/region	Estimated total excluding use for cement 10^6 tpa	Estimated use for cement 10^6 tpa[a]	Total including cement 10^6 tpa
Africa		50	
America – USA	800	70	870
– Other		110	
Austria	12	4	16
Belgium	24	8	32
China		420	
Czech Rep.	2		
Denmark	1	2	3
Germany	33	33	66
Finland	1	1	2
France		21	
Greece		14	
India		56	
Ireland		2	
Italy		34	
Japan	108	100	208
Luxembourg		1	
Mexico		28	
Netherlands		3	
New Zealand	3	2	5
Norway	1	1	2
Poland	10		
Portugal		8	
Sweden	2	3	5
Slovak Rep.	4	28	32
Spain	2	28	30
South Africa	1		
South Korea		50	
Thailand		28	
Turkey	28	31	59
UK	110	10	120
Former USSR		55	
Estimated World Total	3,080	1,420	4,500

[a] The "estimated use for cement" is an approximate value, derived from Cembureau data for 1995, on an assumed conversion rate of 1.0 tonne of calcium carbonate per tonne of clinker, and on ILA estimates (1994).

Table 7.3. Sales of limestone to the main market segments

Market segment	Germany 10^6 tpa	%	Japan 10^6 tpa	%	U.K. 10^6 tpa	%
Construction and building						
Building materials [a]	5.6	8.5	34.2	16.4	13.2	11.0
Building construction	1.6	2.4			39.1	32.5
Roads & below grade	16.0	24.4	27.3	13.1	47.0	39.0
Cement	33.3	51.1	100.5	48.3	10.0	8.3
Agriculture	1.4	2.1	1.3	0.6	2.2	1.8
Iron and steel	3.1	4.7	22.5	10.8	2.7	2.2
Environmental prot'n [b]	1.3	2.0	3.8	1.8	0	0
Other	3.1	4.7	18.5	8.9	6.3	5.2
Total	65.4		208.1		120.5	

[a] Excluding cement.
[b] The major use in environmental protection is flue gas desulfurisation.
In 1994, the U.K. had not commissioned any major units.

- *Gaseous Effluents* (section 12.5). Increasingly, significant quantities of limestone are being used in large-scale flue gas desulfurisation plants. The by-product, gypsum, can be sold although the quantities could exceed the local demand.

7.3   Economic Aspects

Limestone competes with other hard rocks in the major market of aggregate for concrete and roadstone. In most countries, such rocks are widespread and there is intense competition between suppliers, based more on the delivered prices than on the properties of the aggregate.

The *production cost* of limestone depends on a number of factors. The nature of the deposit can be important; massive deposits with little overburden, horizontal strata and consistent physical/chemical properties favour low extraction costs, particularly if linked with a large-scale operation. Selection of appropriate equipment to keep the combined costs of labour, capital charges and other operating costs to a minimum is important to ensure a strong competitive position (see chapters 4 and 5).

The *cost of haulage* is an important factor, as it can equal the ex-works cost of aggregate at a distance of 50 km. For road transport by tipper wagon, the economic distance may be as low as 40 to 60 km. Rail transport may extend that distance to 200 km and water transport can extend the distance considerably further.

Wastage can markedly affect the economics of a quarry operation. It can be reduced by having a balanced portfolio of customers, thereby obtaining sales for most of the grades of limestone produced. Wastage can range from as low as 2 % in some integrated quarries, to a typical value of 20 to 30 % and to 40 to 50 % in quarries supplying only lime kilns.

Ex-works prices for limestone used as aggregate in construction in the UK are in the range £2 to £5/t (1997). Chemical quality limestone for lime-burning, glass manufacture and flue gas desulfurisation commands prices in the range £5 to £10/t. Specialist ground limestone products, such as whiting, can command prices of £25 to £250/t, depending on grade and quality. Precipitated calcium carbonate (PCC) is a specialist product, which commands prices of £250 to £1,000/t (N.B. the prices of limestone products vary considerably from one country to another). While the values quoted above will rapidly become out-of-date, the relative levels are likely to remain valid.

7.4 References

[7.1]   "Crushed Stone", US Bureau of Mines' Mineral Industry Survey, Annual Review, 1994.
[7.2]   "BAT Note — Cement Industry", Dutch Ministry of Environment, Department of Air and Energy, 1997.
[7.3]   "BAT Reference Document — Cement", Cembureau, June 1997.
[7.4]   European Minerals Yearbook, 1994, 1995, 1996.
[7.5]   International Lime Association — Statistics, 1994.
[7.6]   Statistical Year Book, Quarry Products Association (formerly BACMI), London, 1995.

8 Construction and Building

8.1 Introduction

In many countries, the largest market segment for the sale of limestone is as an aggregate in the construction and building industry. The major uses are in concrete and roadstone (both bound and unbound). Other applications include sand for mortar, rip rap, armourstone for sea defence works, land fill, filter media, pebble dash and roofing gravel. Finely divided limestone is used as an inexpensive filler for asphalt concrete.

Aggregates must be hard, durable and clean. They must also meet additional requirements relating to the intended use. Hard, dense limestones are suitable for most applications, while soft porous chalks are generally not sufficiently strong or durable for use as aggregate.

Limestone aggregates compete with crushed igneous rocks, other sedimentary rocks (e.g. sandstones), sand, gravel and "secondary" aggregates (which include artificial aggregates, industrial by-products and waste materials). While technical factors such as particle shape, strength and durability are important, the delivered cost is generally a major factor affecting the choice of aggregate.

The following sections highlight the more important requirements relating to the use of limestone in the construction and building industry. For more detail, the reader should refer to specialist publications (e.g., [8.1]).

8.2 Specifications and Test Methods

As most aggregates are produced and used locally, many countries have their own standards, which relate to the indigenous sources and national uses of aggregate.

In the UK, a general specification for aggregates used in concrete, BS 882 [8.2], contains compliance requirements for particle shape (or flakiness), shell content, grading, cleanliness (or fines content) and mechanical properties. Additionally, BS 63 [8.3] specifies requirements for roadstone, including grading, flakiness and strength. It is often used as a basis for purchasing specifications for asphalt and precast concrete production. BS 812 [8.4] gives sampling and testing methods for these and a wide range of additional properties, but does not provide guidance regarding the interpretation of results.

The results for eight tests relating to strength and durability for thirty-seven limestones used as aggregate in the UK are listed in Table 8.1, together with an indication as to whether high or low values indicate high resistance to the test parameter [8.1].

Table 8.1. Average properties of 37 UK limestones

Parameter	Mean	Standard deviation	Increased resistance indicated by
Aggregate impact value	20.0	3.6	Lower values
Aggregate crushing value	22.3	4.7	Lower values
Aggregate abrasion value	9.6	2.9	Lower values
10% fines value (kN)	182	32	Higher values
Polished stone value	40.5	4.4	Higher values
Water absorption (%)	0.8	0.8	Lower values
$MgSO_4$ soundness value	99.6	2.2	Higher values
Los Angeles abrasion value	26.3	2.9	Lower values

CEN Standards for specifications and test methods for aggregates are being prepared. Details are given in section 8.8.

8.3   Aggregates for Concrete

8.3.1   Introduction

The properties of an aggregate influence both the characteristics of the fresh mix and the strength and durability of the hardened concrete.

In the production of limestone aggregate, the selection of appropriate crushers (see section 5.2) helps to ensure an acceptable particle shape. Well designed screens help to ensure that the aggregate is of the correct size and that it has a low content of fines and clay, all of which have a marked effect on the water demand for a given workability. The strength and durability of the hardened concrete is generally related to the water demand.

Other properties of the aggregate can affect the strength and durability of the hardened concrete.

8.3.2   Grading

BS 882 specifies limits for the particle size distributions of the following products:

a) 40 to 5 mm, 20 to 5 mm and 14 to 5 mm graded aggregates,
b) 40, 20, 14, 10 and 5 mm "single-sized" aggregate,
c) coarse, medium and fine sand and
d) 40, 20, 10 and 5 mm "all-in" aggregate.

8.3.3   Particle Shape

"Flaky" or "elongated" particles are undesirable for several reasons [8.1]:

a) they adversely affect the workability and mobility of wet concrete,
b) they can cause blockages in pump pipelines,
c) they increase water demand which can lead to bleeding of the fresh mix and segregation of particles which can result in reduced strength and durability of the hardened concrete,
d) they can adversely affect finishing of the surface, and
e) they tend to reduce the strength (particularly flexural) of the hardened concrete.

BS 882 specifies that the flakiness index shall not exceed 40 for crushed rock.

8.3.4   Water Absorption

Water absorption is an indirect measure of the porosity of a stone, which can correlate with other physical properties such as mechanical strength, shrinkage, soundness, durability and the resistance of particles to frost damage.

Absorption limits are specified in BS 882 [8.2]. BS 8007 [8.5] recommends that absorption should not "generally" exceed 3 %. A maximum value of 2.5 % is sometimes specified. In general, if the absorption is less than 1 %, the aggregate can be regarded as non-frost susceptible [8.1], although values of 2 % and above may be acceptable.

8.3.5   Mechanical Properties

BS 882 [8.4] specifies requirements for the mechanical properties of coarse aggregates, as measured by the "ten percent fines value" test or the aggregate impact value test (Table 8.2).

Other mechanical test procedures are given in BS 812, namely the aggregate crushing value, aggregate abrasion value and the polished stone value. Other tests are widely used, such as the Los Angeles test for impact resistance [8.6].

The mechanical properties of aggregate should be adequate to prevent excessive breakage during handling, transport, mixing of concrete or compaction, and to enable the required compressive strength and durability of the concrete to be

Table 8.2. Requirements for mechanical properties of aggregates

Type of concrete	10% Fines value (kN)	Aggregate impact value (%)
Heavy duty concrete floor finishes	≥ 150	≤ 25
Pavement wearing courses	≥ 100	≤ 30
Others	≥ 50	≤ 45

achieved. For most applications, the requirements in BS 882 [8.2] are adequate, but for concrete road pavements and abrasion resistant floors, additional tests, such as abrasion value and polished stone value, are necessary.

As can be seen by comparing the requirements in Table 8.2 with the values in Table 8.1, aggregates produced from hard, dense limestones generally meet the mechanical property requirements of BS 882.

8.3.6 Durability

The durability of concrete depends on both the long-term stability of the aggregate and on the quality of the cement paste system. The main aspects of aggregate durability are soundness, alkali reactivity and frost susceptibility.

Soundness is a measure of the resistance of an aggregate to degradation or disintegration as a result of:

a) crystallisation of salts within the pores of the aggregate,
b) wetting — drying cycles, and
c) heating — cooling cycles.

The soundness test adopted in BS 812 [8.4] uses magnesium sulfate. The Highways Agency' Specification [8.7] gives soundness limits for coarse and fine aggregate of 65 and 75 % respectively.

Alkali reactivity results from sodium and potassium hydroxides present in cement reacting with silica, silicates, or carbonates in aggregates. Under certain conditions, the reaction can result in expansion of the aggregate and cracking of the concrete. This can lead to mis-alignment of components within structures and corrosion of reinforcement.

Limestones generally do not contain sufficient reactive silica or silicates to cause expansion, and damaging alkali-carbonate reaction has rarely been reported. The reactions involving carbonate rocks can be either expansive or non-expansive and are more likely to occur when the limestone contains appreciable quantities of dolomite and clay minerals [8.1]. ASTM C586 [8.8] gives a test method for determining the potential alkali reactivity of carbonate rock aggregates.

Frost susceptibility has two meanings. It refers to frost heave of unbound sub-base mixtures and also to the resistance of aggregate particles. In a UK specification [8.7], if the water absorption of the aggregate exceeds 2 %, additional magnesium sulfate soundness tests are required. The frost susceptibility of individual particles also correlates with the magnesium sulfate soundness value (MSSV). As mentioned in section 8.3.4, it is generally recognised that aggregates with a porosity of less than 1 % are rarely susceptible to frost damage.

The proposed CEN standard for aggregates will specify soundness and frost resistance.

8.3.7 Impurities

The impurities which may adversely affect the strength and durability of concrete are fines (clay, silt and dust), chlorides, shell debris, organic matter, sulfates, chalk, mica, pyrites and other metallic impurities.

Clay, silt and *dust* in aggregates are defined in BS 882 as any solid material passing through a 75 µm sieve. The limits are 4 % by mass in crushed rock and 16 % in sand produced by crushing rock (9 % if the sand is for use in heavy duty floor finishes).

The ratio between clay-sized and silt-sized particles (less than 2 µm, and 2 to 75 µm, respectively) can influence the properties of the concrete [8.1]. Fines can have two detrimental effects. They may coat the surface of the aggregate, thereby weakening the bond between the matrix and the aggregate. They also tend to increase the water demand at constant workability. Both effects tend to reduce the strength of the hardened concrete.

Chlorides, generally in the form of the sodium salt are found in sedimentary deposits, particularly in marine and coastal areas. In reinforced concrete, they can increase the corrosion rate of the steel. Chlorides can also adversely affect the performance of sulfate-resisting Portland cements. BS 5328 [8.9] specifies chloride contents in concrete for various types and uses. BS 882 (Appendix C) [8.2] provides guidance on limits for chloride in aggregates "when it is required to limit the chloride ion content", ranging from 0.01 to 0.05 %.

Shell debris in marine and coastal aggregate deposits is acceptable in concrete, providing it does not contain excessive numbers of whole shells with inaccessible voids.

Some types of *organic matter* can retard the hardening of concrete Others are mechanically weak and can adversely affect the mechanical integrity of the concrete. Yet others can cause staining on the surface of the concrete. Qualitative colour tests are given in BS 812 [8.4] and ASTM C40 [8.10], which are indicative of some, but not all organic material. A test for fulvo acid (a component of humic acid) is being included in EN 1744, Part 1 (see section 6.5). ASTM C33 [8.11] specifies limits of 0.5 % for coal and lignite in coarse aggregates and 1.0 % in fine aggregates.

Contamination of gravels and sands by *chalk* can cause problems owing to its mechanical weakness, softness and high porosity. In particular, under freeze — thaw conditions, chalk particles can expand and cause "popping" (i.e., lifting of the surface) [8.12].

Other impurities which may have adverse effects on concrete are *alkalis, sulfates* and *mica* [8.1]. They are unlikely to be present to a significant extent in hard, dense limestones.

8.3.8 Other Parameters

Other aggregate parameters which can affect the performance of cement include surface texture, bulk density, particle density, drying shrinkage, thermal movement and fire resistance. The properties of hard, dense limestones are such that they are generally suitable for use in concrete [8.1].

8.4   Sand for Mortars

8.4.1   Introduction

Although most sand used for mortars is naturally occurring siliceous sand, crushing hard rocks such as dense limestones produces fines which, if correctly processed are suitable for use as building sands.

Limestone sand is generally produced by wet screening, although dry screening is also used. In wet systems, hydrocyclones may be used to separate sand from finer particles.

The main factors affecting the quality of a sand used for mortar are:

a) average particle size,
b) range of particle size,
c) particle shape, and
d) impurities and particularly the clay content.

They are discussed below.

The crushing strength must be adequate for the duty, but this is not an issue for sands produced from dense limestones.

8.4.2   Specifications

Most countries have their own specifications for sand used in mortars. In Europe, work is well in hand to produce a standard (see section 8.8.1). In the UK, the relevant standards are BS 1199 (for plastering and rendering mortars) [8.13] and BS 1200 (for masonry mortars) [8.14]. Sands for floor screeds are specified in BS 882 [8.2] and in BS 1199 [8.13].

8.4.3   Particle Size

As mentioned above, the quality of a sand depends on the average particle size and the range of particle sizes.

As an example, BS 1200 [8.14] specifies wide ranges for Type S (special) and Type G (general) (Table 8.3). It should be noted that these ranges were designed to encompass most of the sands which have proved, in practice, to be satisfactory. An "ideal" grading for bricklaying mortars [8.15] is also given.

8.4.4   Particle Shape

The particle shape of sands produced by crushing dense limestones tends to be angular and cubical. This is satisfactory from the viewpoint of the properties of the hardened mortar, but tends to produce a somewhat harsher and less workable fresh mortar than many naturally occurring silica sands, which generally have a more rounded shape.

Table 8.3. Gradings of sands for bricklaying mortars

BS sieve	% by mass passing BS sieve		
	Type S	Type G	"Ideal"
5.00 mm	98–100	98–100	100
2.36 mm	90–100	90–100	97–100
1.18 mm	70–100	70–100	85–99
600 μm	40–100	40–100	70–95
300 μm	5–70	20–90	25–65
150 μm	0–15	0–25	5–15
75 μm	0–5	0–8	

8.4.5   Impurities

BS 1200 [8.14] refers to "harmful materials such as iron pyrites, salts, coal or other organic impurities, mica, shale and similar laminated materials, and flaky or elongated particles".

For most limestone sands, the most significant impurity is likely to be clay and precautions may need to be taken to reduce clay contamination (e.g. by rinsing the sand, or by pre-screening stone to remove the clay-rich fine fraction before crushing).

Testing for fines and clay is described in [8.1] and EN 993 (in preparation, see section 8.8.2).

8.5   Unbound Aggregates for Roads

8.5.1   Introduction

Unbound (or uncoated) aggregate is widely used in the construction of roads. It is mainly used in the UK for the subbase (see Fig. 8.1), but may also be used in the capping layer. Uncoated roadstone may also be used for the road-base and, in the case of low volume roads, for the complete structure.

The role of the uncoated roadstone layer(s) may be one or more of the following:

a) a working platform on which cement- or bitumen-bound layers are placed,
b) a structural layer to spread loads and prevent rutting,
c) a replacement for a frost-susceptible sub-grade,
d) a drainage layer.

The principle features which determine the suitability of an aggregate for use in the unbound state are its resistance to:

a) crushing and impact (i.e., fragmentation) during handling and placing,
b) attrition (i.e., wear) during repeated loading by traffic,
c) breakdown due to weathering and frost (soundness), and
d) frost heave in the layer.

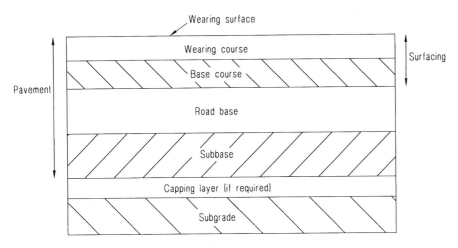

Figure 8.1. Typical section through a flexible pavement

8.5.2 Specifications

Specifications vary widely from country to country and reflect the local climate, traffic conditions and economic constraints.

In the UK, BS 63 [8.3] specifies requirements for single-sized aggregates (50, 40, 28, 20, 14, 10, 6 and 3 mm) for general purposes and for surface dressing (although limestone is rarely used for surface dressing). The single sizes may be recombined to produce the particle size distribution required for a particular application.

The UK Highways Agency's specification [8.7] refers to unbound material for subbase and road-base (Table 8.4). A CEN specification is in preparation (see section 8.8.1). Specifications for use in tropical and sub-tropical countries are described in [8.16].

Table 8.4. A specification for subbase

Sieve size (mm)	75 mm	37.5 mm	10 mm	5 mm	600 µm	75 µm
Limits (% m/m)	100	85–100	40–70	25–45	8–22	0–10

Parameter	Plasticity index	10% Fines value	Soundness (mag. sulf.)	Other
Limits	Non-plastic	≥ 50 kN	≥ 75	(a), (b)

(a) plus sulfate limit if within 500 mm of ordinary Portland cement elements.
(b) plus frost heave limit if within 450 mm of surface.

8.5.3   Grading

Where the aggregate is used as a subbase, an appropriate particle size distribution is required so that it consolidates well. Traditionally subbase has been a crusher-run product, with less than 15 % above 37.5 mm and including some 10 to 20 % less than 0.6 mm, although it is increasingly being produced as a screened product.

8.5.4   Particle Shape

As "flaky" particles are prone to breakage when loaded, a low flakiness index is sometimes specified.

8.5.5   Durability

In the context of durability, with respect to weathering, wetting-drying cycles and heating–cooling cycles, the properties of hard, dense limestones are generally fully adequate.

8.5.6   Mechanical Properties

As most limestones have a low resistance to polishing, they are not particularly suitable for use in the surface layer. However, hard, dense limestones usually have fully adequate resistance to crushing and impact loads and are, therefore, eminently suitable for lower layers.

8.5.7   Water Migration

Good drainage of the pavement is essential [8.17]. Use of a grading such as that specified by the UK's Highway Agency (Table 8.5 in [8.7]) ensures adequate permeability. The weakening effects of frost and thaw cycles on the pavement is a complex issue [8.18]. The effects of moisture movement and salt on road foundations can also be significant. Limits of 0.2 to 1.0 % for soluble salts and 0.05 to 0.25 % for sulfate (as SO_3) have been recommended for cohesive lime- and cement-treated aggregate [8.1].

8.6 Aggregates for Asphalts

8.6.1 Introduction

Dense limestones are widely used in asphalts (also called bituminous macadam). The properties which determine the suitability of an aggregate are its:

a) strength — resistance to crushing and impact,
b) hardness — resistance to abrasion and attrition,
c) resistance to polishing (for wearing course materials),
d) ability to adhere to the bituminous binder,
e) resistance to weathering, and
f) ability to produce a strong and stiff bound mix.

8.6.2 Specifications

The requirements for aggregate in asphalts vary widely and depend on the application [8.1].

In the UK, BS 63 specifies limits for grading and flakiness index for eight single-sized grades (50, 40, 28, 20, 14, 10, 6 and 3 mm). These are blended to give the required particle size distribution as specified in [8.19, 8.20].

The Highways Agency's Specification 1991 [8.7] identifies strength, durability and cleanliness as being important factors, but places the responsibility on the specifying engineer to assess the requirements for a particular application. A CEN Standard is in preparation (section 8.8.1).

8.6.3 Strength

Dense limestones (including dolomites) are relatively strong, as measured by the aggregate crushing test and ten percent fines test [8.4]. Their resistance to impact, as measured by the aggregate impact test, modified aggregate impact test [8.4] and Los Angeles test [8.6], is generally fully acceptable.

Porous limestones are seldom used because they have a high bitumen demand, which increases cost.

8.6.4 Hardness

The abrasion/attrition resistance of dense limestones as measured by the aggregate abrasion test, are lower than many competitive aggregates, but are still acceptable for most applications.

8.6.5   Resistance to Polishing

The low resistance to polishing of most limestones means that they are seldom used at the surface layer of heavily used roads. Limestones have, however, given good service in car park and footway surfaces.

8.6.6   Adhesion to the Binder

Generally, bituminous binders adhere well to limestones and are resistant to stripping [8.1]. Section 26.5 describes the use of hydrated lime as an anti-stripping agent.

8.6.7   Weathering

Dense limestones weather well and easily meet the requirement in the Highways Agency's Specification [8.7] of a magnesium sulfate soundness value [8.4] greater than 75 %.

8.6.8   Mix Strength and Stiffness

The strength and stiffness of the bound mix reflects the strength of the aggregate, and the way in which the aggregate particles interlock. Crushed limestones tend to be cubical in shape and perform well in this context.

8.7   Other Applications

8.7.1   Drainage and Filtration

Aggregates are used in a variety of applications relating to drainage of water from land and structures, and to filtration of water and effluent. The key factors are the permeability of the aggregate, ease of construction, stability and durability and these are governed by the particle size distribution and the strength/durability of the aggregate [8.1].

Crushed, dense limestone aggregates are generally suitable for such applications.

8.7.2   Railway Track Ballast

The main requirement is for high strength and resistance to abrasion (see section 8.8.1). As limestone generally has a lower abrasion resistance than many other aggregates, its use tends to be restricted to areas of low traffic density, such as sidings, and regions where there is no suitable local alternative.

8.7.3 Armourstone

Limestone blocks, weighing up to several tonnes, are used as armourstone for coastal and shoreline engineering [8.21]. A CEN product specification standard is in preparation (section 8.8.1).

8.8 CEN Standards for Aggregates

8.8.1 Proposed European Standards for Aggregate Specifications

– Aggregates for mortars
– Aggregates for concrete including those for use in roads and pavements (prEN 12620)
– Aggregates for bituminous mixtures and surface dressings for roads, airfields and other trafficked areas (prEN 13043)
– Hydraulic bound and unbound aggregates for road layer construction
– Armourstone
– Aggregates for railway track ballast

8.8.2 CEN Test Methods for Properties of Aggregates
(see section 6.5 for details)

– EN 1097: Tests for Mechanical and Physical Properties of Aggregates
– EN 1367: Tests for Thermal and weathering Properties of Aggregates
– EN 1744: Tests for Chemical Properties of Aggregates
– EN 932: Tests for General Properties of Aggregates
– EN 993: Tests for Geometrical Properties of Aggregates
– EN XXX: Tests for Bituminous Bound Fillers

8.9 References

[8.1] M.R. Smith, L. Collis, "Aggregates", The Geological Society, Bath, 1993, ISBN 0-903317-89-3.
[8.2] BS 882: "Specification for Aggregates from Natural Sources for Concrete", 1992.
[8.3] BS 63: "Road aggregates", 1997.
[8.4] BS 812, Parts 100–124, "Testing Aggregates" — see section 6.6 for details.
[8.5] BS 8007: "Code of Practice for Design of Concrete Structures for retaining Aqueous Liquids", 1987.

[8.6]   ASTM C 131-89: "Test methods for resistance to abrasion of small size coarse aggregate by use of the Los Angeles machine", 1989.

[8.7]   Highways Agency, "Specification for Highway Works, HMSO, London, 1991.

[8.8]   ASTM C586-92: "Test method for potential alkali-reactivity of carbonate rocks for concrete aggregates (rock cylinder method)", 1992.

[8.9]   BS 5328: "Concrete", 1990.

[8.10] ASTM C40-92: "Test method for organic impurities in fine aggregates for concrete", 1992.

[8.11] ASTM C33-93, "Specification for Concrete Aggregates", 1993.

[8.12] T.P. Lees, "Impurities in concreting aggregates", British Cement Association, 1987, ref 45.016.

[8.13] BS 1199: "Specifications for building sands from natural sources — sands for mortars for plastering and rendering", 1986.

[8.14] BS 1200: "Specifications for building sands from natural sources — sands for mortars for bricklaying", 1984.

[8.15] N. Benningfield, personal communication.

[8.16] Transport and Road Research Laboratory, "Guide to the Structural Design of Bitumen Surfaced Roads in Tropical and Sub-tropical Countries", Road Note 31, HMSO, London, 1977.

[8.17] S.F. Brown, E.T. Selig, "The design of pavement and track foundations". In: "Cyclic Loading of Soil", Eds. M.P. O'Reilly, S.F. Brown, Blackie, Glasgow, 1991.

[8.18] Jones, "Developments in the British Approach to prevention of frost heave in pavements", Transportation Research Record 1146, Transportation Research Board, Washington, 1987.

[8.19] BS 594, Part 1: "Hot rolled asphalt for roads and other paved areas — Specification for constituent materials and asphalt mixtures", 1992.

[8.20] BS 4987, Part 1: "Coated macadam for roads and other paved areas — Specification for constituent materials and for mixtres", 1993.

[8.21] "Manual on the use of rock in coastal and shoreline engineering", Construction Industry Research and Information Association (CIRIA), Special Publication 83.

9 Use of Limestone in Cement Production

9.1 Introduction

In many countries, cement production represents the second largest outlet for limestone and is exceeded only by the use of limestone for aggregate in construction and building. The world production of Portland cement is in the order of 1420 million tonnes per year [9.1]. The amount of limestone required varies with the grade of cement and the raw materials used, but is estimated to be approximately 1.0 tonne per tonne of cement. Thus the global cement industry uses roughly 1,400 million tonnes per year of limestone, or about 32 % of the extracted limestone.

Many lime producers make Portland cement as a co-product to maximise the use of a limestone deposit and minimise tipping of low grade material. For this reason, and because cement is such an important user of limestone, the process of cement production is described briefly in this chapter.

Portland limestone cements are a relatively recent development. They consist of intimate blends of Portland cement and limestone, in which the limestone acts as an extender [9.2, 9.3]. About 1 million tonnes of limestone was used in 1990 in this application in the European Union [9.1], and, following the publication of a European Prestandard for the product in 1992 [9.2], the amount is likely to grow rapidly.

Masonry cement [9.4, 9.5] is a traditional product which consists of Portland cement as the primary strength-developing constituent and an inorganic component which may be ground limestone.

High-alumina cement is produced by reacting limestone with bauxite. The use of limestone for this process is in the order of 250,000 tonnes per year.

9.2 Portland Cement Production

The production of Portland cement is a complex subject. This section, therefore concentrates on the information relevant to a limestone supplier. For more detailed information, the reader is referred to the standard texts e.g., [9.6, 9.7].

9.2.1 Raw Materials

The production of Portland cement is essentially a chemical process in which the four main components, namely calcium oxide, silica, alumina and iron oxide are combined chemically to produce cement clinker.

The raw materials are blended and milled to produce an intimate mixture of the correct chemical composition. An analysis of a typical cement kiln feed is given in Table 9.1 [9.6].

The process is remarkably flexible in terms of the raw materials used to achieve the required chemical composition. The presence of silica, alumina and iron oxide (e.g., in the form of clay) in a calcium oxide-containing material is not a problem, as the producer requires some of those materials in the feed to the kiln. Table 9.2 lists some of the CaO-containing raw materials that are used [9.8].

Table 9.1. Analysis of a typical cement kiln feed

Component	Analysis % m/m
CaO	44.4
SiO_2	14.3
Al_2O_3	3.0
Fe_2O_3	1.1
Loss on ignition	35.9

Table 9.2. Calcareous raw materials used in making Portland cement

Material	Main components
Cementstone	CaO, SiO_2, Al_2O_3
Limestone	CaO
Marble	CaO
Chalk	CaO
Marl	CaO
Shell deposits	CaO
Alkali waste	CaO
Anhydrite	CaO
Blast furnace slag	CaO, SiO_2, Al_2O_3
Other slags	CaO, SiO_2, Al_2O_3
Carbide lime	CaO

Magnesium carbonate is the main undesirable impurity which is generally present in naturally occurring calcarious materials. The level of MgO in the clinker should not exceed 5 % [9.2]. Indeed, many cement producers favour an upper limit of 3 %. For this reason, dolomitic and magnesian limestones are unsuitable for cement making.

9.2.2   Production of Cement Clinker

There are essentially four types of cement process:

a) the dry process, in which the raw materials are ground together to produce a dry powder, which is fed into the kiln system,
b) the semi-dry process, in which the raw materials are ground, nodulised using water, and fed into the kiln as damp pellets, containing 10 to 15 % by weight of water,
c) the semi-wet process, in which the raw materials are ground as a slurry, filtered, and fed to the kiln as a filter cake, containing 15 to 20 % of water, and
d) the wet process, in which the raw materials are ground and fed as a slurry containing 30 to 35 % of water.

Modern cement plants employ rotary kilns with various types of pre-heating systems. Cement is also produced in shaft kilns [9.6, 9.7].

The process in cement kilns can be divided into six stages.

a) the feed is heated to 100 °C by the kiln gases and any water is driven off,
b) the material is heated to the decomposition temperature of the calcareous component (900 °C in the case of CaCO3),
c) the calcareous component decomposes (with evolution of CO_2 in the case of CaCO3),
d) the calcined product is heated to clinkering temperature (1300–1350 °C),
e) about 20 % of the mix forms a molten phase, which acts as a reaction medium, with the components dissolving and reacting, and the reaction products being precipitated from the melt,
f) the partially molten clinker is cooled rapidly to quench the glassy phases.

9.2.3 Production of Cement

Cement is produced by grinding cement clinker to the required degree of fineness. Gypsum is added at the grinding stage to control the setting characteristics of the cement (see section 9.2.4). Other materials, including limestone, may be co-ground with the clinker and gypsum (see section 9.3). Table 9.3 gives a typical chemical composition of Portland cement [9.8].

Table 9.3. Typical composition of Portland cement

Compound	Oxide composition	Approx content % m/m
Tricalcium silicate	$(CaO)_3SiO_2$	45
Dicalcium silicate	$(CaO)_2SiO_2$	27
Tricalcium aluminate	$(CaO)_3Al_2O_3$	11
Tetracalcium alumino-ferrite	$(CaO)_4(Al_2O_3)(Fe_2O_3)$	8
Calcium sulfate	$CaSO_4$	2.5

9.2.4 Hydration of Cement

When the compounds formed in the kiln are mixed with water, they hydrate to form complex silicates and aluminates and cause the cement to set. The interlocking crystals give the cement its strength.

If the feed to the kiln contains too much calcium oxide, or if the reaction in the clinkering zone is insufficiently complete, the cement contains free lime which slakes, after the cement has set. The reaction is accompanied by an increase in volume, causing blistering of the concrete. This effect is called "unsoundness".

Conversely, insufficient calcium oxide in the kiln feed reduces the formation of tricalcium silicate, with a consequent reduction in strength.

Magnesium carbonate in the kiln feed decomposes to magnesium oxide in the clinkering zone. The temperatures in the kiln are too low for the formation of magnesium silicates and the magnesium oxide sinters into dense crystallites, which do not react in the clinkering zone. The resulting cement contains magnesium oxide which, although it has a very low reactivity to water, eventually hydrates long after the cement has set. As with calcium oxide, the hydration reaction causes swelling and unsoundness. For this reason, the magnesium oxide level in the clinker must be kept below 5 %.

The role of the gypsum is to retard the initial hydration reactions and thereby to avoid a premature, or "false set".

9.3   Composite Cements

Composite cements contain mineral additions, which may be inert (such as limestone), or possess pozzolanic, or hydraulic properties (such as pozzolana, fly ash and burnt shale).

EN 197-1 [9.2] permits 6 to 20 % of limestone to be inter-ground with the clinker in Type II/A-L and 21 to 35 % in Type II/B-L Portland limestone cement. The limestone is required to contain at least 75 % by weight of calcium carbonate, with less than 1.2 % of clay and less than 0.2 % of organic material. It notes that limestone with organic material contents of up to 0.5 % may also be acceptable.

The expected loss of 28-day strength arising from the presence of an "inert" component can be offset by finer grinding. For equal 28-day strengths, the 1-day strength of a Portland limestone cement can be greater than that of the Portland cement.

The effects of ground limestone are partly physical and partly chemical. The limestone acts as a filler between the clinker grains and gives a more dense end-product.

The use of composite cements offers financial advantages, as the additives are less expensive than clinker. They have been widely used for many years in some countries (eg. France and Spain). With the publication of a European Prestandard [9.2], the use of composite cements is likely to grow rapidly.

9.4   Masonry Cements

As described in section 26.5.1, the traditional mortar material for building work was based on lime and sand. When cement became more widely available, it was used to replace lime. However, workable mixtures, containing one volume of cement to three volumes of sand, gave too strong a mortar for general use. Increasing the sand content to give lower final strengths resulted in a harsh and unworkable mortar. The use of air-entrainment agents improved the handling properties, but excessive amounts of air can reduce the strength of the bond between the mortar and the masonry units.

Masonry cement was developed to overcome the above problems. It is specified in ENV 413-1 [9.4]. Essentially it consists of Portland cement (at least 25 % in Type MC 5 and at least 40 % in the other Types) and an inorganic material (such as limestone), finely ground to more than 85 % passing 90 microns. Two of the four Types contain an air-entraining agent, while the other Types do not. The Standard specifies the characteristics of the cement, the fresh mortar and the compressive strengths. The limestone content of masonry cement can be over 70 % in Type MC 5 and over 55 % in the other Types.

It should be noted that mortars using masonry cement compete with cement-lime mortars. Their relative merits are discussed further in section 26.5.

9.5   Calcium Aluminate Cements

Calcium aluminate cements (otherwise known as Ciment Fondu, and high-alumina, or aluminous cement) have high resistances to chemical attack, including sulfates. They set normally, but develop useful strength in 6 hours. Their 1-day strengths are equivalent to those produced after 28 days by ordinary Portland cement. Calcium aluminate cements also have a good abrasion resistance at low temperatures [9.9]. They are made by co-grinding limestone with bauxite (or other aluminous material) and calcining the resulting mixture at temperatures ranging from 1000 to 1600 °C. Both materials should be low in silica, as otherwise the rapid-setting properties are adversely affected.

A variety of furnaces are used for the production of calcium aluminate cements [9.6], including rotary kilns [9.10]. The calcium aluminate, which may be in the form of a clinker, or a solidified melt is ground in tube mills to 94 to 98 % less than 90 µm.

9.6   References

[9.1]   Cembureau 1995 statistics.
[9.2]   ENV 197-1: "Cement — Composition, specifications and conformity criteria — Common cements", 1993.
[9.3]   BS 7583: "Specification for Portland limestone cement", 1996.
[9.4]   ENV 413-1: "Masonry cement, Part 1. Specification", 1995.
[9.5]   BS 5224: "Specification for masonry cement", 1995.
[9.6]   "Lea's Chemistry of Cement and Concrete", 4th ed., Arnold, 1997, ISBN 0-340-56589-6.
[9.7]   H.F.W.Taylor, "Cement Chemistry", Academic Press, 1990, ISBN 0-12-683900-X.
[9.8]   R.S. Boynton, "Chemistry and Technology of Lime and Limestone", John Wiley & Sons, 1980, ISBN 0-471-02771-5.
[9.9]   "Calcium aluminate cements in construction — a re-assessment", The Concrete Society Technical Report 46, 1997, ISBN 0-946-69151-7.
[9.10] British Patent 250246, "Aluminous Cements ", Soc. Anon. des Chaux et Ciments de la Farge et du Teil, 1926.

10 The Use of Limestone in Agriculture

10.1 Introduction

In many countries, agriculture accounts for 1 to 2 % of the limestone market. While this may appear to be a relatively small proportion, the tonnages involved can be large (e.g., in the U.S.A. they exceed 13 million tpa [10.1].

Most of the limestone used in agriculture is applied directly to the soil. Smaller amounts are used as a filler and conditioner in fertiliser formulations. It is also used in animal feeds, and as poultry grit. Furthermore, limestone is increasingly being employed to combat the adverse effects of acid rainfall.

10.2 Arable Land and Pasture

10.2.1 Historical [10.2]

The long-term benefits of applying marls and soft chalks to pastures and to land used for cereals, was known by the Romans in the first century A.D. Pliny reported that the Ubians north of Mainz used "white earth", a calcarious marl, to fertilise their fields. However, it was not until the middle of the 19th century that Liebig and Rothamsted demonstrated the interdependent roles of the major nutrients and growth factors on plant development. The deficiency of any one of them could limit growth. One of those factors was soil pH, which affects the availability of nitrogen, phosphorous and potash, as well as calcium and magnesium.

As the importance of soil came to be understood, the application of "agricultural lime" increased. Until the middle of the 20th century, most "liming" of agricultural land was done using quicklime and hydrated lime products (see section 30.2). These produced fine particles of calcium hydroxide in the soil, which initially raised the pH and then reacted with atmospheric carbon dioxide to form finely divided calcium carbonate. The carbonate was sufficiently soluble to maintain the pH at the required level for a prolonged period.

The production of large tonnages of finely divided limestone and chalk products only became economic with the development of dry grinding equipment in the middle of the 20th century. Such products are commonly called "agricultural lime". To avoid confusion with quicklime and hydrated lime products, the term "agricultural limestone" will be used.

10.2.2   Crop Requirements

A considerable amount of research has been done into the optimum pH range for various crops [10.2]. The results are summarised in Table 10.1.

At pH levels above and below the optimum range for a particular crop, overall nutrient availability decreases, with a consequent adverse effect on quality and yield. It also follows that the effectiveness of fertilisers is maximised when the optimum pH is maintained.

In practice, as the pH of soils tends to decrease with time, it is usual to advise liming to maintain pH values above 6.0 for grass and above 6.5 for most arable crops (the main exception being potatoes).

Table 10.1. Optimum pH ranges for various crops

Crop	Optimum pH for crop growth
Potatoes	5.0–6.5
Oats	5.4–7.0
Linseed	5.5–7.0
Ryegrass mixture	5.5–7.1
Permanent pasture	5.6–6.8
Oilseed rape	5.8–7.5
Wheat	5.8–7.5
Maize	5.8–7.5
Peas	6.1–7.5
Beans	6.1–7.5
Sugar beet	6.3–8.0
Barley	6.4–7.8
Field vegetables	6.4–7.8
Field brassica	6.4–7.7

10.2.3   Lime Losses

The acidity of soils tends to increase as a result of five factors [10.2].

a) Rainwater naturally contains dissolved carbon dioxide and oxides of nitrogen (e.g. from thunder storms) which are acidic.
b) Pollution of the atmosphere with oxides of sulfur and of nitrogen also causes acidity.
c) The use of fertilisers rich in NH_4^+ contributes to soil acidity as a result of its nitrification to produce dilute nitric acid (conversely, NO_3^- results in the release of OH^- ions).
d) Plant roots extrude hydrogen ions, resulting in increased acidity.
e) When organic matter decomposes in the soil, microbial oxidation of organic N and S, produces strong acids, such as nitric and sulfuric acids.

The acids arising from the above processes react with calcium and magnesium carbonates in the soil and form soluble salts. These are then leached out of the soil and reduce both the pH and the availability of calcium and magnesium.

10.2.4   Identification of Liming Requirements

The amount, type and frequency of limestone applications, required to achieve and maintain the optimum pH, depend on six factors:

a) the initial pH,
b) the required pH,
c) the nature of the soil,
d) the rate of lime loss,
e) the magnesium level, and
f) the effectiveness of the limestone.

There are several methods of testing the pH of soils and of assessing the lime requirements. Professional advice should be sought for each situation. It is necessary to distinguish between *actual acidity*, which is measured by the dissolved hydrogen ion concentration, or pH, and the *potential acidity*, which is measured by titration and includes; the H^+ of the soil solution, the adsorbed H^+, and the soluble aluminium species (which also consume OH^-).

Soil samples should be taken systematically across each field and each sample should be tested individually, as lime requirements in adjacent areas can vary significantly. The results should be plotted on a map of the field to enable the optimum additions to be applied.

Heavy soils, particularly clays, have a greater ability to retain calcium and magnesium ions than lighter, sandy soils. It is, therefore, common practice to apply limestone less frequently to heavy than to light soils. The frequency of application also depends on the rate of lime loss, which can only be determined by monitoring over successive years.

As magnesium is an important nutrient, its concentration should be maintained at an acceptable level. This is most easily achieved by liming with ground dolomitic limestone.

Typical recommended application rates for various soils at an initial pH of 5.5 are given in Table 10.2 [10.2].

10.2.5   Limestone Effectiveness

The effectiveness of an agricultural limestone in raising the pH of the soil depends on

a) its neutralising value,
b) its fineness,
c) the hardness of the original rock, and
d) the ease with which it is spread.

Table 10.2. Example of typical application rates

Type of soil	Use	pH		Application rate (te/ha) [a]
		Initial	Target	
Sands/loamy sands	Arable	5.5	7.0	9
	Grass	5.5	6.5	5
Sandy loams/silt loams	Arable	5.5	7.0	11
	Grass	5.5	6.5	5
Clay loams/clays	Arable	5.5	7.0	12
	Grass	5.5	6.5	6
Organic soils [b]	Arable	5.5	6.7	12
	Grass	5.5	6.3	5
Peaty soils [c]	Arable	5.5	6.2	13
	Grass	5.5	5.8	4

[a] Based on soil depth, arable of 200 mm, pasture of 150 mm, neutralising value of 54% with 40% passing 150 µm.
[b] With 10 to 25% of organic matter.
[c] With greater than 25% of organic matter.

The neutralising value is a measure of the calcium plus magnesium carbonate content of the limestone (expressed in terms of CaO — see glossary).

Neutralising values of above 50 % CaO are obtained with high purity limestones, but lower values can be offset by increasing the quantity added per unit area.

The fineness of the limestone affects the rate at which it can react in the soil. Particles of hard limestones do not break down in the soil as a result of the action of frost, or of mechanical cultivation, and it is generally considered that particles larger than 600 µm react too slowly to be effective.

Particles of soft, porous chalks may break down in the soil, enabling coarser products to be effective.

Mechanical spreading equipment is generally designed to distribute dry powders with a maximum particle size of less than 5 mm. Some waste products containing calcium carbonate are very fine, but contain moisture (e.g. the sludge from the processing of sugar beet), which necessitates the use of special equipment to obtain an accurate and even spread.

10.2.6 Application

Agricultural limestone can be spread at any time when ground conditions permit. It is usually applied to arable land after cropping. Because it can take several weeks for the limestone to react sufficiently to raise the pH into the required range, the application should be made well in advance of sowing.

When limestone is applied as a top dressing, it initially affects only the surface layer of soil. Thus, when a soil is acidic throughout its depth, or when lower soil layers are more acidic than the surface layer, it is generally advisable to apply half of the limestone before ploughing and to spread the remainder as a surface dressing.

10.2.7   Limestone Specifications

The requirements for some of the quarry-produced liming materials are summarised in Table 10.3 [10.3].

Table 10.3. Requirements for carbonate liming materials

Material	Form	% MgO	Grading (% passing) 3.35 mm	150 µm
Limestone (Calcium)	Ground	< 15	> 95	> 40
	Screened/dust	< 15	> 95	> 20
	Coarse	< 15	> 95	> 15
Limestone (Magnesian)	Ground	< 15	> 95	> 40
	Screened/dust	< 15	> 95	> 20
	Coarse	< 15	> 95	> 15
Chalk	Ground	–	a)	–
	Screened	–	b)	–

a) More than 98% passing 6.3 mm.
b) More than 98% passing 45 mm.

In the UK, the Fertiliser Regulations [10.3] require the declaration of the neutralising value and, in the case of limestones other than chalks, the percentage passing a 150 µm sieve.

10.3   Fertilisers

Coarsely pulverised dolomitic limestone in the size range 0.2 to 1 mm is frequently added to compound fertilisers at levels of up to about 10 %. It confers several benefits:

a) it "conditions" the fertiliser, making it less prone to caking,
b) it helps to neutralise acidity arising from nitrogenous compounds (ammonium sulfate/nitrate and urea),
c) it neutralises any free acid in the superphosphate component, and
d) it contributes small amounts of Ca and Mg as plant nutrients.

High-calcium limestone is not widely used in fertiliser formulations, as it can release ammonia from nitrogenous compounds.

10.4 Animal Feedstuffs

Animal feedstuffs frequently contain finely ground calcium limestone as a source of calcium. In general, high purity products with low levels of toxic impurities (e.g. with lead levels below 10 mg/kg) are specified [10.4].

10.5 Poultry Grits

Poultry require grit in their gizzard to help them grind their food. Closely graded limestone and shell products in the size range 3 to 10 mm are generally specified.

10.6 Neutralising Acid Rainfall

10.6.1 Forestry

Trees, and particularly conifers, do not require liming as they prefer acid soils and can tolerate pH levels below 4. They are, however, more effective at removing acidic pollutants from the atmosphere than shorter crops and are frequently grown on acid and poorly buffered soils. As a result, water draining from forested areas is frequently very acidic (see section 10.6.3.1).

Applications of agricultural limestone are used to raise the pH of streams draining forested areas. The limestone is spread in a very finely divided form (e.g. 50 % less than 10 µm), at about 15 tonnes per hectare [10.2], over the boggy areas in the headwater regions of sensitive catchment areas. Such areas are generally left unplanted so the limestone does not adversely affect tree growth. Regular repeat applications are necessary.

While spreading limestone may improve stream water quality, it can have an adverse effect on certain plant species (e.g., mosses and lichens) in the areas in which it is applied. In some areas, therefore, conservation requirements may prevent its use.

10.6.2 Upland Catchment Areas

In some upland areas, the combination of high rainfall and acid rain have removed Ca and Mg from soils and lowered the pH to below 4. Under these conditions, aluminium accounts for over 70 % of the exchangeable cations and calcium for only a few percent [10.1]. Such conditions increase the mobility of many heavy metals, such as cadmium, zinc, manganese and nickel. Soluble aluminium is toxic to many plants (e.g. perennial ryegrass, clover, fine fescues and other acid-intolerant herbaceous species) and also to fish. It also locks up phosphates. Dissolved manganese hampers root development and plant growth.

Raising the pH to above 5.5 by liming renders aluminium insoluble, reduces the levels of manganese and other heavy metals It also ensures that calcium is the predominant exchangeable cation.

Selective liming not only increases botanical diversity, but also biological diversity including bacteria, earthworms, birds, badgers, algae, insects and fish. The view has been expressed that, while there is no agricultural need to spread limestone in upland catchment areas, there are strong environmental grounds for doing so [10.2], subject always to any conservation requirements (see section 10.6.1).

10.6.3   Acid Lakes and Streams

10.6.3.1   General

The effects of acid rain have been shown to be most pronounced in areas where the soil is largely composed of weathered particles of granite. Large areas of Norway, Sweden, Finland, Scotland and North America are particularly affected by the problem [10.5]. In addition, the discharge of industrial effluents can cause acidification [10.6].

In Sweden alone, some 7,500 lakes and over 11,000 km of streams are limed repeatedly in an attempt to keep the pH above 6.0. The cost in 1994 was about US$ 25 million [10.7].

Liming of lakes or streams is the only large scale measure available to counter some of the adverse effects of acid rainfall. It will need to be continued until the emission of acid gases is substantially reduced.

10.6.3.2   Lakes

The utilisation of limestone applied to lakes [10.5] increases with:

a) increasing fineness (a median particle size of 6 to 12 μm is often used),
b) decreasing pH of the water,
c) the depth of the water, and
d) the degree of dispersion (powders are generally slurried with water).

Solution rates decrease at high dosage levels (in excess of 25 g/m³), and with increasing $MgCO_3$ content in the limestone (a limit of 5 % is generally advised). The geological origin of the limestone and the carbon dioxide content of the water do not have a significant effect.

Limestone is generally applied from boats, which use the lake water to prepare a slurry containing up to 30 % by weight of calcium carbonate. Some systems prepare the slurry either on the lake shore, or adjacent to a nearby road, and pump the slurry via floating pipelines to a boat which disperses the slurry over the surface of the lake. Aerial application from fixed wing aircraft and helicopters is also used for remote lakes, using either powder, or pre-mixed slurry.

10.6.3.3   Streams

Several techniques have been developed for liming acidified streams [10.5].

- In the *diversion well*, relatively coarsely pulverised limestone is fluidised in a chamber fed with stream water. The fine particles dissolve and somewhat larger particles are maintained in suspension within the well. Typically, about 5 % of the limestone dissolves each day and utilisation levels of 80 to 90 % are obtained.

- In the *powder doser*, limestone with a median particle size of about 12 μm is used. The flow rate of the stream is metered and used to control the limestone feed rate into a slurrying chamber. The resulting suspension is then discharged into the stream. Limestone utilisation rates, however, are only about 50 %.

- The *slurry doser* operates on a similar principle to the powder doser. It uses a pre-mixed slurry containing 70 % by weight of limestone (with a median particle size of about 1 μm), stabilised with a dispensing agent. Utilisation rates approaching 100 % are achieved.

The above techniques can raise the pH value to over 6.5.

10.7   References

[10.1] "Crushed Stone — Annual Review — 1994", US Bureau of Mines' Mineral Industry Survey.
[10.2] "Agricultural Lime and the Environment", The Agricultural Lime Producers' Council, published by the British Aggregate and Concrete Materials Industries, 1994.
[10.3] "The U.K. Fertiliser (Sampling and Analysis) Regulations", HMSO, London, 1991.
[10.4] "Food Chemicals Codex", National Academy of Science, Washington.
[10.5] H.U. Sverdrup, "Liming of Lakes and Streams", Proc. 6th International Lime Congress, London, 1986.
[10.6] A. Calderoni et al., "Recovery of Lake Orta by Liming", Proc. 7th International Lime Congress, Rome, 1990.
[10.7] T. Svenson et al., "The Swedish Liming Programme", Water, Air and Soil Pollution, **85**, 1995, 1003–1008.

11 Use of Limestone in Refining Metals

11.1 The Production of Iron

11.1.1 Introduction

Iron is produced by the reduction of iron ore (oxides of iron), by coke in blast furnaces. Limestone is used as a fluxing agent. It dissociates into quicklime in the blast furnace, and helps to remove impurities in the iron ore (mainly silica and alumina) by reacting with them to form a molten slag. The slag also assists in the removal of other impurities. Traditionally, all of the limestone was charged with the other raw materials into the blast furnace. In current practice, limestone is also used as a component of the feed to sinter strands, which are used to agglomerate finely divided iron ore before it is fed into the blast furnace.

11.1.2 Burden Preparation [11.1]

The blast furnace requires the feed materials, including the iron ore, to be in a granular form. All of the materials must, therefore, have adequate strength to resist degradation during transport, storage and handling.

The ore may be supplied in the following forms:

a) granular lumps of ore,
b) pellets (granules of ore),
c) fines (produced by screening the ore), and
d) concentrate (produced by benefication of low grade ore e.g., by froth flotation).

A typical analysis of a high-grade iron ore is given in Table 11.1 [11.2].

Table 11.1. Typical analysis of a high-grade ion ore

Element	Concentration % (m/m dry)
Fe	89.4
Si	9.6
Al	0.55
Mn	0.41
P	0.04

The proportion of fines and concentrate used varies with the quality of ore and with local economic factors. On average, in Europe, the sinter process is used to prepare about 60 % of the iron ore used, with pellets accounting for 30 % and lump ores for 10 %.

In the *sinter strand* process, the finely divided iron ore is mixed with coke, limestone, and dolomite. The strand consists of a travelling grate which conveys the mixture through a tunnel kiln. Typically 120 kg of limestone is added per tonne of sinter to give a CaO ÷ SiO$_2$ molar ratio of about 1.7. The amount of dolomite used depends on the desired slag composition in the blast furnace. Increasing the MgO content of the slag helps to prolong the life of the refractory lining of the blast furnace, but increases the viscosity of the slag and reduces its capacity for removing sulphur. Olivine [(Mg·Fe)$_2$·SiO$_2$] is increasingly being used as a replacement for dolomite. After being blended, the raw mix is moistened with water and agglomerated into pellets.

The pellets are then fed onto the sinter strand and are heated by burning blast furnace gas. This ignites the coke and raises the temperature of the solids to about 1280 °C. The limestone is calcined and reacts with silica and alumina in the ore to produce a molten calcium silicate-aluminate-ferrite system, which bonds the iron ore particles and produces strong pellets.

Some producers add 1 to 2 % of quicklime to the raw mix, to improve agglomeration and produce stronger "green" pellets. This increases the output of the sinter strand and reduces heat usage (see section 27.2).

11.1.3   The Blast Furnace [11.1, 11.3]

A typical blast furnace has a residence time of 1 day and produces 10,000 tonnes per day of molten iron (Fig. 11.1). Coke, limestone and iron ore plus additives are introduced in layers some 400 to 700 mm thick.

The additives are introduced to control the metallurgy of the slag and may vary from 10 to 200 kg per tonne of pig iron (i.e., of molten iron discharged from the furnace). During the past 30 years, the use of iron-rich ores and of better quality sinter and pellets (particularly with lower silica contents), have reduced the production of slag from more than 700 kg per tonne to an average of 250 kg per tonne of pig iron. It has also reduced the fuel consumption of the blast furnace.

Sufficient limestone is added to produce a molten slag with a "basicity" (CaO ÷ SiO$_2$) of about 1.2, corresponding approximately to the production of monocalcium silicate (equation 11.1). At a slag production level of 250 kg per tonne, this corresponds to a total calcium carbonate addition of about 160 kg per tonne pig iron.

$$CaO + SiO_2 \rightarrow CaSiO_3 \tag{11.1}$$

Oxygen is blown into the furnace near the base. It reacts with the coke to produce carbon monoxide and temperatures of up to 2,300 °C in the "active coke zone". The carbon monoxide progressively reduces the various forms of iron oxide to iron (equations 11.2, 11.3 and 11.4). It also reacts with water from combustion (the water gas shift reaction) to produce hydrogen (equation 11.5), which also reduces the iron oxides via analogous reactions.

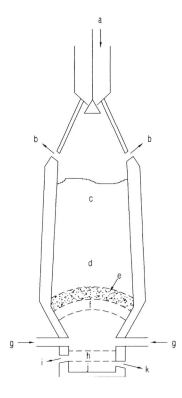

Figure 11.1. Diagram of a blast furnace
(a) ore, coke and limestone; (b) exhaust gas;
(c) alternating layers of ore and coke; (d) solid zone;
(e) alternating layers of permeable coke and cohesive
sinter; (f) active coke zone; (g) oxygen; (h) layer of
molten slag; (i) slag discharge; (j) layer of molten iron;
(k) iron discharge

$$3Fe_2O_3 + CO \xrightarrow{\text{at 200 to 300 °C}} 2Fe_3O_4 + CO_2 \tag{11.2}$$

$$2Fe_3O_4 + 2CO \xrightarrow{\text{at 400 to 700 °C}} 6FeO + 2CO_2 \tag{11.3}$$

$$FeO + CO \xrightarrow{\text{at > 1000 °C}} Fe + CO_2 \tag{11.4}$$

$$CO + H_2O \rightarrow H_2 + CO_2 \tag{11.5}$$

Much of the sulfur in the feed to the blast furnace, which mainly originates from the coke, and virtually all of the phosphorous is absorbed by the iron.

The outputs from the blast furnace are:

a) gas containing carbon monoxide and hydrogen, which is cleaned and used as a low calorific-value fuel,
b) pig iron (see Table 11.2 [11.3]),
c) slag (see Table 11.3 [11.3]).

The use of lime products in the treatment of pig iron and its conversion into steel are described in chapter 27.

Table 11.2. Typical analysis of pig iron

Element	Concentration % (m/m)
Fe	94
C	4.7
P	0.09
Si	0.3–0.7
S	0.02
Mn	0.3

Table 11.3. Typical analysis of blast furnace slag

Component	Concentration % (m/m)
CaO	40
SiO_2	36
Al_2O_3	< 14
MgO	9

11.1.4 Limestone Specification

The high calcium limestone and dolomite should both be high in calcium plus magnesium carbonate content and low in silica and phosphorous (Table 11.4 [11.2]). They should be resistant to decrepitation on heating, and produce limes which are strong enough to resist being crushed in the blast furnace.

Table 11.4. Typical chemical specification for limestone

Component	Concentration % (m/m)
$CaCO_3 + MgCO_3$	> 97.0
S	< 0.1
P	< 0.02

11.2 Open Hearth Steelmaking

The open hearth steelmaking process was invented in 1863 and up to the early 1960s was the principal means of producing steel in the U.S.A. and the U.K. [11.2]. Since then, it has been displaced by the basic oxygen steelmaking processes (see sections 27.4 and 27.5).

Originally, the process was operated with a molten acid slag, which only permitted the removal of Si, Mn and C. Small amounts of limestone were often added towards the end of the refining process to produce a "boil", caused by the calcination of the limestone and the evolution of carbon dioxide, but the amounts were insufficient to affect significantly the chemistry of the slag.

Subsequently, the furnace was operated with a molten basic slag, which enabled the removal of P and S. Limestone, or a mixture of 80 to 90 % of limestone and 10 to 20 % of quicklime, were added to give a basicity ($CaO \div SiO_2$) of 2.5 to 3. Typically this required about 25 kg of limestone per tonne of hot metal [11.4, 11.5]. Limestone with low silica and sulfur contents and a particle size of up to 250 mm was preferred.

11.3   Smelting

Limestone is widely used as a flux in the smelting of copper, lead, zinc and antimony from their ores. It is charged into the smelters with the concentrated ores. The limestone calcines and the resulting lime reacts with silica, alumina and other impurities to form a molten slag. The slag chemistry is similar to that in the blast furnace (section 11.1.3).

The limestone should be high in calcium carbonate and low in silica and alumina. As with limestones for use in blast furnaces, it should be resistant to decrepitation on heating and produce lime which is strong enough to resist crushing in the smelter.

11.4   The Production of Alumina

Low grade bauxite typically contains 50 % alumina and about 12 % of silica. Limestone is used to convert the silica into an insoluble silicate, allowing the alumina to be extracted as a soluble aluminate.

In the combination-Bayer process, the bauxite is finely ground with sodium carbonate and sufficient limestone to give a CaO ÷ SiO$_2$ ratio of 2.0 to 2.2 [11.6]. The mixture is sintered in a rotary kiln at 1100 °C to calcine the sodium carbonate to sodium oxide (equation 11.6) and the limestone to quicklime (11.7). The sodium oxide reacts with the alumina to form sodium aluminate (11.8) and the quicklime reacts with the silica to produce dicalcium silicate (11.9).

The cooled product is ground and extracted with water. The soluble sodium aluminate is removed as a filtrate from the dicalcium silicate and insoluble matter. Injection of carbon dioxide into the filtrate produces a precipitate of aluminium hydroxide and sodium carbonate liquor (11.10), which is recycled to the initial stage. The aluminium hydroxide is heated to above 300 °C to produce pure alumina which is dissolved in fused cryolite (NaAlF$_4$) at about 900 °C and electrolysed between carbon electrodes to produce molten aluminium.

$$Na_2CO_3 \rightarrow Na_2O + CO_2\uparrow \tag{11.6}$$

$$CaCO_3 \rightarrow CaO + CO_2\uparrow \tag{11.7}$$

$$Al_2O_3 + Na_2O \rightarrow 2NaAlO_2 \tag{11.8}$$

$$SiO_2 + 2CaO \rightarrow Ca_2SiO_4 \tag{11.9}$$

$$2NaAlO_2 + CO_2 + 3H_2O \rightarrow 2Al(OH)_3\downarrow + Na_2CO_3 \tag{11.10}$$

11.5 References

[11.1] D. Springorum, "Development of the Specific CaO Consumption in the Iron and Steel Industry", Proc. European Lime Association's Technical Conference, Cologne, Oct. 1992.

[11.2] R.S. Boynton, "Chemistry and Technology of Lime and Limestone", John Wiley & Sons, 1980.

[11.3] Author's notes (confirmed) of a seminar given by S. Millman, T. Fray and G. Thornton, British Steel Technical, Buxton, June 1993.

[11.4] A. Jackson, "Modern Steelmaking for Steelmakers", G. Newnes Ltd, London, 1967.

[11.5] A. Jackson, "Oxygen Steelmaking for Steelmakers", Butterworth, London, 1969.

[11.6] D. Blue, "Raw Materials for Aluminium Reduction", U.S. Bureau of Mines, I.C. 7675, Mar. 1954.

12 Other Uses of Limestone

12.1 Introduction

Limestone and calcium carbonate products are used in a wide range of industries other than those mentioned in chapters 7 to 11. While the quantities used in each of these market segments are small in relation to that quarried (in most cases less than 0.3 % of the total), the products play an important role in those industries.

The products can largely be divided into three categories:

a) screened or pulverised limestone with a nominal top size in excess of 50 μm,
b) whiting, with a top size below 50 μm, (see section 5.3), and
c) precipitated calcium carbonate (see section 31.2).

The chemical and physical properties which are exploited include:

a) its CaO, or CaO plus MgO content in the manufacture of glass, ceramics and mineral wool,
b) its neutralising capacity in the removal of acids from flue gases, the sulfite process for making paper pulp, neutralisation of acidic water, and in the preparation of a wide range of organic compounds,
c) its inertness, as a coal mine dust, and its whiteness as a filler for plastics, paints, rubber and many other products, and
d) its stability as an aggregate for filter beds.

The applications described in this chapter are summarised in Table 12.1.

12.2 Glass Manufacture

The major components of glass are silica sand, soda ash (Na_2CO_3) and limestone. Both high-calcium limestone and dolomite are used (see also the use of quicklime in glass manufacture, section 32.6).

The raw materials are blended and charged into the glass furnace, which is a refractory lined bath covered by a refractory roof. The furnace is heated to about 2000 °C by burners in the space above the molten glass. As the raw materials enter the bath, the carbonates decompose and evolve carbon dioxide, which helps to agitate the bath and to disperse the solids in the melt. The solids react to produce the sodium/calcium/magnesium silicate which is the principle constituent of most glasses.

Typical ranges for the components of soda-lime glass (by far the most common form) are given in Table 12.2 [12.1].

Table 12.1. Other uses of limestone/calcium carbonate

Section	Application	Section	Application
12.2	Glass	12.9	Fillers and extenders
12.3	Ceramics		– floor coverings
12.4	Mineral wool		– paints & coatings
12.5	Acid gas removal		– paper
12.5.2	FGD — wet processes		– plastics
12.5.3	FGD — dry processes		– rubber
12.5.4	HF removal		– sealing mastics
12.6	Paper and pulp		– flame retardants
12.7	Organic chemicals production		– pesticides
12.8	Dust suppressant for mines		– printing inks
12.10.1	Drinking water treatment		– toothpaste & powder
12.10.2	Sewage treatment	12.12	Calcium zirconate production
12.10.3	Acid neutralisation	12.13	Strontium carbonate production
12.11	Sodium dichromate		

Table 12.2. Composition of soda-lime glass

Component	Typical level %
SiO_2	68–75
Na_2O	10–18
CaO	5–10
MgO	0–10

It is important that the particle size distributions of the raw materials are controlled within the specified range as:

a) this helps to limit segregation of the blend,
b) oversize (e.g. over 1 or 2 mm) dissolves slowly and can cause imperfections in the glass, and
c) dust (e.g. less than 75 µm) tends to increase gaseous inclusions in the glass and may result in the formation of a foam on the molten glass, which reduces heat transfer into the melt.

The preferred size range varies from one producer to another. Fibreglass producers, who are especially sensitive to undissolved particles, generally specify products which are predominantly less than 75 µm.

Dolomitic limestone is usually specified for the production of containers and tumblers, as the MgO content increases resistance to etching by acids and other solvents. High-calcium limestone is generally favoured for flat glass.

Most glass producers require high purity limestone. Iron is particularly important with less than 0.1 % by weight of Fe_2O_3 being widely specified: for colourless glasses, levels of 0.02 % and below may be specified [12.2]. Levels of organic carbon, sulfur and phosphorous are also specified.

12.3   Ceramics

There are three categories of earthenware: clay, lime and feldspathic. Lime earthenware contains 5 to 35 % of lime, introduced as marl, finely pulverised limestone or whiting. When the green earthenware is fired at temperatures up to 1200 °C, the carbonate components decompose and react with the clay to produce anorthite ($CaAl_2Si_2O_8$) and diopside ($CaMg(SiO_3)_2$).

12.4   Mineral Wool

Mineral wool, also known as rock and slag wool, is made by heating basalt, rhyolite, diabase, or blast furnace slag with limestone. The limestone decomposes and the calcium oxide fluxes the other component to produce a melt. The fluid melt is then discharged into the path of a high pressure jet of steam, which breaks it up into fine threads of mineral fibre.

12.5   Acid Gas Removal

12.5.1   Introduction

Increasingly, industry is being required to limit its emissions of acidic gases. Limestone reacts with the most common acidic gases (i.e. SO_2, SO_3, HCl and HF), and is considerably less expensive than alternative alkaline materials, such as lime (see chapter 29), sodium carbonate/bicarbonate and caustic soda. It is, therefore, not surprising that considerable effort has been put into developing processes using limestone as an absorbent.

The processes which have been developed may be divided into wet and dry scrubbing. The choice of process depends on several factors, including:

a) the acid gases,
b) the proportion to be removed,
c) the operating and capital costs, and
d) the nature of the waste products.

Commercially, the most important application is the use of wet scrubbing with limestone for flue gas desulfurisation (FGD) at electricity generating stations. For example, a 2000 MW station operating on a 2 % sulfur coal, at a 70 % load factor, requires about 300,000 tpa of limestone to remove 90 % of the oxides of sulfur. The majority of generating stations fitted with FGD use limestone, and about half of those oxidise the calcium sulfite produced to gypsum ($CaSO_4 \cdot 2H_2O$) [12.18]. Some stations use lime (see chapter 29), while others use a variety of other processes [12.3].

Dry scrubbing with limestone is also used to remove hydrogen fluoride from the exhaust gases from kilns in which heavy clay goods and ceramic products are fired. It is also employed for FGD on small to medium sized boilers.

12.5.2 Flue Gas Desulfurisation Using Wet Scrubbing

12.5.2.1 Chemistry

Coals typically contain 0.5 to 4 % by weight of sulfur in the form of iron sulfide (FeS_2), organic sulfur compounds and inorganic sulfates. Many other fuels contain significant amounts of sulfur (e.g., heavy fuel oil and PET coke). During combustion the sulfur is oxidised to sulfur dioxide, with smaller amounts of sulfur trioxide also being produced. Sulfur dioxide is only moderately soluble in water. It dissolves more rapidly in a suspension of limestone (and still more rapidly in alkaline solutions/suspensions — see chapter 29).

The reactions between sulfur dioxide and trioxide with limestone are:

$$CaCO_3 + SO_2 + 0.5H_2O \rightarrow CaSO_3 \cdot 0.5H_2O \downarrow + CO_2\uparrow \qquad (12.1)$$

$$CaCO_3 + SO_3 + 2H_2O \rightarrow CaSO_4 \cdot 2H_2O \downarrow + CO_2\uparrow \qquad (12.2)$$

The mixture of calcium sulfite hemihydrate and calcium sulfate dihydrate (gypsum) comes out of solution as a finely divided precipitate. The product is not saleable and has to be disposed to tip. A further disadvantage of the process is that, because the crystallisation is not controlled, the fine particles form a thixotropic sludge which is difficult to de-water for disposal [12.3].

In a development of the original process, which is widely used for new installations, the precipitated calcium sulfite is oxidised to gypsum by compressed air and the gypsum crystals are grown to a size which can readily be de-watered. The resulting product may be sold for the manufacture of gypsum plaster, or may be used for landfill.

$$2CaSO_3 \cdot 0.5H_2O + O_2 + 3H_2O \rightarrow 2CaSO_4 \cdot 2H_2O \qquad (12.3)$$

12.5.2.2 The Process

Coal can contain up to 0.7 % of chloride in addition to sulfur. When burnt, it produces hydrogen chloride and oxides of sulfur. Before removing the sulfur oxides in an absorber, the gases are cooled and the hydrogen chloride gas is removed by water scrubbing and neutralised with limestone.

Finely divided limestone (with a typical specification of 90 % passing 45 µm) is slurried in water and pumped into the absorber. The scrubbed gases, which are cooled to about 65 °C in the lower part of the absorber, pass upwards through a spray of recycled gypsum/limestone slurry (Fig. 12.1).

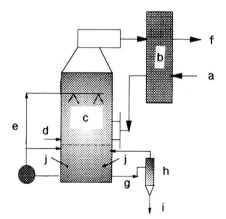

Figure 12.1. Diagram of an FGD absorber system
(a) flue gases; (b) gas-gas heat exchanger;
(c) absorber; (d) limestone slurry;
(e) recycled slurry; (f) de-sulfurised gases;
(g) slurry from sump; (h) gypsum hydroclone;
(i) gypsum slurry to filter; (j) air

Although the contact time between the gas and the spray is only about 5 to 10 sec., 90 % or more of the oxides of sulfur are removed [12.4].

The un-reacted limestone, together with calcium sulfite and some calcium sulfate, collects in a sump at the base of the absorber which gives a liquid retention time of about 5 min. The solids are kept in suspension by agitators. Compressed air is blown into the slurry to oxidise the calcium sulfite. The slurry is then recycled to the spray section to absorb more oxides of sulfur and to ensure that most of the limestone is reacted.

Slurry is drawn from the sump of the absorber and passed through hydrocyclones, which concentrate gypsum crystals that have grown to the required size. The gypsum slurry is fed to filters, which remove the liquor and rinse the solids. The gypsum so produced contains about 10 % of water.

The slurry containing under-sized crystals of gypsum and particles of partially reacted limestone is recycled to the absorber.

Part of the liquor removed by the filters is recycled and part is purged from the system via the effluent treatment plant. This uses lime to precipitate metals (extracted from the gases, the residual fly ash and the limestone), which are removed as a sludge.

12.5.2.3   Limestone Specification [12.4]

The key requirements are that the limestone should have an acceptable "reactivity" with respect to the reaction with sulfur dioxide and a high % $CaCO_3$.

The factors which determine the reactivity of the limestone are not fully understood, but include the magnesium content, the aluminium and fluoride contents, and the crystal structure. The rate of reaction also depends on the particle size of the limestone. Thus, while a lower reactivity can be offset by finer grinding, it raises both operating and capital costs. Details of reactivity tests have been published, see [12.5, 12.6].

The percentage $CaCO_3$ in the limestone should be at least 95 % — not only to provide the neutralising value to react with the oxides of sulfur, but also to minimise the level of impurities. Insoluble impurities tend to contaminate the gypsum, while soluble impurities pass into the effluent treatment plant and may give problems with meeting consent limits for the discharge.

The $MgCO_3$ content should be below 2 % as, although it reacts with the oxides of sulfur, higher levels may reduce the reactivity of the limestone and will raise the level of magnesium compounds in the gypsum to above the specified limits for the gypsum.

The level of SiO_2 in limestone is constrained by limits specified for the gypsum. In addition, silica could increase abrasion of equipment in the FGD process and in the gypsum processing plant, as well as introducing a potential respirable dust problem during the processing of the gypsum.

Aluminium and fluoride should be as low as practicable, because they can combine to produce a phenomenon called "limestone binding". This is believed to be caused by the formation of insoluble compounds on the surface of the limestone particles, which inhibit the neutralisation of the oxides of sulfur.

The permissible level of iron in the gypsum is limited, as iron can cause discolouration of gypsum-based products. This, in turn, imposes a limit on the amount of Fe_2O_3 in the limestone. Nevertheless, some iron is desirable as it accelerates the rate of oxidation of the calcium sulfite.

Manganese has also been found to assist the oxidation of sulfite. Cobalt, nickel and copper have been reported to have a minor beneficial effect on the oxidation process, although, under certain conditions, copper can have an inhibiting effect.

As the limestone must be ground to about 90 % less than 40 µm, it is important that the operating and capital costs of the grinding equipment should not be excessive. The Bond Work Index test [12.7] was developed as a measure of the ease with which a substance can be ground. The value of the index for many limestones is between 8 and 10 kW · hr/short ton.

Where the limestone is ground by the power generator, the particle size distribution and moisture content should be specified. The top size of the limestone should be selected to suit both the grinding equipment (usually ball mills) and the limestone producer's plant. Typically, a washed and screened product with a top size of 15 mm is selected. While a 15 mm to dust product would reduce the grinding costs by perhaps 20 %, it is generally preferable to remove the fines by screening at, say, 5 mm to avoid the production of excessive airborne dust, when transporting and handling the product. Some moisture also helps to control dust.

Table 12.3 summarises a typical specification for limestone [12.4, 12.8].

12.5.3 FGD Using Dry Processes

12.5.3.1 Introduction

The reaction between sulfur dioxide and dry limestone is too slow to be of commercial interest. However, the reaction with quicklime is more rapid. Investigations have, therefore, been made into the in situ production of quicklime, by the injection of finely divided limestone powder into the boiler.

Table 12.3. A typical limestone specification for FGD

Parameter	Requirement
$CaCO_3$ (as CaO)	$\geq 53.2\%$
CO_2	$\geq 41.8\%$
Total Mg (as MgO)	$\leq 0.5\%$
Insoluble Si (as SiO_2)	$\leq 0.65\%$
Insoluble Fe (as Fe_2O_3)	$\leq 0.25\%$
Total insolubles	$\leq 1.0\%$
Hardness (Bond Work Index)	≤ 10
Organic constituents	$\leq 0.2\%$
Whiteness (% absolute)	≥ 80
COD[a] level	≤ 150 mg/l

[a] COD = chemical oxygen demand

12.5.3.2   Injection Process

This consists of injecting very finely divided, ground calcium carbonate into the boiler at temperatures in excess of 900 °C. The quicklime so produced reacts with sulfur oxides to form calcium sulfite and sulfate and with hydrogen chloride to form calcium chloride.

$$CaO + SO_2 \rightarrow CaSO_3 \tag{12.4}$$

$$CaO + SO_3 \rightarrow CaSO_4 \tag{12.5}$$

$$CaO + 2HCl \rightarrow CaCl_2 + H_2O \tag{12.6}$$

The dust (a mixture of CaO, $CaCO_3$, $CaSO_3$, $CaSO_4$, $CaCl_2$ and fly ash) is collected either in an electrostatic precipitator, or a bag filter.

[12.9] compares the use of finely ground limestone, with a mean particle size of 10 to 20 µm, with that of hydrated lime (median particle size 2.5 to 4 µm). To remove 60 % of the sulfur dioxide produced by burning bituminous coal containing 0.66 % S, 4.0 times the stoichiometric quantity of limestone needs to be added. This compares with 2.0 times the stoichiometric quantity of hydrated lime. [12.10] comes to a similar conclusion, reporting that at a Ca ÷ S stoichiometric ratio of 2.0, limestone calcines in about 150 milli sec. at 1200 °C and removes 30 to 40 % of the sulfur, compared with 50 to 80 % removal with hydrated lime.

Although finely ground limestone costs less than hydrated lime per ton of sulfur removed, the latter has generally been favoured by operators of small to medium boilers, because of its greater reactivity and because it produces about 40 % less residue than limestone.

12.5.3.3 Fluidised Bed Combustion

Sulfur dioxide emissions from fluidised bed boilers are routinely reduced by adding limestone to the bed (e.g., [12.11]). The favoured top size of limestone for the "bubbling" and "turbulent" bed systems is less than 3 mm, with a bottom size of about 250 μm. More finely divided limestone is injected into re-circulating bed systems, as the fine fraction is recycled to the bed.

The reactivity of limestones with respect to the reaction with sulfur dioxide varies markedly. For example, for a given fluidised bed combustor, the Ca ÷ S stoichiometric ratio required to achieve a 90 % reduction in sulfur emission at atmospheric pressure, varies from 2 to 5. The reasons for such a variation are not understood, but are likely to include decrepitation, catalytic effects of minor components such as iron, and the structure of the limestone and lime [12.12]. Laboratory test methods have been developed for predicting the performance of sorbents [12.13, 12.14].

The optimum temperature for operation at atmospheric pressure is about 850°C for both high-calcium limestone and dolomite (N.B., it is understood that the magnesium oxide component does not react with the oxides of sulfur).

12.5.4 Hydrogen Fluoride Removal

12.5.4.1 Introduction

The clays used in many ceramic and earthenware products contain appreciable amounts of fluoride. During calcination, part of the fluorine is emitted as hydrogen fluoride, giving typical emission levels of 20 to 30 mg/m^3. To meet modern emission limits, the concentration needs to be reduced to below 10 mg/m^3. The technique of injecting powdered calcium carbonate or hydrated lime, however, is not particularly appropriate as such furnaces do not emit significant levels of dust and are, therefore, not fitted with dust arrestment equipment.

The favoured absorbent in this situation is granular limestone [12.15], although an absorbent based on calcium hydroxide is also used (see section 29.5). Both absorbents also remove oxides of sulfur.

12.5.4.2 The Process

Information on the process is sketchy, but at least two approaches have been adopted.

One, which is available commercially [12.16] appears to use a cross-flow reactor containing sand-sized limestone, believed to be 2 to 6 mm in size. With a contact time of about 0.5 sec., at least 85 % of the hydrogen fluoride is removed. The pressure drop is reported to be about 0.3 kPa.

By withdrawing part of the limestone from the base of the absorber and re-injecting it into the top, the bed of limestone is kept in a porous condition and fresh surfaces are exposed to the gases. The absorption efficiency is understood to remain at above 85 % until roughly 5 % of the limestone has reacted, which, with the design proposed, would require the limestone to be replaced every 15 months.

Another approach [12.10] uses the technology developed for the gravel bed filter. It operates at a pressure drop of 1.5 kPa and has given absorption efficiencies of 97 to 99 %.

The above processes offer two advantages over the hydrated lime injection process, namely:

a) they do not require a bag filter to capture the absorbent and
b) the spent absorbent consists mainly of calcium carbonate and calcium fluoride, both of which are insoluble and do not present any disposal problems (compare with section 29.5).

12.6   Sulfite Process for Paper Pulp

In the sulfite process for making paper pulp, lumps of high-calcium limestone may be used to react with sulfur dioxide to produce a $Ca(HSO_3)_2$ liquor. This liquor is used to digest the pulp to remove components of the pulp other than the cellulose. Increasingly, however, limestone is being replaced by alternative alkalis (magnesia, ammonia or soda ash), which can be recycled more readily.

12.7   Production of Organic Chemicals

Finely divided calcium carbonate is widely used in the production of organic chemicals, either to regulate the pH in reaction vessels or to neutralise the excess acid required in certain reactions (e.g. sulfonation or nitration). It is also used as a component of hydrogenation catalysts.

12.8   "Rock Dust" for Mines

A significant tonnage of pulverised limestone is used in mines, primarily to dilute coal dust and thereby to reduce the explosion risk. A typical specification is for a product which is less than 1 mm, with 70 % less than 75 μm [12.1]. The content of silica and combustible matter should be low. The dust is sprayed, generally as a dust, but also as a slurry over the roof, walls and floor of the mine.

Limestone dust is also used in metallic mines to control the airborne concentrations of silica and heavy metals [12.1].

12.9   Fillers and Extenders

Whiting and precipitated calcium carbonate (PCC) are incorporated into a wide range of materials and products to improve their performance, reduce their cost and/or increase their density. Many grades of filler are made with differing particle size distributions, particle shapes and (optionally) coatings to meet the requirements of different specifications. It should be emphasised that whiting and especially PCC are high added value products which are frequently tailored to precise customer requirements. Table 12.4 lists some of the applications [12.17], the more important of which are described in the following paragraphs.

a) *Floor Coverings*. Whiting/PCC are the most widely used fillers in PVC resins. They are also used in the backings of carpets, where the ratio of $CaCO_3$ to latex can range from 3 to 1 to 7 to 1 [12.19, 12.20].
b) *Paints and Coatings*. Whiting/PCC make up 80 to 90 % of the extenders used in paints in the EU. They are used in both oil-based and water-based paints, where they reduce shrinkage on hardening and improve adhesion. They also act as an optical brightener and reduce the cost of the paint [12.21].
c) *Paper*. Whiting is used to increase brightness, and opacity as well as reducing costs [12.22, 12.23].
d) *Plastics*. Whiting/PCC are the main fillers used in a number of plastics [12.24–12.27].
e) *Rubber*. Whiting reinforces rubber, making it harder, increasing its density and acting as an extender [12.28–12.30].
f) *Sealing Materials*. Whiting is used to reinforce and extend many sealants. Varying the particle size distribution enables the rheology of the sealant to be controlled [12.31].
g) *Flame Retarder*. Calcium carbonate filler in rubber, plastics and paints also acts as a flame retarder as a result of its decomposing into carbon dioxide and quicklime [12.32].
h) *Pesticides*. Whiting is used as a dispersant for powders that are applied as an aqueous suspension in pest control.
i) *Printing Inks*. Both whiting and PCC (coated with waxes/fatty acids) are used in white ink formulations.
j) *Toothpaste and Powder*. PCC is frequently used as a mild abrasive and filler in formulations [12.33].

12.10   Water Treatment

12.10.1   Drinking Water

Calcium limestones are used as a closely graded granular filter medium (e.g., 0.71 to 1.25 mm and 1.6 to 2.8 mm) [12.34]. The physical requirements are described in [12.35]. As a filter medium, it has the advantage that it can neutralise acidic water

Table 12.4. Uses of whiting and PCC

Absorbent	– of oil/chemicals	Fireworks	
Abrasive (mild)	– detergent cleaners	Floor coverings & tiles	– asphalt
	– cleaner-polishes		– linoleum
Adhesives	– rheology modifier		– vinyl
Adsorber	– surface degreaser		– vinyl
	– of emulsion		asbestos (white)
	stabilising	French chalk	
	surfactants	Mastics	– emulsifier
Agriculture	– adjustment		– stiffener in
	of soil pH		putties/mastics
Antacids (alkalisers)		Medicines	– anti-acid powders
Baking	– flour additive	Neutralisation of acid pickling baths etc.	
Brewing	– control of acidity	Paint	– filler & extender
Bituminous products			– rheology modifier
	– filler & stiffener		– hardener
Carrier	– insecticides &	Paper	– coating pigment
	fertilisers		– filler
Caulking compounds			– pH control
Ceramic glazes		Pharmaceuticals	
Cosmetics	– tint carrier	Plastics	– pvc extrusions
Dusting powder	– baby powder		– cable insulation
	– french chalk	Printing	– improvement of
	– anti-blocking		set of inks
	powder	Putties	
	– seeds	Rubber	– latex-based
Dentifrices			products
Digestion aids			– foam rubber
Emulsifier	– synthetic rubber		– polyurethanes
	– paper coatings		– mechanical
	– mastics		compounds
Explosives		Textiles	– weighter
Extender	– cost reduction	Toothpastes	– mild abrasive

(see section 12.10.3) and produce a controlled pH that approaches neutrality. However, where the water contains high levels of sulfate, or humic acid, the build-up of insoluble deposits on the surface of the grains can inhibit the neutralisation reaction.

As the limestone dissolves, the particle size decreases and the porosity of the packed bed decreases. Therefore, at intervals, the fines need to be removed and replaced with fresh material.

A European Standard is in preparation [12.34], which specifies the qualities of limestones used for drinking water treatment. Table 12.5 shows how the current proposals classify calcium carbonates into:

a) "non-porous", which is divided into three Types,
b) "porous", which is divided into two Types.

It sub-divides the porous and non-porous groups into Types A and B, depending on the levels of "toxic substances" (Table 12.6). (N.B., proposed revisions of the EU Directive for drinking water may reduce the limits for lead and arsenic in water, which could affect the limit values given in Table 12.6. Further information about the skew distributions of the concentrations of trace elements is given in section 28.1.8.)

Table 12.5. Calcium carbonate for use in water treatment – proposed major and minor components

Parameter	Non-porous calcium carbonate			Porous calcium carbonate	
	Type 1	Type 2	Type 3	Type 1	Type 2
	%	%	%	%	%
$CaCO_3$	> 98	> 94	> 88	> 95	> 85
$(CaCO_3 + MgCO_3)$	> 98	> 94	> 88	> 99	> 95
Insoluble in HCl	≤ 2	≤ 6	≤ 12	≤ 1	≤ 5

Table 12.6. Calcium carbonate for use in water treatment — proposed toxic substances

Parameter	Upper limits	
	Type A mg/kg	Type B mg/kg
Arsenic	3	5
Cadmium	2	2
Chromium	10	20
Nickel	10	20
Lead	10	20
Antimony	3	5
Selenium	3	5
Mercury	0.5	1

12.10.2   Sewage Treatment

Limestone aggregate (e.g. 40 to 60 mm) has been used in sewage plants as a "filter" bed through which plant effluent is passed. The aggregate provides a surface on which bacteria can become established and feed on other bacteria in the effluent [12.36]. The stone should pass a severe soundness test.

12.10.3   Acid Neutralisation

Limestone is undoubtedly the lowest cost reagent for the neutralisation of acids. It suffers from certain disadvantages, which may have prevented it from being used to its full potential.

a) It can only neutralise to pH 6.5, or perhaps pH 7.0 with aeration to displace the dissolved carbon dioxide.
b) Its rate of reaction is slow relative to milk of lime and other alkalis.
c) With the exception of nitric and hydrochloric acids, the reaction products are either insoluble or sparingly soluble and inhibit the reaction. This effect can be overcome by using finely divided calcium carbonate. Such a product is likely to be expensive, unless it is a waste product.
d) The reaction generates carbon dioxide, which may cause frothing problems.
e) The volume and weight of sludge tends to be greater than with other alkalis, resulting in higher disposal costs.

Nevertheless, it is believed that a significant quantity of high-calcium limestone/calcium carbonate is used for this purpose. A two-stage neutralisation process may offer the best overall economics. In the first stage, calcium carbonate is used to neutralise the greater part of the acid under acidic conditions. In the second stage, an alkali is used to neutralise the residual acid and to raise the pH to the required level. For more details of the chemistry involved, see section 28.3.

Dolomitic limestone is unsuitable for acid neutralisation, as its reaction rates are too slow.

12.11   Sodium Dichromate

Sodium dichromate is produced by roasting chromite ores with soda ash. High-calcium limestone or dolomite is incorporated into the mixture for two reasons. Firstly it acts as a source of lime, which lowers the melting point to below 1000 °C. Secondly, it releases carbon dioxide, which gasifies the melt and, after cooling, produces a porous product. The quantity of limestone/dolomite added is controlled to ensure that the product corresponds to the compound $5Na_2CrO_4 \cdot CaCrO_4$ [12.37]. The product is ground and the sodium dichromate extracted for use in the production of chromium oxides (including the magnetic chromium dioxide, used in video and audio tapes), some forms of which are used as green pigments.

12.12   Calcium Zirconate

One of the processes for extracting zirconium from the mineral zircon ($ZrSiO_4$) is to fuse the mineral with limestone or dolomite. The reaction product disintegrates on cooling into powdered calcium silicate and coarse crystals of calcium zirconate (equation 12.7). The zirconate is dissolved in acid and converted into zirconium salts or zirconium oxide, much of which are converted into the corrosion-resistant zirconium metal [12.38].

$$2ZrSiO_4 + 5CaCO_3 \rightarrow 2CaZrO_3 + (CaO)_3(SiO_2)_2 + 5CO_2 \qquad (12.7)$$

12.13 **References**

[12.1]   R.S. Boynton, "Chemistry and Technology of Lime and Limestone", John Wiley & Sons, 1980.

[12.2]   BS 3108: "Specification for Limestone for Making Colourless Glasses", 1980 (1993), ISBN 0-580-11395-7.

[12.3]   J.R. Cooper, I.A. Johnston, "The Engineering Requirements of the CEGB's Programme for Flue Gas Desulfurisation". In: IMechE Seminar on "Fossil Fired Emissions", 6 Dec. 1988.

[12.4]   I.A. Johnston, S. Westaway, "A Review of the Factors Influencing the Selection of Desulfurisation Process of Large Fossil-fired Plant", Proc. International Lime Congress, Berlin, 1994.

[12.5]   J.W. Morse, American Journal of Science, Vol. 274, Feb. 1974, 97–107.

[12.6]   A.R. Ellis, "An Investigation of UK Limestone Reactivity for the Limestone-gypsum FGD Process", from "Desulfurisation in Coal Combustion Systems", Institution of Chemical Engineers Series **106**, April 1989, 19–21.

[12.7]   "Test Method for Limestone Grindability and Calculation of Bond Work Index", Method 7. In: "FGD Chemistry and Analytical Methods Handbook", Vol. 2, EPRI Report CS 3612, July 1984.

[12.8]   B. Oppermann, M. Mehlmann, N. Peschen, "Products from the Lime Industry for Environmental Protection", Zement Kalk Gips **6**, 1991, 265; **8**, 1991, 159.

[12.9]   K. Schneider, "Mopping-up Pollution", Gypsum, Lime and Building Products, April, 1996, 23.

[12.10] P. Flament, M. Morgan, "Fundamental and Technical Aspects of SO_2 Capture by Ca-based Sorbents in Pulverised Coal Combustion", International Flame Research Foundation, Doc. No. F 138/a/8, 1987.

[12.11] M.J. Cooke, I. Highley, "The use of Limestone for Sulfur Capture in Fluidised Bed Combustion", Proc. 1st European Technical Conference of the European Lime Association, Cologne, Oct. 1992.

[12.12] J.E. Stanton, "Sulfur Retention in FBC". In: "Fluidised Beds — Combustion and Applications", ed. J.R. Howard, Applied Science Publishers, London, 1983, chapter 5.

[12.13] D.C. Fee et al., "Sulfur control in Fluidised Bed Combustors: Methodology for Predicting the Performance of limestone and Dolomite Sorbents", Argonne National Laboratory, Illinois, 1980.

[12.14] R.A. Newby et al., "A Technique to Project Sulfur Removal Performance of Fluidised Bed Combustors", Proc. Sixth International Conference of Fluidized Bed Combustion, Atlanta, April 1980.

[12.15] "Ceramic Processes", IPC Guidance Note S2 3.04, 1996.

[12.16] Brochure by E.I. Tec. GmbH of D-95463, Bindlach, Germany.

[12.17] R.R. Davidson, "Natural Chalk Whiting — Nature and Uses", Information Paper, Welyn Hall Research Association, 1969.

[12.18] J.R. Cooper et al., "Sulfur Removal from Flue Gases in the Utility Sector: Practicalities and Economics", Proc. Institution of Mechanical Engineers, Vol. 211, Part A, 1997, 11–26.

[12.19] F. Werny et al., "Floor coverings", Ullmann's*.

[12.20] K. Fischer et al., "Textile Auxiliaries", Ullmann's*.

[12.21] D. Stoye et al., "Paints and Coatings", Ullmann's*.

[12.22] P. Platt, "Paper and Pulp", Ullmann's*.

[12.23] H. Heine et al., "Pigments, Inorganic", Ullmann's*.

[12.24] R. Wolf, B.L. Kaul, "Plastics, Additives", Ullmann's*.

[12.25] H.-G. Elias, "Plastics, General Survey", Ullmann's*.

[12.26] M.W. Allsopp, G. Vianello, "Poly(vinyl chloride)", Ullmann's*.

[12.27] G.W. Ehrenstein, J. Kabelka, "Reinforced Plastics", Ullmann's*.

[12.28] H.-H. Greve, "Rubber, 2 Natural", Ullmann's*.

[12.29] H.-W. Engels, "Rubber, 4 Chemicals and Additives", Ullmann's*.

[12.30] W. von Langenthal, J. Schnetzer, "Rubber, 5 Technology", Ullmann's*.

[12.31] R.A. Palmer, J.M. Klosowski, "Sealing Materials", Ullmann's*.
[12.32] B.J. Sutker, "Flame retardants", Ullmann's*.
[12.33] W. Weinert, "Oral Hygiene Products", Ullmann's*.
[12.34] prEN 1018: "Chemicals used for treatment of water intended for human consumption — Calcium carbonate".
[12.35] M.R. Smith, L. Collis, "Aggregates", The Geological Society, Bath, 1993.
[12.36] "Filtering Materials for Sewage Treatment Plants", American Society of Civil Engineers, Manual Eng. Prac. **13**, 1937.
[12.37] G. Anger et al., "Chromium Compounds", Ullmann's*.
[12.38] R. Nielsen, "Zirconium", Ullmann's*.

---

* Ullmann's Encyclopedia of Industrial Chemistry, 5th ed. on CD-ROM, WILEY-VCH, Weinheim, 1997.

Part 3   Production of Quicklime

13 Physical and Chemical Properties of Quicklime

The Chemical Abstracts Service (CAS) registry numbers for CaO and MgO are 1305-78-8 and 1309-48-4. The corresponding reference numbers in the European Inventory of Existing Commercial Substances (EINECS) are 215-138-9 and 215-171-9.

13.1 Physical Properties

- *Molecular weight.* The values for calcium and magnesium oxides are 56.08 and 40.31 respectively.

- *Colour.* Most quicklimes are white. Impurities, particularly iron and manganese, can result in grey, brown, or yellow tints. When quicklime is burned using a solid fuel, the ash can produce a brown or grey surface coating.

- *Odour.* Quicklime has a slight earthy odour.

- *Texture.* All quicklimes are micro-crystalline, but appear to be amorphous to the naked eye.

- *Crystal structure.* The crystal structures of both calcium oxide and magnesium oxide are cubic. The sum of the ionic radii of CaO is 2.4 Å while that of MgO is 2.1 Å, which accounts for doloma (CaO · MgO) having a greater density than calcium oxide [13.1].

- *Porosity.* Part of the porosity of particles of commercial quicklime arises from the porosity of the limestone, and part from the decomposition process. The porosity of commercially produced quicklime can be as high as 55 % (by volume), when a porous limestone is lightly burned. Exposure to elevated temperatures results in sintering (see sections 15.4 and 15.5), which can reduce the porosity to below 25 %. Dead-burned dolomite has a porosity of about 10 %.

- *Specific gravity.* Reported values for the specific gravity of calcium oxide (with zero porosity) range from 3.25 to 3.38 g/cc [13.2]. That for calcined dolomite is 3.5 to 3.6 g/cc [13.1].

- *Apparent density.* Light-burned quicklimes produced from porous limestones can have apparent densities below 1.4 g/cc. Sintering, caused by prolonged exposure to high temperatures progressively raises the apparent density (see sections 15.4 and 15.5). Calcined dolomite generally has a higher apparent density than high-calcium quicklime, if given the same heat treatment. Dead-burned dolomite, with a porosity of 10 %, has an apparent density of about 3.2 g/cc [13.1].

- *Mean apparent density.* As the heat treatment of individual lumps in a commercial lime kiln can vary widely, their apparent densities also vary. The mean apparent density is used to characterise the average degree of sintering of a particular quicklime. For a particular high-calcium quicklime, the mean apparent density correlates with the reactivity to water (see Fig. 13.2).

- *Bulk density.* This depends on the mean apparent density of the particles and on the voidage between them. The latter is related to the particle size distribution and particle shape. Most commercial screened quicklime products have bulk densities in the range 900 to 1200 kg/m^3. Inclusion of fines, which fill the interstices, can increase the bulk density by 30 %.

- *Angle of repose.* The angle of repose for cubical, well-graded "pebble" quicklime is about 35°. The angle of repose increases with increasing fines content.

- *Hardness.* Most commercial quicklime products have a hardness of 2 to 3 Mohs. The value for dead-burned dolomite is in the range 3 to 5 Mohs [13.1].

- *Coefficient of thermal expansion.* Values of about 140×10^{-7} have been reported [13.3].

- *Thermal conductivity.* A value of 0.0015 to 0.002 cal/cm$^2 \cdot$ sec $\cdot$ °C is reported [13.1].

- *Melting point.* The melting point temperatures for calcium oxide and magnesium oxide are 2580 °C and 2800 °C respectively [13.2]. The value for calcined dolomite is reported to be about 2400 °C, with a eutectic mixture of 67 % CaO and 33 % MgO melting at 2370 °C [13.4].

- *Boiling point.* Values of 2850 °C and 3600 °C have been reported for calcium oxide and magnesium oxide respectively [13.2].

- *Specific heat.* The mean specific heats at 20 °C for CaO and doloma are 0.182 and 0.198 cal/g $\cdot$ °C respectively. The values at higher temperatures for CaO are given in Table 13.1, with integrated values over temperature ranges being given in Fig. 16.18 [13.5].

Table 13.1. Specific heat of calcium oxide

Temperature °C	Specific heat cal/g · °C
0	0.182
200	0.209
400	0.230
600	0.252
800	0.270
1000	0.288
1200	0.307

- *Heat of formation.* Values of 151,900 cal/mole and 143,800 cal/ mole have been quoted for calcium oxide and magnesium oxide, respectively [13.1].

- *Solubility.* The equilibrium solubility in water is that of hydrated lime, i.e., 1.85 g $Ca(OH)_2$/l at 0 °C — for more details see section 19.1.

- *Electrical resistivity.* A value of 71×10^8 ohms/cm at 15 °C has been reported [13.1].

13.2 Chemical Properties

- *Stability.* Both quicklime and calcined dolomite are refractory substances, as evidenced by their high melting points. They react with water in both the liquid and vapour state (see below). Many reactions apparently involving quicklime and calcined dolomite proceed via the hydrated products. Such reactions are described in section 19.2, but see also the reactions with carbon dioxide below.

- *Heat of hydration/slaking.* The heat liberated by the reaction of quicklime with water is 1140 kJ/kg of CaO. The value for fully slaked calcined dolomite is 880 kJ/kg of CaO · MgO [13.1].

- *Reactivity* to water is measured by the rate of release of the heat of hydration [13.3, 13.5, 13.6], or by the rate at which an aqueous suspension produces hydroxyl ions (which are neutralised at pH 9.2 by hydrochloric acid) [13.4]. A large number of reactivity tests have been used [13.7–13.11]. The relationships between some of those test methods (for a high-calcium quicklime) are given in Fig. 13.1 (N.B., the relationships for other limes may differ significantly from those given).

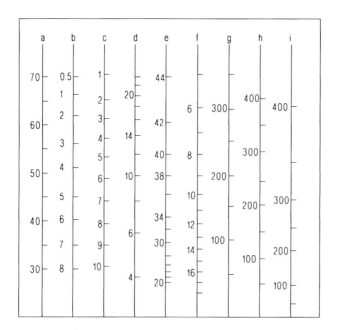

Figure 13.1. Approximate relationships between reactivity tests for a high-calcium quicklime. (N.B. results vary with grading of quicklime and source of limestone)
(a) BS 6463, °C after 2 min. [13.7]; (b) EN 459-2, t_{60} (min.) — time to reach 60 °C [13.11]; (c) EN 459-2, t_u (min.) — time for 80 % slaking [13.11]; (d) ASTM C110, temperature rise (°C) after 30 sec [13.8]; (e) ASTM C110, maximum temperature rise (°C) [13.8]; (f) ASTM C110, time to maximum temperature (min.) [13.8]; (g) acid titration, (ml) at 3 min. [13.9]; (h) acid titration, (ml) at 5 min. [13.9]; (i) acid titration, (ml) at 10 min. [13.9]

The terms "low", "moderate", "high" and "very high" reactivity are used throughout this book and may be taken as corresponding to the following approximate ranges (expressed in terms of the BS 6463 reactivity test [13.7]):

a) low — below 40 °C,
b) moderate — 40 to 55 °C
c) high — above 55 °C
d) very high — above 65 °C

Figure 13.2 shows how the reactivity of a dense high-calcium quicklime is related to its mean apparent density.

It should be noted that the result of a reactivity test can be markedly affected by impurities in the water. For this reason, distilled water is used as the reference standard. The reactivity of a quicklime can also be depressed by absorption of water and carbon dioxide from the atmosphere during storage or sample preparation (see affinity for water and section 24.2).

The low reactivities of calcined dolomite and dolomitic quicklimes arise from the inhibiting effect on the rate of hydration of the magnesium oxide component (which is much more prone to over-burning, hydrates more slowly and also has a lower solubility).

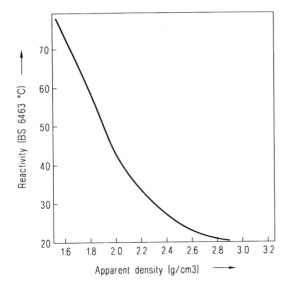

Figure 13.2. Relationship between reactivity and mean apparent density for a dense, high-calcium quicklime

- *Affinity for water.* Quicklime has a high affinity for water and is a more efficient desiccant than silica gel. When quicklime must be placed in a desiccator, phosphorous pentoxide should be used as the drying agent. Laboratory samples should, therefore, be stored in sealed containers and not in desiccators. Because of its affinity for water and, after partial hydration, for carbon dioxide, care should be taken to minimise exposure of quicklime to the atmosphere.

Relatively low levels of air slaking can reduce the reactivity significantly. For example, 1.5 % of combined water, coupled with a smaller, but un-measured amount of re-combined carbon dioxide has been observed to reduce the BS 6463 reactivity from 60 to 50 °C. As a general rule, reactivity results for quicklime samples with combined water contents in excess of 1.0 % may not be valid.

The reaction of quicklime with water is associated with an increase in volume of at least a 2.5 times factor.

- *Reaction with carbon dioxide.* In practice, quicklime reacts with carbon dioxide under ambient conditions. This is because commercially-produced quicklime always contains sufficient calcium hydroxide (produced by reaction with atmospheric water vapour in the cooling zone of the lime kiln) for carbon dioxide to be absorbed by the hydroxide (equation 13.1). That reaction releases water, which is available to react with more quicklime to produce more hydroxide etc, until the quicklime is fully converted into calcium carbonate (equation 13.2), see section 19.2.

$$Ca(OH)_2 + CO_2 \rightarrow CaCO_3 + H_2O \qquad (13.1)$$

$$H_2O + CaO \rightarrow Ca(OH)_2 \qquad (13.2)$$

In the absence of calcium hydroxide, quicklime only reacts with carbon dioxide at temperatures above about 300 °C and below 800 °C. This reaction can occur in the cooling zone of lime kilns under abnormal conditions. It results in carbonation of quicklime at the surface of the pores, and can give abnormally low reactivities for a given mean apparent density.

- *Impurities and trace elements*

    a) In most high-calcium quicklimes, *calcium carbonate* is the second largest component after CaO. In some applications calcium carbonate can be regarded as an inert substance, while in others it is an active component.

    b) *Magnesium oxide* is not regarded as an impurity in quicklime used for steel-making (see section 27.4.2), but levels in excess of 2 to 5 % can render a quicklime unsuitable for other applications (e. g., plaster, render, sand-lime bricks, autoclaved aerated concrete, soda lime and the ammonia-soda process).

    c) *Silicon, aluminum, iron and sulfur* are generally present in the limestone, but the levels may be increased significantly when solid fuel is used to calcine the limestone: such impurities are generally higher in the finer fractions of run-of-kiln product (see section 17.1.3). Limes containing over 8 % of silica plus alumina may possess hydraulic properties (see section 16.10.1).

    d) The levels of *trace elements* such as lead, antimony, arsenic, cadmium, chromium, molybdenum, nickel and selenium are largely determined by their concentrations in the limestone from which the lime is made (N. B. some trace elements, such as boron and mercury are volatile and may be removed by calcination). High levels of "toxic substances" can render a quicklime unsuitable for applications such as drinking water treatment (see section 28.1.8) [13.12, 13.13].

- *Reactions with acid gases.* In practice, quicklime reacts with acidic gases under ambient conditions. This is because commercially-produced quicklime always contains sufficient calcium hydroxide for the gases to react with the hydroxide (compare with reaction with carbon dioxide).

Dry acid gases do not generally react readily with CaO at ambient temperatures, although it is reported that dry hydrogen chloride reacts with quicklime above 80 °C [13.1].

At temperatures above about 350 °C (at which calcium hydroxide begins to dissociate), hydrogen fluoride and hydrogen chloride react readily with calcium oxide. The reactions with the oxides of sulfur are slower.

The reactions of calcium oxide with sulfur trioxide at temperatures below 1300 °C and with sulfur dioxide in the range below 980 °C are commercially important in the context of both the sulfur cycle in lime kilns (e.g., see sections 16.4.9 and 16.4.10), and the desulfurisation of flue gases (see section 29.4). Sulfur dioxide reacts with calcium oxide at temperatures below 980 °C to form calcium sulfite (equation 13.3). In the presence of excess oxygen, the sulfite is oxidised to sulfate (13.4), which is stable up to 1370 °C [13.14]. Calcium sulfate is also produced by the reaction of CaO with SO_3 (13.5). In the presence of carbon monoxide, however, less sulfite is oxidised to sulfate. Moreover, the carbon monoxide reacts with the oxygen produced by the decomposition of the sulfate, thereby reducing the effective decomposition temperature by up to 100 °C (13.6).

$$CaO + SO_2 \rightleftarrows CaSO_3 \qquad (13.3)$$

$$CaSO_3 + 0.5O_2 \rightleftarrows CaSO_4 \qquad (13.4)$$

$$CaO + SO_3 \rightleftarrows CaSO_4 \qquad (13.5)$$

$$CaSO_4 + CO \rightarrow CaO + SO_2 + CO_2 \qquad (13.6)$$

- *Reactions with metals.* In the absence of water, quicklime does not react with metals. However, see section 19.2 for the reaction of hydrated lime with the amphoteric metals (namely aluminium, lead, tin, brass and zinc). Iron and steel are not attacked and are widely used as materials of construction in lime handling plants.

- *Reaction with carbon.* Calcium carbide is produced by reacting quicklime with carbon, at temperatures of 1800 to 2100 °C, to produce calcium carbide (see section 31.4).

13.3   References

[13.1]   R.S. Boynton, "Chemistry and Technology of Lime and Limestone", John Wiley & Sons, 1980.
[13.2]   "CRC Handbook of Chemistry and Physics", 77th ed., 1996–97.
[13.3]   Norton, J. Am. Ceram. Soc. **8**, 1925, 799.
[13.4]   R. Doman et al., J. Am. Ceram. Soc. **46**, July 1963, 313.
[13.5]   J. Murray, "Specific Heat Data for evaluation of lime kiln performance", Rock Prod., Aug. 1947, 148.
[13.6]   F. Smyth, L. Adams, J. Am. Chem. Soc. **45**, 1926, 785.
[13.7]   BS 6463: "Quicklime, Hydrated Lime and Natural Calcium Carbonate", Part 3, 1987.
[13.8]   ASTM C 110-95: "Test Methods for Physical Testing of Quicklime, Hydrated Lime and Limestone", 1995.
[13.9]   N.E. Rogers, Cem. Lime and Gravel, June 1970, 149–153.
[13.10] EN 459-2: "Building Lime: Test methods", 1995.
[13.11] EN 459: "Building lime — Part 2: Test methods", 1994 (being revised).
[13.12] prEN 12518: "Chemicals used for the treatment of water intended for human consumption — high calcium lime", in preparation.
[13.13] prEN 1017: "Chemicals used for the treatment of water intended for human consumption — half-burnt dolomite", in preparation.
[13.14] F. Schwarzkopf, "Lime Burning Technology — a Manual for Lime Plant Operators", 3rd ed., Svedala Industries, Kennedy Van Saun, 1994.

14 Raw Materials for Lime Burning (Limestone, Fuel and Refractories)

14.1 General

The limestone, fuel and refractory linings used for lime burning should be selected with care in the context of the kiln design and the required quicklime quality. For a new installation, the kiln supplier should be able to test and assess the suitability of the raw materials. For an existing lime works, the potential effects of proposed changes in the limestone, fuel or refractories should be assessed with care.

14.2 Limestone

14.2.1 Introduction

When selecting a limestone for the production of quicklime, the necessary tests should be done to ensure that it is suitable in all respects. Indeed, in the context of present day requirements for quicklime, only a small proportion of limestone deposits are suitable for lime burning.

The following requirements should be taken into consideration when assessing the quality of a limestone deposit.

a) strength and abrasion resistance of the stone,
b) resistance of the limestone to thermal degradation,
c) strength and abrasion resistance of the quicklime,
d) rate of calcination,
e) reactivity of product,
f) content of $CaCO_3$, $MgCO_3$, and
g) content of impurities that might affect kiln operation, or the acceptability of the lime.

In addition, the quality of the selected grade of processed limestone should be assessed, in terms of:

a) particle size distribution,
b) shape,
c) contamination with clay particles,
d) surface cleanliness, and
e) consistency.

Limestones that are relatively soft, are subject to thermal degradation, or produce a soft lime may not be suitable for calcining in shaft kilns. They may, however, be calcined in certain types of rotary or other kilns (see section 16.4.11).

14.2.2 Strength and Abrasion Resistance of the Limestone

The strength and abrasion resistance of the limestone should be adequate to withstand the physical forces to which they are exposed in the stone handling system and in the kiln. The production of fines by crushing and abrasion tends to reduce the porosity of the limestone bed in the kiln, thereby increasing pressure drops in shaft kilns and reducing the rate of heat transfer in rotary kilns. Where breakage presents problems, re-screening, to remove undersized particles immediately before charging into the kiln, is good practice.

14.2.3 Thermal Degradation of Limestone

Thermal degradation can occur for three reasons.

a) Some limestones decrepitate at temperatures well below 900 °C.
b) Some limestones are prone to break up if exposed to thermal cycling (e.g. as in the parallel flow regenerative kiln).
c) Some limestones degrade on calcination.

Such degradation reduces the porosity of the bed and can adversely affect calcination and cooling of the lime. Moreover, the quantity of fines (e.g., 5 mm to dust) increases, reducing the yield of the premium screened products.

The causes of decrepitation are related to the effects of heat on the crystals. Limestones with large crystallites are often more prone to decrepitation than micro-crystalline deposits. The only satisfactory approach is to compare their performance with samples which are known to be suitable for a particular design of kiln [14.1].

14.2.4 Strength and Abrasion Resistance of the Quicklime

The quicklime needs to have an adequate resistance to crushing, impact and abrasion to avoid producing excessive fines within the kiln and in the lime handling system. As mentioned above, fines can reduce the porosity of the bed, increase pressure drop and reduce heat transfer.

Empirical tests have been developed to compare the strength and abrasion resistance of a lime with those of an established product which has proved to be satisfactory.

14.2.5   Rate of Calcination

Rates of calcination of different limestones under identical conditions can vary markedly (see section 15.3). For a given kiln output, a limestone with a slow calcination rate requires either a larger calcining zone, or higher temperatures in that zone, and due allowance should be made when designing the kiln and/or its refractory lining.

14.2.6   Reactivity of Product

For identical calcining conditions, different limestones may produce quicklimes with measurably different reactivities. While the differences in reactivity may not be significant in most operations, they could affect the suitability of the quick-limes for certain applications (e.g. aerated concrete). There may be several causes of such differences, e.g. rate of calcination, effects of impurities and the micro-structure of the lime, (e.g. lime produced from oolitic and coarsely crystalline limestones is more prone to sinter and become less reactive [14.2]).

14.2.7   Content of $CaCO_3$, $MgCO_3$ and Impurities

The requirements of the quicklime market generally dictate the quality of lime-stone required. Although a high $CaCO_3$ content of > 97 % is an advantage in some markets, it is by no means essential, as shown by the following two exam-ples.

a) Class CL70 building lime (ENV 459-1 [4.3]) specifies a $CaO + MgO$ content of at least 70 % with not more than 5 % of MgO. This corresponds approximately to a limestone of at least 81 % $(CaCO_3 + MgCO_3)$ and not more than 6.7 % of $MgCO_3$.
b) Type 3 high-calcium lime for water treatment (prEN 12518 [14.4]) requires a CaO content of at least 80 %, corresponding to at least 87.7 % $CaCO_3$ in the limestone.

For some applications, magnesium carbonate in the limestone can be regarded as an active ingredient, although tests should be made to ensure that the resulting MgO in the lime is reactive. For example, limestone used for producing quicklime for basic oxygen steelmaking may be judged by its total $(CaCO_3 + MgCO_3)$ con-tent, although over-burning of the MgO may depress the reactivity to water.

There are many applications where magnesium is regarded as an undesirable impurity. These include the production of hydrated lime (at atmospheric pres-sure), aerated concrete, sandlime bricks, and precipitated calcium carbonate, for which $MgCO_3$ levels should preferably be less than 2 % and, ideally, less than 1 % in the limestone.

The required levels of impurities tend to be specific to particular industries. Perhaps the most comprehensive set of requirements are those for water treatment (prEN 12518 [14.4]), which includes limits for SiO_2, Al_2O_3, Fe_2O_3, MnO_2, Sb, As, Cd, Cr, Pb, Hg, Ni and Se (see section 28.1.8).

High levels of sodium and potassium can volatilise and contribute to the formation of rings in rotary kilns (see section 16.4.9). High levels of silica, alumina and iron oxide (e.g., 5 to 10 %) can cause fusion leading to "crotches" (also known as "scaffolds" and "bears") in shaft kilns [14.5]. They can also contribute to rings and clinkering in rotary kilns. Where the lime is to be supplied to the steel industry, sulfur levels above 0.01 % S in the stone may either restrict the fuel/kiln options, or render the lime unsuitable (see section 27.4.3).

14.2.8 Particle Size Distribution

The particle size distribution must be compatible with the requirements of the kiln. This generally requires the stone to be positively screened to give a size ratio of, ideally 2 to 1 or at least 3 to 1. While some kilns might tolerate a wider size range, the consequence is almost certainly a reduction in product quality, output and/or heat usage per tonne product, caused by (a) the range of calcining times and (b) reduced voidage in the bed.

The top size of the feedstone to shaft kilns is generally in the range of 75 to 150 mm, although some kilns have recently been designed to accept stone with top sizes down to 30 mm [14.6, 14.7]. Rotary lime kilns can accept feedstones with top sizes from dust up to 50 mm, depending on the design. (N.B., as mentioned in section 5.3.2.1, a 100 mm square mesh screen will allow slabby particles with maximum dimensions approaching 200 mm to pass.)

14.2.9 Particle Shape

Stone for lime-burning can be slightly slabby. This allows gyratory, rolls and jaw crushers to be used, which produce a lower proportion of undersized product and consume less power per tonne of stone.

Such slabby stone does result in slightly higher resistance to gas flow than a similar size of more cubical stone. This necessitates the use of fans with increased suction, but the extra power requirement is more than offset by the above benefits.

14.2.10 Cleanliness

It is important that the feedstone to lime kilns should be clean. Cleanliness refers to low levels of:

a) undersized lumps of stone, which tend to reduce voidage,
b) contamination of the surface of the stone particularly with clay, which can fuse one lump to another, and
c) lumps of clay which also can cause fusion.

Undersized lumps may be removed by screening prior to charging the stone into the kiln. Surface contamination may be removed by rinsing the stone with water on a screen. The presence of lumps of clay are best avoided by a combination of selective quarrying, screening and washing.

14.2.11   Consistency

The importance of consistency in the quality of feedstone, particularly to shaft kilns cannot be over-emphasised. It is a frequent source of friction between quarry and kilns personnel.

For example, the author has experience of a sudden and temporary increase in the proportion of large particles within a specified size range (caused, as it was subsequently learned, by "pushing in" a low stockpile of kilnstone). That caused a significant change in the calcining characteristics of the feedstone, which resulted in an increase in lime reactivity and carbonate. The operator attempted to correct the unexpected increase in both reactivity and carbonate by raising the fuel input. However, before he took that action, the grading of the feedstone had returned to normal and his attempted correction resulted in the reactivity of the lime being below specification for some days.

14.3   Fuel

14.3.1   Introduction

The choice of fuel(s) for a lime-burning operation is important as:

a) the cost of fuel per tonne of lime may represent 40 to 50 % of the production cost,
b) an inappropriate fuel can cause major operating problems,
c) the fuel can influence the quality of the lime, notably the $CaCO_3$ level, the reactivity, the sulfur content and impurities such as SiO_2, Al_2O_3 and Fe_2O_3, and
d) different fuels contribute in different ways to the environmental impact of the exhaust gases.

Because of the wide range of fuels used in lime-burning, the variety of kilns used, the various product qualities required and the interactions between them, it is difficult to discuss the subject in a structured way. The approach adopted is to discuss fuels in general and the ramifications for several groups of kiln, in the context of the characteristics of the various fuels used in practice.

Optimising the selection of fuel for a given kiln is a site-specific process, which may well involve trials to ensure that there are no unexpected problems with a particular fuel. The optimum fuel may well change over time, due to variations in fuel prices, or availabilities. Considerable efforts have, therefore, been made by kiln suppliers and operators to ensure that kilns can be fired with alternative fuels.

14.3.2   General

In principle, any fuel capable of producing temperatures in excess of 900 °C could be used to produce lime. Table 14.1 lists many of the fuels used in practice, with an indication of how widely each is used. In practice, the availability of alternative fuels will restrict the options at a particular works.

Table 14.1. Fuels used in lime-burning

Type of fuel	Widely used	Sometimes used	Rarely used
Solid	Bituminous coal Coke	Anthracite Lignite, PET coke	Peat Oil shales
Liquid	Heavy fuel oil	Medium fuel oil	Light fuel oil, SLF [a]
Gaseous	Natural gas	Butane/propane Producer gas	Towns gas
Vegetation		Wood/sawdust	Biomass

[a] SLF = secondary liquid fuels

The cost of fuel per tonne of lime depends on:

a) the unit cost of the fuel,
b) the *net* calorific value (see section 14.3.6), and
c) the kiln *net* heat usage per tonne lime, which can be affected by the characteristics of the fuel.

The operational acceptability of a particular fuel depends on:

a) the ease of handling and firing it,
b) the level of control over the combustion process, and
c) any operating problems caused by the ash fusing on to the lime or refractories, and the sulfur in the fuel causing calcium sulfate deposits.

The acceptability of a fuel from the viewpoint of product quality is largely related to:

a) the level of ash and its fusion temperature,
b) the sulfur content,
c) the extent to which the fuel contaminates the product, and
d) the extent to which that contamination can be tolerated by the customers.

In addition, the choice of fuel can affect the emission of carbon dioxide, carbon monoxide, smoke, dust, sulfur dioxide and oxides of nitrogen, all of which have an environmental impact and may have financial implications.

For a particular fuel, the importance of many of the above factors is specific to the kilns used and/or to the works where they are situated. Nevertheless, it is useful to make some cautious generalisations.

Table 14.2 presents an analysis [14.8] which compares five of the more commonly used fuels. It suggests that, natural and liquefied petroleum gases are relatively easy to use and have little adverse effect on product quality. The solid fuels are generally more difficult to use and adversely affect product quality. Heavy fuel oil is usually intermediate in terms of ease of use and effect on quality.

Table 14.2. Characteristics of fuels for lime-burning

Type of fuel	Ease of handling and feeding	Control over combustion	Impurities sulfur	ash
Natural gas	✓✓	✓✓	✓✓	✓✓
Liquified petroleum gas	✓✓	✓✓	✓✓	✓✓
Heavy fuel oil	0	✓	×	✓
Pulverised coal	×	✓	×	×
Coke and anthracite [a]	0	×	×	×

[a] For mixed feed shaft kilns only.
✓✓ very favourable, ✓ favourable, 0 moderately favourable, × unfavourable.

In a logically structured fuel market, where bituminous coal, heavy fuel oil and natural gas are all available, it would be expected that the cost per unit of net heat would increase progressively from coal to oil to natural gas. This would reflect the different capital and operating costs of the preparation and control equipment, their decreasing levels of impurity and environmental impact. However, market forces often distort the expected price differentials and the astute lime producer will seek to take advantage of those distortions to produce the required quality of lime at minimum total cost.

14.3.3   Kilns — Fuels Used for Various Types of Kiln

14.3.3.1   Early Designs of Kiln

The early "pot" or "field" kilns (see section 16.3.2 for details) were generally wood-fired. Many operated on a batch basis, in which alternate layers of wood and stone were placed in the kiln before setting light to the wood at the base of the kiln. Others operated continuously, with alternate layers of wood and limestone being charged into the top of the kiln.

Wood is still a favoured fuel for the basic kilns which are used in many developing countries [14.2, 14.9].

Green wood contains a large proportion of water and is generally air-dried for at least twelve months before use. It contains a high proportion of volatile matter, but a low level of ash.

Wood has been described as the "ideal" fuel for lime-burning as it burns with a long, lazy flame which is relatively cool. This helps to ensure that the quicklime is reactive.

From the viewpoint of thermal efficiency, however, wood is far from being an ideal fuel. Much of the volatile matter distils from the calcining zone and its calorific value is lost from the viewpoint of calcination (see section 16.7.4 for the discussion of "high-grade" and "low-grade" heat). Most of the volatiles emerge as smoke in the exhaust gases.

As wood became scarcer and coal became more readily available, the early types of kiln were increasingly fired with bituminous coal in lump form. Kiln control became more difficult because coal burns with a shorter, fiercer flame than wood and is more prone to over-burn the lime. In addition, the greater ash content, coupled with the higher temperatures, increased the risk of the ash fusing and causing "crotching" of the lumps of lime (see sections 14.2.7 and 16.4.1.2).

As coal generally contains less volatile matter (and less water) than wood, more of the heat released when it burns is "high-grade". The resulting kiln heat usages per tonne of lime are significantly lower than with wood. The emission of volatiles to atmosphere, however, still results in the emission of copious quantities of smoke.

In areas where wood and coal were not available, peat and dried vegetable matter have been used, but neither makes a satisfactory fuel for lime burning.

14.3.3.2 Traditional Shaft Kilns

"Traditional" shaft kilns operate continuously and are fired with fuel introduced into the calcining zone (see section 16.2.2). Various fuels have been used, including bituminous coal, producer gas, fuel oil and natural gas.

In the case of *bituminous coal*, the relatively inexpensive "smalls" grade was widely used. The coal was introduced to the kiln at the top of the calcining zone, either by hand-shovelling or by mechanical projection, into a space created by a narrowing of the shaft (see section 16.4.2 for details). After charging, the kiln was drawn to cover the bed of coal with pre-heated limestone.

The coal had to have an adequate ash fusion temperature, preferably over 1200 °C and an acceptably low ash content. A high ash fusion temperature reduced the risk of crotching, caused by the ash fusing and bonding lumps of lime to each other and to the refractories. It also ensured that most of the ash did not adhere to the lime and could be removed by screening at (say) 6 mm. The ash-contaminated fine lime could be sold as a low-grade product.

The coal used had to have adequate coking properties, so that it formed cohesive agglomerates, which burned relatively slowly and became trapped between the lumps of calcining limestone.

Most of the volatile matter in the coal distilled shortly after firing and either burned in the pre-heating zone, or was emitted to atmosphere as dark smoke.

Most of the sulfur in the coal was absorbed on the surface of the lime, forming calcium sulfate. Because of their greater surface area per unit mass, the concentration of sulfur was greatest in the smaller fractions of lime.

Producer gas has been used for lime burning. It is made by the partial combustion of coal (or oil) in a deficiency of air in "gas producers". Steam may also be added. The main fuel components of the gas are carbon monoxide and hydrogen with a minor amount of methane. Inert gases make up most of the volume of the gas. They consist of nitrogen with small amounts of carbon dioxide, steam and oxygen. The sulfur in the fuel is converted into either hydrogen sulfide or sulfur dioxide. Most of the coal ash is retained in the producer, so that the lime is not heavily contaminated with ash.

As a low calorific value "lean" gas (ca. 1,500 kcal/Nm^3), producer gas is a convenient fuel for lime-burning. Its long, lazy flames produce a reactive lime, which is contaminated with sulfur from the coal.

There are, however, two major disadvantages of using producer gas. The first is the high total heat usage per tonne of lime, as a result of heat losses in the gas producer and of the low proportion of high-grade heat produced by the gas. The second is the additional capital and operating costs associated with the gas producer.

Fuel oil is generally either gassified in refractory-lined chambers attached to the kiln, or vaporised by spraying on to the surface of the lime. It is generally burned with a low level of excess air to produce the relatively long, lazy flames required for controlled lime-burning. Nevertheless, the lime produced tends to be of moderate to low reactivity. In shaft kilns, most of the sulfur in the oil is retained as calcium sulfate on the surface of the lumps of lime. Fuel oil is ash-free.

Operation with a limited excess of air results in the emission of some volatiles as dark smoke. This, coupled with the relatively low reactivity of the lime has caused many oil-fired shaft kilns to be replaced by modern designs.

The heat usages of oil-fired shaft kilns can be quite low (1,100 to 1,250 kcal/kg), although where the gasifiers are water-cooled, the usage is about 10 % higher.

Natural gas (calorific value about 8,600 kcal/Nm^3) is introduced via ports in the walls of the kiln. It burns with a longer, lazier flame than oil or coal and more readily produces medium reactivity lime. As it contains insignificant amounts of sulfur and produces no ash, the resulting lime is not contaminated by the fuel in any way. Moreover, natural gas does not readily produce smoke, so that the exhaust gases have a low opacity. The heat usages of gas-fired shaft kilns are generally about 1,150 kcal/kg.

14.3.3.3   Mixed Feed Kilns

Mixed feed kilns currently in operation generally use coke or anthracite [14.5, 14.10].

Where *coke* is used, it needs to be sufficiently large to be trapped between the lumps of lime/limestone and also needs to be strong enough to resist being crushed and abraded by the burden. It should have a low reactivity with respect to the reduction of carbon dioxide ($C + CO_2 \rightarrow 2CO$), which would result in a loss of effective calorific value and a reduction in yield of carbon dioxide. The latter is particularly important in industries, such as the ammonia soda process, where the carbon dioxide is an important co-product. The ash fusion temperature of the coke should be relatively high to avoid the formation of crotches.

As coke starts to burn at 800 °C, relatively little of its heat is lost as low-grade heat. The net heat usage of coke-fired mixed feed kilns can be as low as 830 kcal/kg [14.11]. The disadvantages of coke are its high cost per unit of heat and the relatively low availability of suitable grades.

Anthracite (see Table 14.3) typically contains 4 to 10 % of volatile matter, which results in higher heat usages than coke and in the emission of dark smoke. It needs to be strong of the correct particle size and have a high ash fusion temperature.

Table 14.3. Dryden's classification of coals [14.9]

Classification	Carbon %	Volatiles %	Hydrogen %	Gross calorific value (kcal/kg)	Rank
1. Anthracites	92–95	10–3.5	4.0–2.9	8870–8353	High
2. Bituminous coals	75–92	50–11	5.6–4.0	6985–8870	
3. Brown coals & lignites	60–75	60–45	5.5–4.5	6653–7207	
4. Peats	45–60	75–45	6.8–3.5	4157–5322	Low

14.3.3.4 Modern Shaft Kilns

The fuels used in the various designs of modern shaft kilns currently available (see section 16.8) are selected to produce medium to light-burned lime, at high thermal efficiencies, without the emission of dark smoke. All designs can use natural gas, most can also use fuel oil and some can operate on pulverised solid fuel.

The solid fuels used include bituminous coal, lignite and PET coke (with a controlled level of volatiles). Some lime producers use a mixture of gas and solid fuel to produce lime with acceptably low sulfur levels at minimum fuel cost.

It is generally important to obtain the correct quality of solid fuel for a specific kiln and the advice of the kiln manufacturer should be sought.

A number of sophisticated handling systems have been developed for pulverised fuels to ensure that the quantities delivered to the burners/lances are equal and controlled accurately (e.g. [14. 12]). This enables the kilns to be operated at high efficiency, with controlled low levels of $CaCO_3$ and high reactivities.

14.3.3.5 Rotary Kilns

From the viewpoint of fuel requirements, rotary kilns are the most flexible of all lime kilns [14.13, 14.16]. They are successfully fired with natural gas, fuel oil and pulverised solid fuels. The solid fuels used include lignite and sawdust.

The correct design of the single burner is critical for all of the fuels. Natural gas tends to produce a long, lazy flame with a low emissivity, so a relatively high degree of mixing is required. In "straight" rotary kilns without a pre-heater, the reduced radiant heat transfer can result in a 10 to 15 % increase in net heat usage relative to oil and coal-firing. This can be partially offset by improved internal fittings to recover heat at the back-end of the kiln.

With oil-firing, the sulfur in the fuel can be expelled in the exhaust gases. The high sulfur recycle load can lead to the formation of "calcium sulfate" rings at the back-end of the kiln, or of calcium sulfate deposits in the pre-heater (see section 16.4.9).

With solid-fuel firing, the ash and sulfur in the fuel can cause rings and/or deposits. Ash deposition can be controlled by purchasing a fuel low in ash (preferably less than 7 %), with a high ash fusion temperature, and by grinding it finely

Table 14.4. Comparison of typical fuel characteristics [14.9, 14.17]

Fuel	Relative capital cost	Relative cost of unit net heat	Ash content %	% volatiles	% sulfur	Net calorific value (kcal/kg) [e]	Relative ease of handling	Level of hazard	Comments
Anthracite (lump)	Low	High	5–10	< 10	1	8,200–8,600	High	Low	Low availability
Bituminous coal (lump)	Low	Low	3–10	11–50	1–5	7,800–8,600	High	Low	High availability
Bituminous coal (pulverised)	High	Low	3–10	11–50	1–5	7,800–8,600	Low	High	High availability
Lignite (pulverised)	High	Low	7–12	40–60	0.5	ca. 7,000	Low	High	Limited availability
Peat	Low	Low	7–12	45–75	0.5	ca. 3,000	High	Low	Not a suitable fuel
Coke (lump)	Low	High	6–12	< 1	1	ca. 8,000	High	Low	Low availability
PET coke (pulverised)	High	Low	< 1	< 5 [b]	3–5	ca. 8,200	Low	High	Moderate availability
Heavy fuel oil	Intermediate	Intermediate	Nil	100	1–4	9,700	High	Low	High availability
Natural gas	Intermediate [a]	Intermediate	Nil	100	Nil	11,400	High	Moderate	Limited availability
Propane	Low	High	Nil	100	Nil	11,400	High	High	
Butane	Low	High	Nil	100	Nil	11,400	High	High	
Producer gas (coal)	High	High	Nil	100	1–5 [c]	ca. 1,300	Moderate	Moderate	See note [d]
Coal gasification	High	High	Nil	100	1–5 [c]	low	High	Low	See note [d]
Wood	Low	Low	< 2	75–80	Low	ca. 2,600	High	Low	See note [d]
Biomass gas	Intermediate	Low	Nil	100	Low	low	High	Low	See note [d]

[a] Excluding the cost of the pipeline.
[b] PET cokes with moderate levels of volatiles are available (and required for certain kilns).
[c] Sulfur levels depend on the coal used and on whether desulfurisation techniques are used.
[d] The low net calorific values of these fuels may markedly increase kiln heat requirements.
[e] Moist, as delivered.

(e.g. to at least 85 % less than 75 µm, if the volatile matter level is 30 %, and the ash content is below 7 %). Higher ash levels require finer grinding [14.13]).

The content of volatile matter in the coal should be at least 18 to 20 %. With such low levels of volatiles, however, finer grinding is necessary to obtain the required combustion characteristics [14.13].

Increasingly, the selection of fuel is being influenced by the need to limit the concentration of sulfur dioxide in the exhaust gases (see section 33.4).

14.3.4 Fuel Characteristics

Table 14.4 summarises some of the more important characteristics of the fuels mentioned in section 14.2.3. In the case of coal, the technology of grinding and handling it is complex and merits detailed consideration [14.13, 14.17].

14.3.5 Monitoring Heat Input

To produce a consistent quality of lime, it is essential to be able to control the quantity of fuel used and to know its calorific value. This is particularly true of shaft kilns which may take two to four days to respond to a change in heat input.

The volume of gaseous fuels used can be measured accurately with turbine meters. However, to measure their mass, it is essential to apply temperature and pressure corrections to the measured volume. Changes have been observed of over 2 % in the calorific value of natural gas — sufficient to cause a significant change in the degree of burning. As a result, many operators have installed calorific value meters to detect any changes as they occur.

The volume of medium and heavy fuel oils can readily be measured accurately by positive displacement meters. However, as they are pre-heated to achieve the required atomization temperature, the temperature at which they are metered should either be kept constant or be monitored so that appropriate corrections can be made.

The quantity of solid fuels used is generally controlled by some form of batch weighing which gives a high consistency. Variations in calorific value are, however, more problematic. They are generally resolved by a combination of blending deliveries and determining the calorific value daily on a representative sample.

14.3.6 Gross and Net Calorific Value

A common cause of confusion arises between the use of the gross (or upper) and the net (or lower) calorific value.

The gross calorific value is determined by measuring the heat released when a sample of fuel is ignited in oxygen. The initial and final temperatures are such that the water vapour produced by the combustion is condensed. Thus the gross calorific value includes the latent heat of condensation of the water.

In practice, however, it is impractical and undesirable to condense the water vapour. Thus, from the viewpoint of determining kiln efficiency, the latent heat of condensation of the water should be subtracted from the gross calorific value. This gives the net calorific value.

Table 14.5 lists the ratio of net to gross calorific values for various fuels.

Table 14.5. Ratio of net to gross calorific values

Fuel	Net ÷ gross calorific value
Bituminous coal — 0% water	0.977
— 10% water	0.960
Heavy fuel oil	0.940
Natural gas	0.904
Propane	0.926
Butane	0.929
Hydrogen	0.846
Carbon monoxide	1.000

14.4   Refractory Linings

14.4.1   Introduction

Selecting the most appropriate refractory lining for a lime kiln is important for the economic viability of the installation. The refractory at the hot face has to provide adequate performance with respect to;

a) abrasion resistance,
b) mechanical strength,
c) resistance to thermal shock, and
d) resistance to chemical attack.

The backing layers of the lining must also provide thermal insulation to reduce heat loss and to protect steelwork from excessive temperatures.

The "optimum" lining varies with the type of kiln, the fuel used and the way in which the kiln is operated. The following examples illustrate just three of the contrasting requirements.

a) The duties of refractories in rotary kilns are very different from those in shaft kilns.
b) Hot face temperatures in modern shaft kilns are generally much higher than those in traditional natural draught kilns.
c) Conversion of a mixed-feed kiln from wood- to anthracite-firing is likely to increase both hot face temperatures and the degree of chemical attack of the lining.

In practice, a number of lining designs may give similar costs per tonne of quicklime produced. The following paragraphs can only indicate some of the issues and practices which are used. Specific advice on the choice of lining for a particular installation can be obtained from the kiln supplier, refractory manufacturers and other lime producers operating similar kilns under similar conditions.

14.4.2 Hot-face Refractories

In natural draught kilns, blue engineering bricks often give adequate performance in the upper part of the preheating zone (say the top quarter of the kiln). Below that, a firebrick with about 40 % of alumina is often used [14.9].

In modern shaft kilns, the preheating zone is often lined with bricks containing 35 to 40 % of alumina, with the calcining and cooling zones being lined with high-alumina refractories (containing about 70 % of Al_2O_3).

The superior performance of magnesite-chrome refractories in the calcining zone (relative to high-alumina) resulted in a rapid increase in their use during the 1970s and 1980s, despite their higher initial cost. More recently, concerns about the environmental effects of chromium (VI) has led to a marked reduction in their use in favour of chrome-free magnesite bricks [14.14]. A relatively recent development is the use of magnesia-zirconia bricks, which have a still higher thermal performance [14.15]. Their use, however, can probably only be justified in high temperature processes such as the production of dead-burned dolomite.

In the cooling zone, a less refractory lining may be used (e.g., 35 to 40 % alumina), although many operators prefer to use the same refractories as in the calcining zone. This is particularly true in the case of mixed-feed kilns, where the base of the calcining zone is not well defined.

Refractories in rotary kilns are subject to cyclic mechanical stresses as a result of flexing of the kiln shell as it rotates. They are also exposed to greater thermal shock when the kiln is shut down or re-lit.

14.4.3 Backing Layers

A considerable amount of technology can go into the design of the backing layers and the selection of materials. A cost-effective design should minimise the total cost per tonne of lime produced, by being relatively thin (to maximise the cross-sectional area of the kiln and hence its potential productive capacity), while maintaining a good level of insulation and a long life for the complete lining.

In the calcining zone of a shaft kiln, the total lining thickness is typically 700 to 800 mm [14.14]. The backing may consist of four or five layers, with the outer layers consisting of insulating mats and bricks. The layer behind the hot-face refractory is generally an intermediate grade of refractory material that, in the event of the inner layer being removed, can withstand kiln operating temperatures for a limited period. That gives time to plan the resulting kiln shut-down for re-lining.

The linings in the preheating and cooling zones have less arduous duties than that in the calcining zone. They are, therefore, generally simpler and use less expensive materials.

In rotary kilns, the linings are, of necessity, much thinner and lighter than in shaft kilns. As a result, the designs of backing layers are less complex, but, arguably, are even more important. In the preheating zone, use is sometimes made of "semi-insulating" refractories, which have a controlled porosity to reduce thermal conductivity and heat loss.

14.5   References

[14.1]   G. Ruckensteiner, J. Burczeck, U. Ludwig, "Factors Affecting the Thermal Stability of the Internal Structure When Burning Limestone", Zement Kalk Gyps **6**, 1995.

[14.2]   G. Bessey, "Production and Use of Lime in Developing Countries", Overseas Building Note 161, Building Research Establishment, Garston, 1975.

[14.3]   ENV 459-1: "Building lime — Definitions, specifications and conformity criteria", 1995.

[14.4]   prEN 12518: "Chemicals Used for Treatment of Water Intended for Human Consumption — High-Calcium Lime", in preparation.

[14.5]   T.-P. Hou, "Manufacture of Soda", Hafner Publishing, 1969.

[14.6]   F. Mereu, "Maerz Vertical Limes Shaft Kiln for the Calcination of Small Size Limestone", Proc. 8th International Lime Congress, Berlin, June, 1994.

[14.7]   G. Gnecchi, "Maerz Vertical Double Shaft Parallel Flow Kiln for the Production of Fine-sized Limestone", Proc. 8th International Lime Congress, Berlin, June, 1994.

[14.8]   H. Ruch, "Possibilities and Prospects for Fuel Supply in the Lime Industry", Zement Kalk Gyps, **6**, 1983.

[14.9]   M. Wingate, "Small-scale Lime-burning", Intermediate Technology Publications, 1985, ISBN 0-946688-01-X.

[14.10] W. Höltje, "Alternatives to coke as fuel for conventional shaft kilns", Zement Kalk Gyps, **4**, Jan. 1989, 21–26; March 1989, 57–60 (English).

[14.11] R.S. Boynton, "Chemistry and Technology of Lime and Limestone", John Wiley & Sons, 1980, ISBN 0-471-02771-5.

[14.12] A. Linssen, "Feed systems for fine-grained fuels", Zement Kalk Gyps, **4**, Febr. 1989, 76–83; April 1989, 86–91 (English).

[14.13] F. Schwarzkopf, "Lime Burning Technology — a Manual for Lime Plant Operators", 3rd ed., Svedala Industries, Kennedy Van Saun, 1994.

[14.14] F. Drobowski, F. Steinwender, "Modern Liming Schemes for Kilns in the Lime Industry, with particular reference to Standard Shaft Kilns", Zement Kalk Gyps, Aug. 1990, 376–380; Oct. 1990, 226–229 (English).

[14.15] M. Naziri-Zadeh, "Use of Magnesia Zirconia Bricks with High Thermal Ratings in Shaft and Rotary Kilns", Zement Kalk Gyps, April 1995, 224–230.

[14.16] B.G. Jenkins, "The effect of the fuel Selection on Kiln Operation and Emissions". In: "Combustion and Gaseous Control III", Inst. of Energy, 1997, ISBN 090 259 7558.

[14.17] P.T. Luckie, L.G. Austin, "Coal Grinding Technology", Kennedy Van Saun, 1979.

15   Calcination of Limestone

15.1   Introduction

The term "calcination of limestone" refers to the process of its thermal decomposition into quicklime and carbon dioxide. It is frequently abbreviated by lime producers to "calcination".

The chemical reactions involved in the decomposition of limestone are deceptively simple. There are, however, complications, particularly with dolomite, which are believed to be caused by variations in crystallography and micro-structure.

The kinetics of the decomposition, especially of granular and lump limestones, are still more complex. Many attempts have been made to produce a unified theory of calcination, but all have had limited validity. It is believed that this is because:

a) several processes are involved in calcination, any one of which may be rate-determining under specific circumstances,
b) differences in the crystallography and micro-structure of limestone, which are difficult to quantify, can have a marked effect, and
c) the micro-structure of the quicklime (and particularly of the surface layer) can have a significant effect. This is dependent on temperature, impurities and the time of exposure after the completion of calcination.

Because of the above complexities, the calcination characteristics of a particular limestone are best determined by practical trials.

The following sections focus on general principles rather than on theory. Those principles are illustrated by reference to a particular limestone to give an understanding of the processes involved. Reference is also made to other sources of information, which, while they may be somewhat contradictory, will be of assistance to those wishing to delve more deeply into the subject.

The calcination of finely divided limestone is addressed in section 15.7.

15.2   The Chemical Reactions

15.2.1   Calcium Carbonate

The reaction for the thermal decomposition of calcium carbonate may be written as:

$$CaCO_3 + heat \rightleftarrows CaO + CO_2\uparrow \qquad (15.1)$$

$$\text{100g} \qquad\qquad \text{56g} \quad \text{44g}$$

The relationship between the dissociation pressure of calcite with temperature is shown in Fig. 15.1. It reaches 1 atmosphere (101.3 kPa) at approximately 900 °C. (For many years, a value of 898 °C has been regarded as the correct value [15.1–15.3], but more recent work [15.4, 15.5] points to a value of 902.5 °C.)

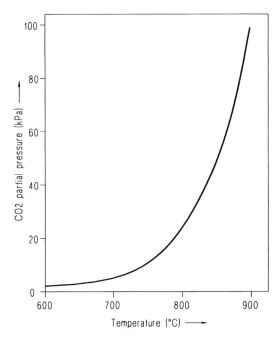

Figure 15.1. Variation of the dissociation pressure of calcite with temperature

The heat of dissociation of calcite relative to 25 °C has variously been reported in the range 695 to 834 kcal/kg of CaO. Boynton [15.1] quotes an average value of 770 kcal/kg, while Schwarzkopf [15.6], which was published subsequently, quotes 754 kcal/kg. The author favours a value of 760 kcal/kg.

The heat of dissociation *at 900 °C*, calculated from the above value of 760 kcal/kg, is 723 kcal/kg CaO (the difference of 37 kcal/kg CaO arises from the fact that 1.786 kg of $CaCO_3$ requires 431 kcal to be heated from 25 to 900 °C, whereas the reaction products — 1.000 kg of CaO and 0.786 kg of CO_2 — release 208 and 186 kcal respectively on cooling from 900 to 25 °C, a total of 394 kcal). Schwarzkopf [15.6] reports a value of 698 kcal/kg at 900 °C.

It should be noted that the reaction is reversible (see section 15.6).

15.2.2   Magnesium Carbonate

The reaction for the thermal decomposition of magnesium carbonate is:

$$MgCO_3 + heat \rightleftarrows MgO + CO_2\uparrow \qquad (15.2)$$
$$84g \qquad\qquad 40g \quad 44g$$

The temperature at which the dissociation pressure of $MgCO_3$ reaches 1 atmosphere has variously been reported to be between 402 and 550 °C [15.1, 15.6]. The heat of dissociation of $MgCO_3$ relative to 25 °C has been reported to be 723 kcal/kg MgO [15.7].

15.2.3 Dolomite and Magnesian/Dolomitic Limestone

The decomposition of dolomites and magnesian/dolomitic limestones is more complex. It is reported that some decompose via two discrete stages, others decompose via a single stage, while others decompose in an intermediate manner [15.1, 15.7].

$$CaCO_3 \cdot MgCO_3 + heat_{(1)} \rightleftarrows CaCO_3 \cdot MgO + CO_2 \uparrow \qquad (15.3)$$
$$184g \qquad\qquad\qquad 140g \qquad 44g$$

$$CaCO_3 \cdot MgO + heat_{(2)} \rightleftarrows CaO \cdot MgO + CO_2 \uparrow \qquad (15.4)$$
$$140g \qquad\qquad\quad 96g \qquad 44g$$

$$CaCO_3 \cdot MgCO_3 + heat_{(1+2)} \rightleftarrows CaO \cdot MgO + 2CO_2 \uparrow \qquad (15.5)$$
$$184g \qquad\qquad\qquad 96g \qquad 88g$$

All dolomites and magnesian/dolomitic limestones decompose at higher temperatures than magnesium carbonate. The onset of calcination can vary from 510 to 750 °C [15.6], and depends on the crystal structure and form of the stone [15.1]. Reaction (15.3) is understood to occur towards the lower end of that range, with reaction (15.4) subsequently occurring at about 900 °C. Reaction (15.5) is reported to occur towards the upper end of the range. This apparent discrepancy may reflect different heat transfer rates, rather than fundamental differences in the physical chemistry of the various limestones.

The heat of dissociation of dolomite, i.e. $heat_{(1+2)}$ is reported to be 723 kcal/kg of $(CaO \cdot MgO)$, relative to 25 °C [15.1]. This is lower than the weighted average of the heats of dissociation of $CaCO_3$ and $MgCO_3$ — 750 kcal/kg $(CaO \cdot MgO)$. This apparent discrepancy may reflect the difference in the heat of formation of dolomite relative to those of calcite and $MgCO_3$.

15.3 Kinetics of Calcination

15.3.1 General

The passage of a limestone particle through a lime kiln can be divided into five stages. The following description refers to high-calcium limestone, but parallels can be drawn with magnesian/dolomitic limestones and dolomite.

a) In the pre-heating zone, the limestone is pre-heated from ambient temperature to about 800 °C by the kiln gases (i.e. products of combustion plus CO_2 from calcination and excess air).
b) At about 800 °C, the pressure of carbon dioxide produced by the dissociation of the limestone equals the partial pressure of the CO_2 in the kiln gases. As the temperature of the limestone rises, the surface layer begins to decompose, so that, when the temperature of the stone reaches 900 °C, the layer of lime may be 0.5 mm thick (corresponding to about 5 % by weight of quicklime for a 25 mm particle).
c) Once the temperature of limestone exceeds the "decomposition temperature" of 900 °C, the partial pressure exceeds 1 atmosphere and the process of dissociation can proceed beyond the surface of the particles.
d) If all of the calcium carbonate dissociates before a given particle leaves the calcining zone, the lime begins to sinter. (This process occurs to a very limited extent during stage (c), but, for most situations, it can be disregarded.)
e) The particles of lime, which may contain residual limestone, leave the calcining zone at 900 °C and are cooled by air used for combustion.

Stages (a), (b) and (e) involve direct heat transfer between a gaseous medium and the particles. As a unit process, the heat transfer mechanism is well understood and not specific to the calcination of limestone. Stages (c) and (d), however, are specific to limestone, are influenced by the design of lime kiln, and influence the properties of the quicklime. They are described in detail below.

15.3.2   Dissociation of High-calcium Limestone

The dissociation of limestone above the decomposition temperature can be regarded as consisting of five processes, as illustrated by Fig. 15.2.

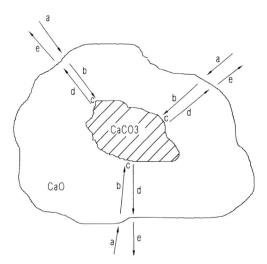

Figure 15.2. Illustration of the processes involved in dissociation of limestone

a) Heat is transferred from the kiln gases to the surface of the decomposing particle.
b) Heat is then conducted from the surface to the reaction interface through the micro-porous layer of lime (see Fig. 15.7a).
c) Heat arriving at the reaction interface causes the dissociation of $CaCO_3$ into CaO and CO_2.
d) The CO_2 produced migrates from the reaction interface, through the lime layer to the surface of the particle, and is simultaneously heated from the temperature of the reaction zone to that of the surface.
e) CO_2 migrates away from the surface into the kiln gases.

The physics and physical chemistry of processes (a), (c) and (e) are relatively well understood, but the effect of the micro-structure of the lime layer on (b) and (d) is complex and depends in part on the characteristics of the limestone. Further complications arise when changes occur in the structure of the surface layers of the lime, caused by sintering, slagging and absorption of sulfur dioxide.

The difficulty in producing a unified theory of the kinetics of calcination is that it would need to account for all of the above processes, any one of which may become rate-determining under particular circumstances.

If the notion that the migration of carbon dioxide from the reaction interface may be the rate-determining process appears unlikely, reference should be made to [15.4]. That suggests that "the fundamental rate controlling step is the solid-state diffusion of CO_2 through CaO in the region of the reaction interface". It contrasts the complex shape of the weight loss curve at atmospheric pressure with the linear relationship under vacuum.

The above generalisations are illustrated by the results of a laboratory calcination of a limestone sphere Fig. 15.3 [15.8]. The sphere had a diameter of 31 mm and had one thermocouple embedded in the surface layer and a second positioned 5 mm below the surface. The sphere was suspended in a furnace at 1020 °C.

The surface temperature rose to above 900 °C within 5 min. and then rose progressively to 1000 °C after 60 min., and reached the furnace temperature after 67 min., by which time the particle was fully calcined. Calcination of the outer part of the sphere kept the temperature 5 mm below the surface below 900 °C until after 45 min., when the reaction interface reached it. The temperature at that point then rose towards the surface temperature. The calculated proportion of the $CaCO_3$ that had dissociated and the calculated radius of the limestone core are also shown in Fig. 15.3.

From the viewpoint of the lime burner, the major variables affecting the rate of dissociation are:

a) the characteristics of the limestone,
b) the particle size distribution,
c) the shape of the particles,
d) the temperature profile in the calcining zone, and
e) the rate of heat exchange between the gases and the particles.

The characteristics of a particular limestone can only be evaluated with confidence by calcining trial quantities and comparing the progress of the calcination

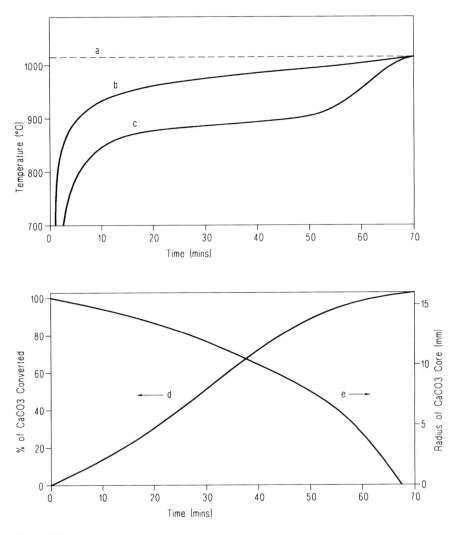

Figure 15.3. Progression of calcination of a 31 mm limestone sphere
(a) furnace temperature; (b) surface temperature; (c) temperature 5 mm below surface;
(d) % of CaCO$_3$ dissociated; (e) radius of CaCO$_3$ core

and the quality of the product with those of a limestone with known, satisfactory characteristics. Such trials should, ideally, be done both in the laboratory and on the full scale. Variations in calcination rates of up to a two times factor are common and even greater factors have been reported [15.1]. Fine-grained limestones generally decompose more rapidly than coarse-grained ones (in the absence of special factors such as decrepitation).

   The progress towards complete dissociation of particles, with a given shape, but of differing sizes, has been found to follow a common curve. Fig. 15.4 shows such a curve for spheres of a dense limestone, the results are summarised in Table 15.1 [15.8].

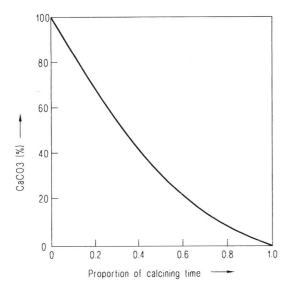

Figure 15.4. Normalised dissociation curve for limestone spheres

Table 15.1 Dissociation of limestone[a]

Proprotion of dissociation time	% dissociation of original $CaCO_3$[b]	% of $CaCO_3$ in sphere[b]
0.16	40	73
0.29	60	54
0.37	70	43
0.48	80	31
0.63	90	17
0.76	95	8.6
0.86	98	3.5
0.90	99	1.8
1.00	100	0

[a] For spheres of dense, high calcium limestone, with a range of diameters.
[b] % values are m/m.

Fig. 15.5 and Table 15.2 [15.8] summarise the combined effects of temperature and size on spheres of the same dense limestone.

Azbe [15.9] concluded that for a given shape factor the calcining time is directly proportional to (thickness)^2 (where thickness is defined as the minimum dimension measured through the centre of gravity of the particle). He also derived a formula for the "calcining effort" of a particular shape. This predicts that the calcining effort of a large rectangular prism of a given thickness is double that of a cube with the same thickness [15.1]. [15.6] gives relative calcining times, for a given thickness, of 1 for a sphere, 1.5 for a cube, 1.7 for a cylinder and 3 for a plate (see also section 14.2.9).

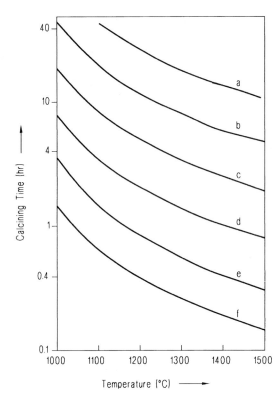

Figure 15.5. Variation of calcining time with size and temperature for spheres of a dense, high-calcium limestone, with diameters of (a) 150 mm; (b) 125 mm; (c) 100 mm; (d) 75 mm; (e) 50 mm; (f) 25 mm

Table 15.2 Variation of calcining time with size and temperature.

Diameter of sphere (mm)	Dissociation time (h) [a]					
	Temperature (°C)					
	1000	1100	1200	1300	1400	1500
12	0.8	0.3	0.2	–	–	–
19	1.1	0.5	0.3	0.2	–	–
25	1.5	0.6	0.4	0.25	0.2	–
37	2.3	1.0	0.6	0.4	0.3	0.2
50	3.5	1.5	0.8	0.6	0.4	0.3
75	7.9	3.8	2.0	1.4	1.0	0.8
100	19	8.5	4.8	3.5	2.6	1.9
125	43	19	11	8.3	6.3	4.6
150	–	45	26	19	15	11

[a] For a dense, high calcium limestone.

15.3.3   Dissociation of Magnesian/Dolomitic Limestones and Dolomite

The processes described in section 15.3.2 also apply to limestones containing significant quantities of magnesium carbonate. From a practical viewpoint, the calcination of such limestones does not differ significantly from that of high-calcium limestones.

15.4   Sintering of High-calcium Quicklime

When a dense calcitic limestone is calcined at temperatures close to 900 °C, the volume of individual particles increases very slightly until about 60 % of the limestone has dissociated, and then decreases slightly to about 98 % of the original volume at the point when calcination is complete [15.8].

Calcination at very high temperatures has been shown to cause significant sintering of the surface layer of lime, while dissociation is still in progress: under extreme conditions this can interfere significantly with the dissociation process [15.1].

The sintering process after the completion of dissociation is of more relevance to practical lime-burning, because of its effect on the reactivity of the lime. Fig. 15.6 shows the effect of temperature and time on sintering, as measured by the variation of apparent density, of a quicklime produced from a dense, high-calcium limestone [15.8].

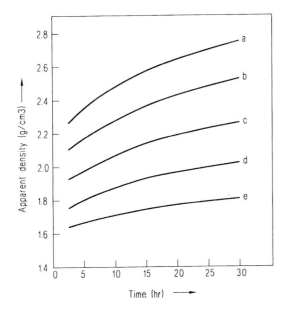

Figure 15.6. Variation of apparent density with temperature and time (a) 1400 °C; (b) 1300 °C; (c) 1200 °C; (d) 1100 °C; (e) 1000 °C

Scanning electron micrographs of quicklime demonstrate the dramatic change in structure resulting from the sintering process, for the same high-calcium quicklime (Fig. 15.7). The effects of sintering on lime reactivity are described in section 13.2.

The implications of the sintering process, coupled with the effects of particle size on calcining time are illustrated by the following example. When a 50 to 150 mm limestone, consisting of spherical particles, is calcined in a furnace at 1300 °C, the times required for complete calcination (from Table 15.2) are as given in Table 15.3.

If the calcining time were restricted to 10 h, the 50, 75 and 100 mm spheres would be fully calcined, and would have sintered to a density about 2.0 g/cm^3 (Fig. 15.7). The 125 mm sphere would be almost fully calcined and might have sintered slightly to a density of 1.6 g/cm^3. The 150 mm sphere would still contain over 40 % of $CaCO_3$ by weight and the density of the lime layer would be about 1.5 g/cm^3.

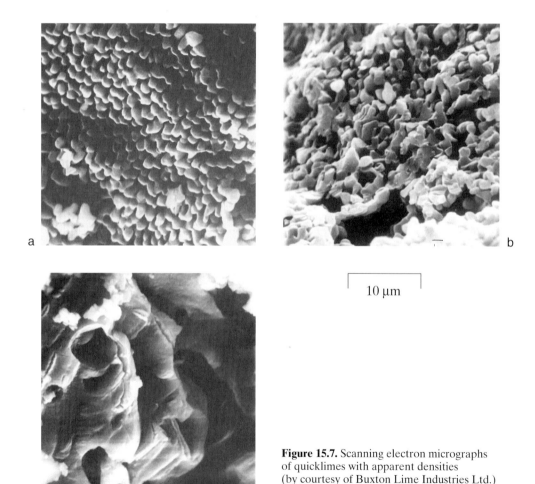

a

b

10 μm

c

Figure 15.7. Scanning electron micrographs of quicklimes with apparent densities (by courtesy of Buxton Lime Industries Ltd.) (a) 1.5 g/mL; (b) 1.9 g/mL; (c) 2.3 g/mL

Table 15.3 Estimated extent of dissociation and sintering after 10 hours at 1200 °C

Diameter of sphere (mm)	Dissociation time (h)	Expected properties after 10 h	
		% $CaCO_3$[a]	Density (g/ml)
50	0.8	0	2.05
75	2.0	0	2.05
100	4.8	0	1.95
125	11	1.8	1.6
150	26	43	1.5

[a] % by mass.

Thus the properties of quicklime from a given kiln reflect the average properties of individual lumps, each of which has experienced a particular time-temperature history. Fig. 15.8 compares the distribution of particle densities for a light-
burned quicklime from an annular shaft kiln, with a mean apparent density of
1.66 g/cm³ with that of a solid-burned quicklime from a coal-fired traditional shaft
kiln with a mean apparent density of 2.15 g/cm³. Table 15.4 presents some typical

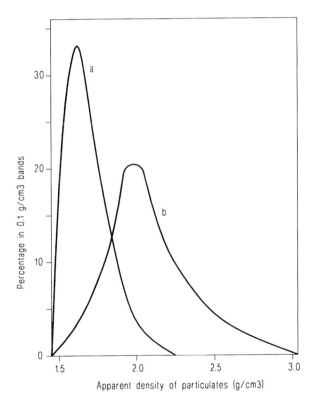

Figure 15.8. Distribution of particle densities of quicklime (a) from an oil-fired annular shaft
kiln; (b) from a coal-fired shaft kiln

Table 15.4 Typical properties of soft medium and hard-burned quicklimes.

Property	Degree of burning		
	Soft	Medium	Hard
Apparent density[a] (g/ml)	1.5–1.8	1.8–2.0	> 2.0
Porosity[a] (% v/v)	55–45	45–39	< 39
Reactivity t_{60}[b] (min)	< 3	3–9	> 9
Calcium carbonate (% m/m)	4–1	1.0–0.5	< 0.5

[a] For quicklime produced from a dense, high-calcium limestone.
[b] EN 459-2.

properties of lime from a gas-fired rotary kiln operated to give "soft", "medium" and "hard" burning.

The sintering characteristics of quicklimes are influenced by impurities. It is reported [15.1] that the presence of high levels of sodium, either in the limestone, or added to the stone as a solution of sodium chloride or sodium carbonate, reduces sintering of some limes [15.1]. The effect on lime reactivity does not appear to have been reported. Finely divided silica, alumina and/or iron oxide dispersed in the limestone may increase sintering (see also section 15.5).

It is also probable that the sintering characteristics of very porous limes produced from micro-porous limestones, such as chalk, differ from those of limes made from dense limestones.

The MgO component of magnesian/dolomitic limes and dolime (or calcined dolomite) calcines more rapidly and is understood to sinter more readily than the CaO component (at least in terms of reactivity to water). Little appears to have been published on this topic.

15.5   Sintering of Calcined Dolomite

The sintering of dolomite is primarily of commercial interest in connection with the manufacture of refractory products with mean apparent densities in excess of $3.0\,g/cm^3$. This involves heating calcined dolomite at temperatures of 1400 to 1800 °C [15.17]. The sintering process is affected by grain size and crystallite size, but the predominant factor is the amount of impurities present with fluxing properties. Iron oxide has a particularly pronounced influence on the rate and extent of sintering [15.18–15.23].

Certain grades of sintered dolomite are produced by blending ground, calcined dolomite with about 0.5 % of finely divided ferric oxide, and heating the mixture at temperatures of up to 1600 °C.

15.6   Steam Injection

It has been reported that steam injection reduces the dissociation temperature of some limestones and increases the rate of dissociation [15.1]. From an academic viewpoint, the steam would be expected to act as an inert gas, reducing the CO_2 partial pressure in the kiln gases, thereby marginally lowering the dissociation temperature and increasing the rate of dissociation.

However, from the viewpoint of lime production, the expected effects would be:

a) moderation of kiln temperatures, particularly if water, rather than steam, were added to the combustion zone, and
b) an increase in heat usage.

It is understood that some producers of dolomite, who operate shaft kilns to make particularly soft-burned dolomite, inject steam into the combustion gases to moderate kiln temperatures. The resulting lime has a higher reactivity, and is particularly suitable for the production of Type S hydrate [15.1].

15.7   Re-carbonation

Quicklime can re-carbonate via two mechanisms.

Above the dissociation temperature of calcium hydroxide (ca 540 °C), and below the dissociation temperature of calcite (ca 800 °C), equation (15.1) (see section 15.2.1) can proceed to the left with evolution of heat. Porous high-calcium limes can absorb over 50 % of the theoretical amount of CO_2, whereas dense limes may only absorb 3 to 5 % of CO_2 [15.8].

The re-carbonation process was demonstrated using a limestone sphere to which thermocouples were fitted (see section 15.3.2). Immediately after calcination had been completed, the furnace was cooled to 700 °C and an atmosphere of 100 % carbon dioxide was established. The temperature of the thermocouple at the surface rose to a maximum value of about 780 °C in about 3 min., while that 5 mm below the surface rose to a maximum of 875 °C in 4 min. After 20 min., 50 % of the calcium oxide originally present had re-carbonated.

High temperature re-carbonation can occur in lime kilns under abnormal conditions in which some kiln gases enter the cooling zone. It has been observed when re-circulating exhaust gases (to moderate calcining zone temperatures), when shaft kiln internal structures, such as arches, have failed and when "crotching" has occurred.

The second mechanism occurs at temperatures below the dissociation temperature of calcium hydroxide, and involves:

a) surface hydration of the quicklime by atmospheric water vapour (equation 15.6), and
b) carbonation of some of the calcium hydroxide to calcium carbonate plus water (15.7), which can then hydrate more quicklime.

$$CaO + H_2O \rightleftharpoons Ca(OH)_2 \tag{15.6}$$

$$Ca(OH)_2 + CO_2 \rightleftharpoons CaCO_3 + H_2O \tag{15.7}$$

Such slaking and re-carbonation occurs routinely, to a very limited extent ($<0.5\%$ by weight of H_2O and $<0.03\%$ of CO_2) in the cooling zone of all lime kilns but does not cause any significant effects. It occurs to a much greater extent when quicklime is exposed to the atmosphere for excessive periods, when it is called "air slaking".

Excessive re-carbonation by either mechanism is generally indicated by an abnormally low reactivity, which is not associated with the expected high mean apparent density.

15.8   Calcination of Finely Divided Limestones

The calcination of finely divided limestones (ranging from 5 µm to 5 mm) is of interest in connection with:

a) the production of quicklime in fluidised bed and flash calciners (section 16.4.11),
b) the production of cement (section 9.2.2),
c) the use of injected limestone to remove oxides of sulfur from boiler flue gases (section 12.5.3), and
d) re-calcination of precipitated calcium carbonate in processes such as the production of wood pulp (sections 16.4.11 and 32.16).

The process has been studied by several investigators (e.g. [15.11, 15.12]), but no consensus has emerged regarding the mechanisms that control the calcination rate. The following paragraphs summarise some of the conclusions drawn by various researchers.

At temperatures below 900 °C, the dissociation of 5 to 10 µm particles appears to be controlled by chemical kinetics [15.13] and an Arrhenius activation energy of 46 kcal/mole has been determined. However, above 900 °C, the apparent activation energy decreases markedly, indicating that chemical kinetics no longer control the rate of dissociation. The time to achieve 80 % calcination for such a particle size varies from 0.55 sec. at 850 °C to 0.03 sec. at 1250 °C [15.13].

Müller and Stark [15.14] reviewed literature relating to particle sizes of 20 to 30 µm and concluded that reactors operating below 900 °C require residence times ranging from 10 sec. to several minutes, whereas reactors with temperatures above 900 °C require residence times ranging from a few tenths of a second to a few seconds for 90 % calcination.

Campbell et al. [15.15] studied the calcination of precipitated calcium carbonate, as produced in the re-causticising cycle of a Kraft (sulfate) pulp mill, and concluded that an activation energy of 50 kcal/mole applies to calcination at temperatures of 1165 and 1240 °C.

Information from Vosteen [15.12] is presented in [15.16], which indicates that, at 1000 °C, the times for complete calcination of 1, 3 and 5 mm particles are 1, 6 and 13 min. respectively.

The above work may be summarised as indicating that, depending on the particle size and temperature, the rate determining process can be;

a) mass transfer,
b) heat transfer,
c) chemical kinetics, or
d) a combination of these factors.

15.9 References

[15.1]   R.S. Boynton, "Chemistry and Technology of Lime and Limestone", John Wiley & Sons, 1980, ISBN 0-471-02771-5.
[15.2]   J. Johnston, J. Am. Chem. Soc. **32**, 1910, 938.
[15.3]   Marc and Simek, Z. Anorg. Chem. **82**, 1913, 17.
[15.4]   J.L. Thompson, "Prediction Lime Burning Rate via New Dynamic Calcination Theory", Pit and Quarry **5**, 1979, 80.
[15.5]   Turkdogan et al., "Calcination of Limestone", SME Transactions, Vol. 254, March 1973.
[15.6]   F. Schwarzkopf, "Lime Burning Technology — a Manual for Lime Plant Operators", 3rd ed., Svedala Industries, Kennedy Van Saun, 1994.
[15.7]   N.V.S. Knibbs, "Lime and Magnesia", Ernest Benn, 1935.
[15.8]   L.C. Anderson, "Resumé of ICI Work on Limestone Calcination, Lime Reactivity and Apparent Density", Oct. 1973 (internal report).
[15.9]   V.J. Azbe, "Theory and Practice of Lime Manufacture", Part II, Rock Products, March 1953, 102–104.
[15.10] EN 459-2: "Building Lime, Part 2 Test methods", 1994 (under revision).
[15.11] G. Flamment, "Direct sulfur capture in flames through the injection of sorbents", IFRR Doc. No. F 09/a/24, 1980.
[15.12] B. Vosteen, "Die physikalische und chemische Kinetik der thermischen Zersetzung von Kalk", Dissertation (in German), T.U. Braunschweig, 1970.
[15.13] P. Flamment et al., "Fundamental and technical aspects of SO_2 capture by Ca based sorbents in pulverised coal combustion", IFRF Doc. No. F 138/a/8, Dec. 1987.
[15.14] A. Müller, J. Stark, "Calcination of limestone powders in suspension", Zement Kalk Gips **2**, 1989, 93 (English translation **4**, 1989, 97).
[15.15] A.J. Campbell et al., "Lime Calcination: time and temperature of calcination expressed as a single variable and the effect of selected impurities on lime properties", Zement Kalk Gips **9**, 1988, 442.
[15.16] J. Bayens et al., "The development, design and operation of a fluidised bed calciner", Zement Kalk Gips **12**, 1989, 620.
[15.17] E.J. Koval et al., "Effects of raw material properties and Fe_2O_3 additions on the sintering of dolomites", Ceramic Bulletin **9**, 1983, 274–277.
[15.18] G.V. Kukolev, G.Z. Doligna, "Study of the properties of difficultly sinterable dolomites in the Abano deposits", Ogneupory **13**, 1948, 17–21.
[15.19] G.V. Kukolev, G.Z. Doligna, "Sintering processes and methods of improving metallurgical dolomite", Ogneupory **15**, 1950, 536–544.
[15.20] G.V. Kukolev, G.Z. Doligna, "Sintering processes and methods of improving the quality of metallurgical dolomite", Ogneupory **16**, 1951, 63–68.
[15.21] P.W. Clark et al., "Further investigations on the sintering of oxides", Trans. Brit. Ceram. Soc. **52**, 1953, 1–49.

[15.22]   G.V. Kukolev, I.A. Kryzhanovsky, "Liquid sintering of alkaline earth oxides of dolomite", J. Appl. Chem. USSR **29**, 1956, 1621–1630.

[15.23]   R. Gonhardt et al., "The sintering of dolomites with different petrographic characteristics", Tonind.-Ztg. Keram. Rundschau **91**, 1967, 121–215.

General References

R.H. Borgwart, "Calcination kinetics and surface area of dispersed limestone particles", AICHE Journal **31**, 1985, 1.

P.K. Gallagher, D.W. Johnson, "The effect of sample size and heating rate on the thermal decomposition of $CaCO_3$", Thermochemical Acta **6**, 1973, 67.

A.W.D. Hills, "The mechanism of the thermal decomposition of calcium carbonate", Chem. Eng. Sci. **23**, 1968, 297.

16   Production of Quicklime

16.1   Introduction

The choice of lime kiln is of paramount importance to a lime producer. It must be suitable for burning the selected feedstone and for producing the required quality of quicklime. It must have sufficiently low capital and operating costs to produce quicklime at a competitive price. Its capacity must also be appropriate for the market requirements.

Over the centuries a large number of designs of lime kiln have been produced and many are still used. As the major item of capital and operating cost, considerable effort is put into optimising and marketing the various designs. The reader involved in selecting and developing kilns would be well advised to study the current literature (see Annex 2), obtain up-to-date information and, where possible, visit other lime producers with relevant up-to-date experience.

The following sections outline the principles underlying the production of quicklime, and describe briefly the main types of kiln being operated and available from suppliers. They also consider how the choice of kiln, feedstone and fuel affects quicklime quality, as well as discussing a number of related subjects. A number of references for further reading is given for most sections. The closely related topics of processing and despatch of quicklime are addressed in chapter 17.

It is frequently stated that every limeworks is unique. This is because of the many permutations of technical, commercial and economic factors. The technical factors include stone, fuel and lime quality (see chapter 14) as well as kiln design.

16.2   Principles of Lime Burning

16.2.1   Heat Requirements

The lime-burning process requires sufficient heat to be transferred to the limestone to preheat it to the dissociation temperature and to decompose the calcium and magnesium carbonates.

The heat requirement for the former process is 442 kcal/kg of CaO [mass × specific heat × temperature difference = $1.79 \times 0.279 \times (900-15)$] while the heat of calcination is 733 kcal/kg CaO (section 15.2.1). It might naively be assumed that the minimum heat requirement for lime burning is $(442 + 733) = 1175$ kcal/kg CaO.

In practice, however, the achieved heat requirement can be much lower than the above value and modern lime kilns achieve net heat usages (see section 14.3.6) below 900 kcal/kg CaO. The above calculation makes two major omissions.

a) Lime kilns make extensive use of heat recovery, which greatly reduces heat requirements.
b) Only part of the heat released by combustion of the fuel is available for calcination (so-called high-grade heat).

These issues are considered further in sections 16.2.2 and 16.7.

Part of the skill of the lime kiln designer is to minimise heat requirements by integrating efficient heat transfer with controlled combustion. This maximises both heat recovery and the amount of high grade heat available for calcination.

16.2.2 Heat Transfer and Recovery

Heat transfer in lime burning can be divided into three stages. To simplify the presentation, the following relates to high-calcium limestone, but the principles also apply to other types of limestone.

a) *Preheating zone.* Limestone is heated from ambient to above 800 °C by direct contact with the gases leaving the calcining zone (i.e. products of combustion, excess air and CO_2 from calcination).
b) *Calcining zone.* Fuel is burnt in preheated air from the cooling zone and (depending on the design) in additional "combustion" air added with the fuel. This produces heat at above 900 °C and causes dissociation of the limestone into quicklime and carbon dioxide.
c) *Cooling zone.* Quicklime leaving the calcining zone at 900 °C is cooled by direct contact with "cooling" air.

These zones are illustrated in Fig. 16.1 for a vertical shaft kiln.

In most designs of lime kiln, there are two major imbalances in the foregoing heat transfer requirements.

a) The kiln gases contain considerably more heat than is required to preheat the limestone to 900 °C.
b) The quicklime does not contain sufficient heat to preheat the air required for combustion (including the necessary level of excess air) to 900 °C.

Both of these imbalances increase the overall heat requirements.

Some modern designs of lime kiln have reduced this imbalance by transferring heat from the exhaust gases to the combustion air.

Heat transfer from gases to solids and vice versa occurs efficiently in packed beds, mainly by conduction and radiation. In some kilns, notably the rotary, heat transfer is less efficient and relies to a greater extent on radiation. This places considerable emphasis on burner design and on the shape and emissivity of the flame [16.1].

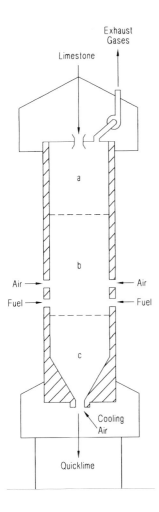

Figure 16.1. Cross-section of a vertical shaft kiln (a) preheating zone; (b) calcining zone; (c) cooling zone

16.2.3   Product Quality

Production of the required quality of quicklime requires the selection of an appropriate limestone, kiln design and fuel. It also involves control of the kiln parameters to ensure that product quality is consistently within specification. These issues are discussed in more detail in later sections.

In making 1 t of quicklime, approximately 1 t of carbon dioxide is also produced — the amount depends on the heat usage per tonne lime and on the fuel used. Most lime producers vent the carbon dioxide to atmosphere as a waste product. Some however use the carbon dioxide as a valuable co-product. Such producers include those making sugar from sugar beet, sodium carbonate via the ammonia soda process, and precipitated calcium carbonate. For them, maximising the % CO_2 helps to minimise gas compression costs and may be achieved by selecting an appropriate kiln, maximising thermal efficiency, minimising excess and "tramp air" (in-leakage) and using a fuel with a high carbon to hydrogen ratio.

16.3   Development of Lime Kilns

16.3.1   Introduction

A large variety of techniques and kiln designs have been used over the centuries
and around the world. In 1935, Searle [16.2] described some 50 types of kiln. Wingate [16.3] described a number of kilns currently used in developing countries for
small scale lime burning. Such kilns are reviewed briefly in the following sections.

Although sales of lime kilns in recent years have been dominated by a relatively small number of designs, many alternatives are available, which may be particularly suitable for specific applications. In addition, as most kilns remain ser-
viceable for several decades, many obsolete designs are still in operation, either
in the original, or a modified form.

The purpose of this section is to give an appreciation of how kilns have developed, which types are in common use and which types are generally being installed.

16.3.2   Intermittent Kilns

The earliest technique used for burning lime probably consisted of an uncovered
heap, containing alternating layers of wood and limestone. The wood was lit and
burnt in an uncontrolled manner for 1 to 2 days. Relatively small amounts of lime
were produced at a very poor thermal efficiency.

Subsequent techniques achieved higher thermal efficiencies by insulating the
heap (e.g. by use of a clay skin) and by controlling the air flow (e.g. by placing the
wood in a pit with an air channel, and building a mound of limestone on top of
the wood).

Flare kilns, pot kilns and field kilns include a number of designs, which may be
roughly cylindrical, rectangular or other suitable shapes (Fig. 16.2). Wood is placed on the grate beneath a crude stone arch (or arches). Progressively smaller
lumps of stone are placed above the arch(es). The wood is lit and the rate of combustion is adjusted by controlling the air flow which enters via the ash pits. Ad-

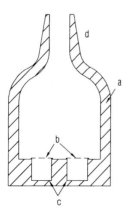

Figure 16.2. Cross-section of a flare kiln
(a) fire-brick lining; (b) twin grates; (c) ash pits; (d) chimney

ditional wood is added via the draw hole until the lime burner judges that the limestone is fully calcined. After an appropriate period (e.g. 3 to 5 days), the kiln is allowed to cool and lime is removed by hand via the draw hole. The kiln is typically built into the side of a hill or an embankment to permit easy access for charging stone into the top of the kiln.

Intermittent mixed-feed kilns are similar in general design, but, instead of additional fuel being added on to the hearth, it is added in layers while the kiln is being charged with limestone.

All intermittent kilns are thermally inefficient because there is little heat recovery in the preheating and cooling zones. In addition, much of the volatile matter in the fuel distils out of the kiln as smoke rather than burning. Judging the amount of fuel required to calcine a charge of limestone is based partly on previous experience and partly by poking rods into the charge to assess the amount of unburned limestone.

16.3.3 Early Continuous Kilns

The earliest continuous kilns (also known as running kilns or draw kilns) were fired using the mixed-feed principle (Fig. 16.3). Alternate layers of limestone and fuel (wood or coal) were charged into the top of the kiln and lime was removed through the drawing door.

Such kilns gave increased outputs as well as considerably improved thermal efficiencies, as effective heat recovery occurred in the preheating and calcining zones. In addition, it was possible to produce a more consistent product by monitoring lime quality and adjusting the fuel input accordingly.

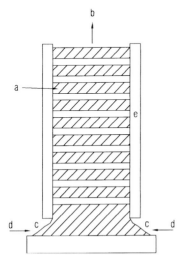

Figure 16.3. Cross-section of a continuous, mixed-feed kiln
(a) alternate layers of limestone and fuel;
(b) exhaust gases; (c) draw holes; (d) combustion air;
(e) kiln lining

16.3.4   Continuous Kilns up to 1935

Searle [16.2] describes a large number of designs of continuous kilns, categorising them into:

a) vertical kilns: mixed feed, furnace-fired and gas-fired
b) horizontal kilns: ring, tunnel and rotary

The mixed feed kilns were developments of those described in section 16.3.3 and the precursors of current designs of kilns. Furnace-fired kilns, often called "Patent" kilns, were heated by wood, coal, gas or oil which was burnt in a furnace adjacent to the kiln. The increases in fuel usage and capital costs were offset by the production of a superior grade of lime that was not contaminated with fuel ash and clinker. Gas-fired kilns generally used producer gas, generated on site from coal. The gas was injected via a number of pipes through the refractory lining towards the base of the calcining zone. It was found that gas firing gave increased control over the kiln and improved product quality.

Horizontal ring kilns (Fig. 16.4), such as the Hoffmann and de Wilt kilns consisted of about 20 interconnected chambers, some of which preheated the limestone with kiln gases, some calcined the limestone using fuel added through trap doors in the roof of the kiln and others served as the cooling zone. After a suitable

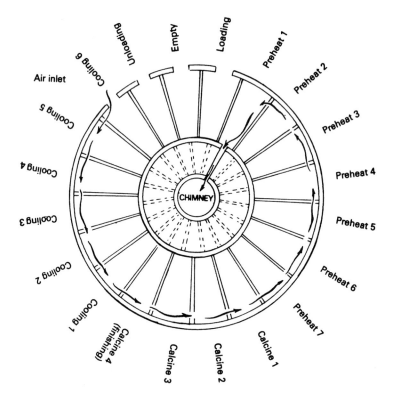

Figure 16.4. Diagram of a ring kiln (by courtesy of Intermediate Technology Publications)

period, the process moved forward by one chamber, with a chamber charged with limestone being added to the front of the preheating zone and the last chamber of the cooling zone being removed from the sequence and the lime in it unloaded.

The concept of tunnel kilns was similar to that of ring kilns. They consisted of a linear, or circular tunnel with fixed preheating, calcining and cooling zones, through which plates or trucks, loaded with limestone, were drawn.

Rotary kilns were developed for the manufacture of cement, and were first used for lime burning in 1885. Their high capital cost, however, limited their use to relatively large productive capacities. They could only calcine small limestone (e.g. with a top size of 60 mm), but, while this was an advantage in terms of the stone balance of the quarry, the small lime produced was not as marketable as lump lime. For many years, rotary kilns were largely used to feed hydrating plants.

Most of the kilns prior to 1935 operated on "natural draught", i.e. the combustion products were drawn through the kiln by the chimney effect of vertical kilns, often supplemented by a small chimney. As a result, output rates were low (e.g. 1.5 t/d per m^2 of cross-section). Horizontal kilns operated on the draught produced by tall chimneys.

16.3.5   Continuous Kilns After 1935

The period from 1935 to about 1965 saw the introduction of increased mechanisation of existing designs of lime kilns. Improvements included mechanical charging and drawing systems and fans to draw kiln gases through the burden (i.e., the limestone and lime) and through simple dust collectors. In some cases, fans were used to blow combustion and/or cooling air into the kiln. As a result, output rates of shaft kilns increased (e.g. to about 7.5 t/d per m^2 of cross-section) and manning levels per tonne of lime decreased markedly.

Much of the lime from shaft kilns was still hand-picked to remove partially calcined lumps (which were tipped), and to sort light-burned from solid-burned lime.

The rapid growth of the Basic Oxygen Steelmaking process in the 1960s required kilns which could produce a 6 to 40 mm quicklime, with a relatively high reactivity, low residual % $CaCO_3$ and low sulfur content. Initially, much of this new demand was met by Calcimatic and rotary kilns.

The Calcimatic kiln can best be described as a circular tunnel kiln, with an annular refractory hearth which rotated, thereby transporting preheated limestone under some 20 or 30 burners, which produced a controlled temperature profile peaking at about 1300 °C. One revolution of the hearth took about 60 min., long enough to calcine 20 to 40 mm limestone. The quicklime was removed from the hearth by a scraper and discharged into the lime cooler, which was also used to preheat the combustion air. Low sulfur quicklime could be made using natural gas and oil containing up to 2 % sulfur (providing reducing conditions were maintained in the kiln). High heat usages and maintenance costs caused most of the kilns to be phased out in the 1980s.

Since the "oil crisis" of 1972, there have been a number of pressures which have caused many lime producers to replace existing kilns, namely:

a) variable and, at times, high fuel prices,
b) fierce competition (arising from spare capacity), which has forced down the real cost of lime,
c) a swing towards quicklime with more consistent quality and particularly with high reactivity, low $CaCO_3$ and low sulfur content, and
d) increasingly stringent environmental standards both for the workplace and for emissions to atmosphere.

Modern kilns fall into the following groups:

a) Vertical shaft – traditional types (including producer gas)
               – mixed feed
               – modern "basic" designs
               – annular
               – parallel-flow regenerative
               – double-inclined
               – multi-chamber
b) Rotary – traditional (without pre-heater)
            – pre-heater
c) Miscellaneous

These are described briefly in the following sections, to give an appreciation of the variety of designs and some of their characteristics. Space constraints mean that it is only possible to include reference to a limited number of designs and the omission of a particular design should not be taken as a comment on its merits. Quoted kiln characteristics are based partly on manufacturers' data and partly on achieved performances. References for further reading are included to assist the reader requiring more detailed information, but reference should also be made to recent editions of the trade journals (Annex 2) and to the manufacturers' literature.

16.4   Modern Kilns

16.4.1   Shaft Kilns — General

Before describing specific designs of shaft kilns, it is appropriate to consider three important features which are common to all designs, namely charging, drawing and combustion.

16.4.1.1   Charging of Limestone

The problems associated with charging are illustrated by consideration of the implications of single point charging on the axis of the kiln. This produces a conical surface to the stone bed (Fig. 16.5).

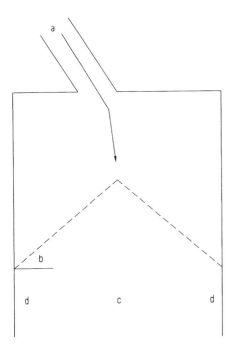

Figure 16.5. Cross-section of a single point charging system
(a) feed chute; (b) angle of repose; (c) finer fraction of stone; (d) coarser fraction of stone

The larger stones tend to roll down the conical heap towards the walls, while the smaller fractions concentrate along the axis of the kiln. As a result, there is a gradation in the resistance to flow of the kiln gases from a high level around the central axis to progressively lower levels towards the walls of the kiln, and to an even lower level at the kiln wall (owing to the "wall effect"). This results in a greatly reduced flow of gases through the central part of the kiln and, in consequence, that part of the burden tends to be under-calcined.

A variety of devices have been developed to mitigate this effect and to minimise the effect of asymmetry in the charging system. In the fixed plate and cone arrangement (Fig. 16.6), the position of the cone and striker plate, relative to the feed chute and to each other, can be adjusted to produce a more-or-less uniform profile around the kiln. Inevitably, fines tend to concentrate in the 60° segments on either side of the feed chute centre line, but the effect on kiln operation is small. The rotating hopper and bell system (Fig. 16.7) is more sophisticated and produces both a more uniform profile and a better dispersion of fines in an annular ring around the kiln.

Mixed feed kilns present the greatest challenge. For them, it is essential that the fuel (generally coke or anthracite) is dispersed uniformly across the kilns. They tend to use cones, similar to that shown in Fig. 16.7, fitted with an extension, which may consist of four quadrants, one deflecting part of the charge towards the centre of the kiln, a second deflecting it further out, and with the third and fourth quadrants deflecting it progressively further away from the axis of the kiln [16.4]. After each charge, the cone and apron are rotated by a small amount (e.g. by $^1/_5$ or $^1/_4$ of a revolution) so that, on average, a uniform distribution is obtained.

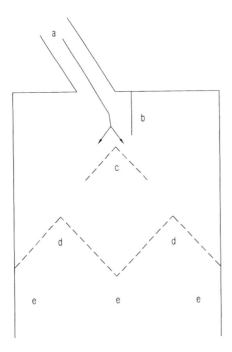

Figure 16.6. Cross-section of a plate-and-cone charging system
(a) feed chute; (b) deflector plate;
(c) cone; (d) finer fraction of stone;
(e) coarser fraction of stone

Figure 16.7. A hopper and bell charging system (by courtesy of Fuel & Combustion Technology International Ltd.)

16.4.1.2   Drawing of Lime

The drawing system should be such as to produce uniform movement of the burden across the kiln. A simple system, using a single discharge point and a conical table is shown in Fig. 16.8. This works satisfactorily while the burden moves freely. However, when there is a tendency for part of the kiln to stick, or when crotched lumps of lime (i.e., lumps that have become fused together) bridge between the table and the wall of the cooling zone, lime is preferentially drawn from the free-flowing parts of the kiln, resulting in further over-heating in the problem area.

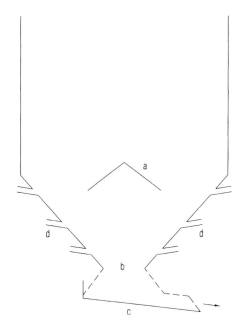

Figure 16.8. Illustration of a basic discharge system
(a) conical table; (b) discharge aperture; (c) feeder; (d) rodding ports

A better system uses four discharge points without a central table. If there is a tendency for part of the kiln to stick, the feeder(s) under that part can be operated at a faster rate than the others to help re-establish free movement. Similarly, if one feeder becomes blocked with crotched material, appropriate action can be taken. Multiple discharge points can also assist with diagnosing problems within a kiln. By operating each in turn, the lime from different segments can be tested separately to identify if a particular segment is under or over-burned.

Still more sophisticated drawing mechanisms are used to obtain mass-flow, such as;

a) hydraulically driven oscillating quadrants,
b) a rotating eccentric plate, and
c) a rotating spiral cone with steps and a slope designed to take lime uniformly across the shaft.

The last mentioned design is used on some mixed-feed kilns [16.4].

16.4.1.3 Combustion

In all combustion processes there is an optimum air to fuel ratio which gives the highest efficiency of combustion. A ratio lower than optimum results in incomplete combustion and increased levels of carbon monoxide, while a higher ratio results in the products of combustion being diluted and cooled by the additional quantities of air.

Combustion within the packed bed in vertical lime kilns is particularly problematical as mixing of gassified fuel and air under those conditions is an inefficient process. From the viewpoint of combustion efficiency, the fuel and air should, ideally, be distributed uniformly across the shaft. However, regardless of the firing system, variations in the air-fuel ratio occur. From the viewpoint of lime-burning, the long, lazy flames resulting from combustion in a deficiency of air are more favourable than short, fierce flames. This is because a deficiency of air reduces the temperatures in the calcining zone, particularly in the "finishing" section, and helps to ensure a high reactivity.

Various techniques have been used to moderate temperatures in the calcining zone. Use of an overall deficiency of air is effective but increases heat usage and can cause the emission of dark smoke. Re-circulation of kiln gases is practised on some kilns (e.g. the West Gas design) to moderate kiln temperatures, particularly at the walls. In the annular shaft and parallel-flow regenerative kilns part or all of the combustion gases pass down the shaft in co-current flow with the lime. This results in a comparatively low temperature in the finishing section of the calcining zone.

16.4.2 "Traditional" Shaft Kilns

When fuel is injected at the wall of a kiln, it usually does not penetrate more than 1 m into the burden. This limits the effective kiln width, or diameter, to about 2 m, and restricts the productive capacity of a 2 m diameter shaft to about 80 t/d, (with an induced draught fan).

Various techniques were used to enable the diameter of the kiln, and hence its productive capacity, to be increased. On producer gas kilns, the large volume of the low calorific value gas favoured greater penetration, and was often assisted by the injection of additional, or "primary" air into the calcining zone (e.g. the Priest design). Some oil-fired kilns used recycled kiln gases to increase the penetration of the vaporised oil (e.g. the West Gas kiln). Others gassified the oil in refractory-lined combustion chambers, using 50 % of stoichiometric air, and injected it 1 m into the burden via water-cooled pipes or tuyères, thereby enabling a 4 m diameter shaft to be used.

Some coal-fired kilns used arches to create voids into which coal batches were projected, either manually or mechanically. Shaft diameters of up to 4.5 m were used, but output rates, even with impressed air and induced draught were only 150 t/d.

Heat usages were generally high at about 1350 kcal/kg net of saleable lime. There was a considerable loss of volatile matter in the exhaust gases, particularly immediately after firing when the volatile matter caused a high excess of fuel over the air supply. This resulted in the emission of dark smoke for at least 6 min. after firing and, in some cases for 20 min.

Frequently, oil firing also resulted in dark smoke emission, owing to the need mentioned above to restrict the levels of excess air to ensure an acceptable lime reactivity.

In many countries, the widespread increase in the availability of natural gas at competitive prices in the 60s and 70s enabled many lime producers to extend the operating lives of their traditional shaft kilns. Natural gas brings three major advantages:

a) it is sulfur free so that the lime meets the sulfur requirements of steelmakers,
b) it burns with a long, lazy flame which has a relatively low emissivity. This enables a more reactive lime to be produced for a given % $CaCO_3$, or enables the % $CaCO_3$ to be reduced for a given reactivity, and
c) it does not produce dark smoke when burnt under normal operating conditions.

The above advantages, coupled with the development of techniques to improve penetration of the burning gas into the centre of the kiln, has enabled gas-fired shaft kilns with diameters of up to 5 m and outputs of over 200 t/d to remain competitive with modern designs of shaft kilns. Such techniques include the addition of beam burners, central burners and the adoption of the asymmetric "wafting" technique.

16.4.3 Mixed-feed Shaft Kilns

Mixed-feed kilns are still widely used in the ammonia soda process [16.4] and elsewhere (e. g. [16.5]).

In the ammonia soda process coke is used to produce exhaust gases with the high CO_2 concentrations of about 40 % by volume (dry) required for the process. However, producers are being forced to consider alternative fuels because of:

a) the increasing cost and reducing availability of suitable cokes, (which must be of a certain size, strength and "reactivity" to the $C + CO_2 \rightarrow 2CO$ reaction),
b) competition from natural sodium carbonate deposits.

Coke-fired mixed-feed kilns can have the lowest heat usage of all kilns. Net heat usages of about 850 kcal/kg are reported [16.4, 16.6], with 950 to 1,100 being more typical of routine operation.

Another feature of mixed-feed kilns is that they can be operated to produce the consistently low reactivity lime favoured by some producers of aircrete. They can also produce higher reactivity lime, but the $CaCO_3$ levels are higher than can be obtained from more modern designs.

Anthracite (see section 14.3.3.3) is widely used in mixed-feed kilns. As much of the volatile organic matter distils at below the dissociation temperature, part of the calorific value is lost and the total heat usage is increased (typically coke usage is in the range 130 to 150 kg/t lime). The volatiles also cause the emission of smoke, which is increasingly becoming environmentally unacceptable.

Coke ignites at about 800 °C, corresponding with the onset of surface calcin-ation, so that there is little wastage of high grade heat. Exhaust gas temperatures of about 100 °C and lime discharge temperatures as low as 50 °C reflect the high thermal efficiency of the kiln.

The operation of mixed-feed kilns is fully described in [16.4] and a design of a 70 t/d kiln is given in [16.5] (see also [16.7, 16.8]).

16.4.4   Modern Shaft Kilns

The concept of the shaft kiln has been modernised in a number of designs, the characteristics of four (Nos. A to D) are summarised in Table 16.1. Some designs are more suitable for low outputs e.g. below 100 t/d, while others can be used for much higher outputs, e.g. up to 800 t/d for the beam-burner design.

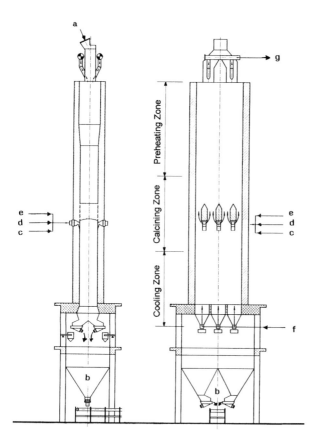

Figure 16.9. Cross-sections of an HPK gas-fired shaft kiln, incorporating internal arches (by courtesy of SiC – Società Impianti Calce S.r.l.)
(a) limestone; (b) quicklime; (c) natural gas; (d) primary air; (e) recycled kiln gas; (f) cooling air; (g) exhaust gases

Table. 16.1. Summary of typical characteristics of some kilns

Ref. no.	Kiln design	Fuels	Output range (t/d)	Range of stone size (mm)	Net heat usage (kcal/kg) [a]	Power usage (kWh/t)
(A)	Central burner	G	40–80	40–150	1,000–1,200	10–15
(B)	External chambers	G, L	40–120	80–350		
(C)	Beam burner	G, L, S	50–800	20–175		
(D)	Triple internal arch [b]	G, L, S	15–250	25–120		
(E)	Annular shaft	G, L, S	80–600	10–250	950–1,100	18–35
(F)	P.F.R. [c] – standard	G, L, S	100–600	25–200	860–1,000	20–40
(G)	P.F.R. [c] – "finelime"	G	100–300	10–30	860–1,000	35–45
(H)	Double-inclined	G, L	10–150	20–100	ca. 1,030	ca. 30
(I)	Multi-chamber	G, L, S	40–225	20–150	950–1,070	20–45
(J)	Long rotary	G, L, S	150–1,500	0–60	1,550–1,800	18–25
(K)	Preheater rotary	G, L, S	150–1,500	0–60 [d]	1,200–1,450 [e]	20–45 [f]
(L)	Mixed-feed	S	10–300	30–150	950–1,120	5–15
(M)	Flash calciner	G, L	300–1,500	<1	1,100–1,200	33–38
(N)	Fluidised bed	G, L	30–150	<2	ca. 1,200	ca. 30
(O)	"Top-shaped" [g]	G, L, S	30–100	5–40	1,100–1,250	ca. 40
(P)	Travelling grate [h]	G, L, S	80–130	15–45	880–1,150	30–38

[a] Average long-term values, for high calcium limes, excluding use for heating or drying fuel.
[b] Or "HPK" kiln.
[c] P.F.R. = parallel-flow regenerative.
[d] 10 to 60 mm with shaft preheater, 10 to 50 mm with grate preheater, 0 to 2 mm with cyclone preheater.
[e] Can be as high as 1800 kcal/kg with cyclone preheater.
[f] Typically 20 to 25 kWh/t with grate preheater.
[g] Or "Chisaki" kiln.
[h] Or "CID" kiln.

Acceptable sizes for the feedstone range from a minimum of 20 mm to a top size of up to 175 mm and even up to 350 mm for design B (although the % $CaCO_3$ for such a large feed stone would probably be relatively high). It should be noted that the feedstone size ratio should generally be 1 to 2 or 1 to 3 and that the dimensions of a 200 t/d kiln designed for a feed of 25 to 50 mm would differ from those of a kiln with the same capacity designed for a feed of 75 to 150 mm.

Reported long-term net heat usages at design output for these types of kiln generally lie in the range 1,000 to 1,200 kcal/kg.

Some kilns are suitable for operation on gaseous, liquid and solid fuels, while the options for others are more restricted. The level of CO_2 in the exhaust gases when burning heavy fuel oil is about 34 %. (See references [16.9–16.17] for more detailed information.)

16.4.5   The Double-inclined Kiln

The double-inclined kiln (design H in Table 16.1) is shown in Fig. 16.10. It is essentially rectangular in cross-section, but incorporates two inclined sections in the calcining zone. Opposite each inclined section, off-set arches create spaces into which fuel and preheated combustion air are fired via three combustion chambers.

Cooling air is drawn into the base of the kiln. Part of it is withdrawn at 350 to 400 °C and is re-injected via the combustion chambers. The tortuous paths for both the gases and the burden, coupled with firing from both sides, ensures an efficient distribution of heat and enables stone as small as 20 to 40 mm to be calcined. The maximum feed size is 100 mm.

The net heat usage is about 1,030 kcal/kg. A range of solid, liquid and gaseous fuels can be used, although they should be selected with care to avoid excessive build-ups caused by fuel ash and calcium sulfate deposits.

By limiting the proportion of fuel fired through the lower burners, the kiln can produce a reactive low carbonate product. Further details are available in the literature [16.18].

16.4.6   The Multi-chamber Kiln

This is a development of the double-inclined kiln (design I in Table 16.1). It consists of 4 or 6 alternately inclined sections in the calcining zone, opposite each of which is an offset arch. The arch serves the same purpose as in the double-inclined kiln.

Cooling air is preheated by lime in the cooling zone and is withdrawn, dedusted and re-injected via the combustion chambers. A feature of the kiln is that the temperature of the lower combustion chambers can be varied to control the reactivity of the lime over a wide range. The kiln can be fired with solid, liquid and gaseous fuels (or a mixture) and the net heat usage is about 1000 kcal/kg. It can accept stone with a minimum size of 20 mm up to a maximum size of 150 mm. Further details are available in references [16.19–16.23].

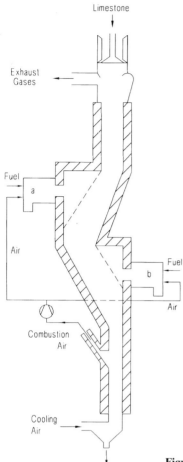

Limestone

Exhaust
Gases

Fuel

a

Air

Fuel

b

Air

Combustion
Air

Cooling
Air

Quicklime

Figure 16.10. Cross-section of a double-inclined shaft kiln
(a) upper combustion chamber; (b) lower combustion chamber

16.4.7   The Annular Shaft Kiln

The annular shaft kiln (design E in Table 16.1) is shown in Figs. 16.11 and 16.12. The characteristic feature of the kiln is a central cylinder, which:

a) restricts the effective thickness of the burden,
b) ensures good heat distribution,
c) enables part of the combustion gases from the lower burners to be drawn down the kiln (creating a relatively low temperature finishing zone in which both the gases and the burden move co-currently),
d) enables kiln gases to be withdrawn into a heat exchanger (where fitted) which preheats part of the combustion air.

The burden is drawn through the annulus between the central cylinder and the walls of the kiln, past the two layers of burners.

Figure 16.11. A Beckenbach annular shaft kiln (by courtesy of Beckenbach Wärmestelle GmbH)

Most of the fuel is fired through the upper burners, together with a sub-stoichiometric quantity of primary air. The remaining oxygen required to burn the fuel efficiently is provided by the kiln gases arising from the lower burners, which operate with an excess of oxygen.

Part of the products of combustion from the lower burners rises up the kiln to the upper burners. The remainder is drawn down the kiln and into the central cylinder, together with the cooling air. The necessary suction for this flow is provided by an air-operated ejector. The resulting mixture of kiln gases and cooling air has a temperature of about 900 °C and serves to moderate the flame temperatures in the lower zone to about 1350 °C.

About 30 % of the kiln gases passing through the preheating zone may be withdrawn at about 750 °C to preheat the primary air in an external recuperator. The % CO_2 in the exhaust gases is about 34 % by volume (dry) when fired with heavy fuel oil.

The kiln accepts a feedstone with a top size in the range 30 to 250 mm and a bottom size as low as 10 mm. It can burn gas, oil or solid fuels. The net heat usage is about 1,000 kcal/kg. (See references [16.24–16.28] for more details.)

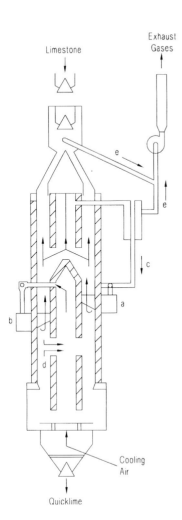

Limestone

Exhaust Gases

Cooling Air

Quicklime

Figure 16.12. Cross-section of an annular shaft kiln
(a) upper burners; (b) lower burners; (c) combustion air
to upper burners; (d) combustion air to lower burners;
(e) kiln exhaust gases

16.4.8   The Parallel-flow Regenerative Kiln

The parallel-flow regenerative kiln (design F in Table 16.1) is shown in Figs. 16.13 and 16.14. Its characteristic feature is that it consists of two interconnected vertical cylindrical shafts (some early designs had three shafts, while others had rectangular shafts, but the operating principles are the same).

Batches of limestone are charged alternately to each shaft. The burden is drawn downwards through a preheating/regenerative heat exchange zone, past the fuel lances and into the calcining zone. From there the quicklime passes into the cooling zone.

The operation of the kiln consists of two equal stages, of 8 to 15 min. duration at full output.

In the first stage, fuel is injected through the lances in shaft 1 and burns in the combustion air blown down that shaft. The heat released is partly absorbed by

Figure 16.13. A pair of Maerz parallel-flow regenerative kilns
(by courtesy of Buxton Lime Industries Ltd.)

the calcination of limestone in shaft 1. Air is blown into the base of each shaft to cool the lime. The cooling air in shaft 1, together with the combustion gases and the carbon dioxide from calcination, passes through the interconnecting cross-duct into shaft 2 at about 1050 °C. In shaft 2, the gases from shaft 1 mix with the cooling air blown into the base of shaft 2 and pass upwards. In so doing, they heat the stone in the preheating zone of that shaft. If this mode of operation were to continue, the exhaust gas temperature would rise to well over 500 °C.

However, after 8 to 15 min., the second stage commences. The fuel and air flows to shaft 1 are stopped, and "reversal" occurs. After charging limestone to shaft 1, fuel and air are injected to shaft 2 and the exhaust gases are vented from the top of shaft 1.

The above method of operation incorporates two key principles.

a) The stone-packed preheating zone in each shaft acts as a regenerative heat exchanger, in addition to preheating the stone to calcining temperature. The surplus heat in the gases (low grade heat — see section 16.7.4) is transferred to the stone in shaft 2 during the first stage of the process. It is then transferred from the stone to the combustion air in the second stage (and, in so doing, becomes high-grade heat). As a result, the combustion air is preheated to about 800 °C.

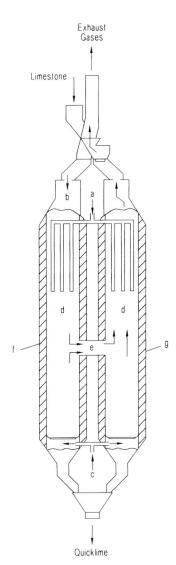

Exhaust
Gases

Limestone

d   d

f

e

g

c

Quicklime

Figure 16.14. Cross-section of a parallel-flow
regenerative kiln
(a) fuel; (b) combustion air; (c) cooling air; (d) kiln gases;
(e) cross-duct; (f) shaft 1; (g) shaft 2

b) The calcination of the quicklime is completed at the level of the cross-duct at
a moderate temperature of about 1100 °C. This favours the production of a
highly reactive quicklime, which may, if required, be produced with a low
$CaCO_3$ content.

The standard kiln can be designed to accept feedstones in the range of 25 to
200 mm. It can be fired with gas, oil or solid fuel (in the case of solid fuel, its
characteristics must be carefully selected). The net heat usage is low at about
900 kcal/kg. A modified design (the "finelime" kiln — design G in Table 16.1) is
able to accept a feedstone in the range 10 to 30 mm.

Because the kiln is designed to operate with a high level of excess air (none of the cooling air is required for combustion), the level of CO_2 in the exhaust gases is low at about 20 % by volume (dry). For further details see [16.29–16.38].

16.4.9   Traditional Rotary Kiln [16.1]

The traditional/long rotary kiln (type J in Table 16.1 and Fig. 16.15) consists of a rotating cylinder (110 to 140 m long) inclined at an angle of 3 to 4° to the horizontal. Limestone is fed into the upper "back-end" and fuel plus combustion air is fired into the lower "front-end". Quicklime is discharged from the kiln into a lime cooler, where it is used to pre-heat the combustion air.

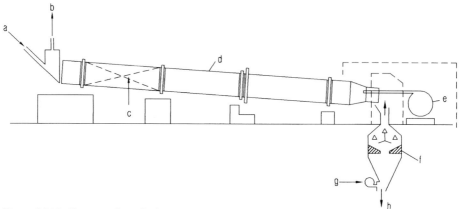

Figure 16.15. Cross-section of a long rotary kiln
(a) limestone; (b) exhaust gases; (c) refractory trefoils; (d) kiln shell; (e) fuel plus secondary air; (f) lime cooler; (g) cooling air; (h) quicklime

Various designs of lime cooler are used, and include "planetary" units mounted around the kiln shell, travelling grates and various types using louvres to contain the lime, while permitting the passage of air.

Many kilns have internal features to recover heat from the kiln gases and to preheat the limestone. These include

a) chains (in kilns fed with calcium carbonate sludge),
b) metal dividers and refractory trefoils, which effectively divide the kiln into smaller tubes,
c) lifters which cause the stone to cascade through the gases, and
d) internal refractory dams, which increase the residence time of the burden.

The design of burner is important for the efficient and reliable operation of the kiln. The flame should be of the correct length — too short and it causes excessive temperatures and refractory failure, too long and it does not transfer sufficient radiant heat in the calcining zone with the result that the back-end temperature rises and thermal efficiency decreases. The flame should not impinge on the refractory. Oxygen enrichment of the combustion air, and particularly of that under the flame, is used to raise flame temperatures and increase radiant heat transfer. It can increase output by 20 % and reduce net heat usage per tonne lime by 10 % [16.62].

Rotary kilns can accept a wide range of sizes from 60 mm down to dust. An interesting feature of the tumbling bed in the kiln is that larger stones migrate towards the outside of the bed, while smaller ones concentrate at the centre of the bed. This results in the larger stones being exposed to higher temperatures and avoids over-calcination of the finer fractions. Indeed, it is frequently necessary to incorporate "mixers" or steps into the refractory lining to mix the bed and to ensure that the finer fractions are fully calcined.

Because of the ease with which they can be controlled, rotary kilns can produce a wider range of reactivities and lower $CaCO_3$ levels than shaft kilns. The variability of reactivity, however, tends to be greater than that of shaft kilns.

Relatively weak feedstones, such as shell deposits, and limestones that decrepitate, are unsuitable as feed to shaft kilns but may prove to be acceptable for rotary kilns. Long rotary kilns can also accept a wet feed, i.e., a sludge or filter cake of finely divided carbonate.

Rotary kilns can be fired with a wide range of fuels. As heat transfer in the calcining zone is largely by radiation, and as the infra-red emissivities increase in the sequence gas, oil and solid fuel, the choice of fuel can have a marked effect on heat usage. Values as high as 2,200 kcal/kg have been observed for simple gas-fired kilns, while a similar coal-fired kiln may have a heat usage of 1800 kcal/kg. The use of internal fittings can reduce those heat usages to below 1600 kcal/kg. Radiation and convection losses from the kiln are high relative to other designs of lime kiln.

A feature of rotary kilns is the formation of "rings". These consist of an accumulation of material on the refractory in a part of the kiln which has the appropriate temperature for a semi-liquid phase to form. Such rings can form from ash in coal-fired kilns and from calcium sulfate deposits. Alkalis (sodium and potassium oxides), clay and lime can contribute to the build-ups, which can be troublesome. In the case of coal-firing, fine grinding of the fuel can significantly reduce the rate of build-up.

Another feature of rotary kilns is that sulfur from the fuel, and, to a lesser extent from the limestone, can be expelled from the kiln in the kiln gases, without over-burning the lime, by a combination of controlling the temperature and the % CO in the calcining zone (see section 13.2). Thus high reactivity, low sulfur limes can be produced using relatively inexpensive high sulfur fuels — subject to any emission limits for SO_2 in the exhaust gases.

16.4.10   The Pre-heater Rotary Kiln [16.1]

Rotary kilns fitted with pre-heaters (type K in Table 16.1; Figs. 16.16 and 16.17) are generally considerably shorter than the conventional rotary kiln (e.g. 40 to 60 m). The heat usage decreases because of reduced radiation and convection losses as well as the increased heat recovery from the exhaust gases. Thus, with coal firing, net heat usages below 1250 kcal/kg are reported.

Figure 16.16. A KVS preheater rotary kiln (by courtesy of Svedala Industries Inc.)

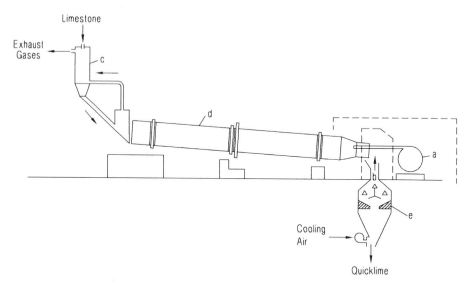

Figure 16.17. Cross-section of a preheater rotary kiln
(a) fuel; (b) preheated combustion air; (c) preheater; (d) rotary kiln; (e) cooler

A number of pre-heater designs have been developed, including vertical shafts and travelling grates. The pre-heater should be selected on the basis of the size and properties of the feedstone. Most can accept a bottom size of 10 mm; some cannot tolerate weak stones, or stone that is prone to decrepitate. Semi-wet, powdered feeds, consisting of partially dried filter cake or granules, containing about 20 % of water, can be calcined using grate preheater systems.

While the elimination of sulfur is more difficult with pre-heater kilns, it can be achieved by various means, e.g.

a) establishing a purge of SO_2 by taking some of the kiln gases around the pre-heater (at the cost of increased heat usage),
b) operating the kiln under reducing conditions and introducing additional air at the back end (only applicable with grate pre-heaters), and
c) adding sufficient finely divided limestone in the feed that it preferentially absorbs SO_2 and is either collected in the back-end dust collector, or is screened out of the lime discharged from the cooler.

For further information on rotary kilns see [16.39–16.47].

16.4.11   Miscellaneous Kilns

Several interesting designs of kiln have been developed, but few appear to have found a wide application.

One of the driving forces for developing new designs of kiln is that substantial quantities of calcium carbonate are available in a finely divided form. The sugar and paper/wood pulp industries, for example, produce a calcium carbonate sludge (see sections 30.4 and 32.16), which can be calcined and recycled. Many plants use rotary kilns for this purpose, while some use fluidised bed kilns [16.6, 16.48, 16.49].

Many limestone quarries produce a surplus of fine stone, which, in principle, would be suitable for calcination (although it is often contaminated with clay). Moreover, in some countries (e.g., Western Australia) the limestone deposits are sandy, or friable, and are unsuitable for calcining in shaft kilns. Several companies operate various designs of fluidised bed kilns. It is understood that problems have been experienced with operation and product quality, particularly with larger units (e.g., over 100 t/d). At the time of writing, (mid-1997), a flash calciner is being built. It is linked to a short rotary kiln, and is reported to have a design capacity in excess of 1200 t/d [16.51]. While such kilns are suitable for calcining fine limestone, their capital plus operating costs tend to be unacceptably high for small installations.

For limestone in the size range 15 to 45 mm, a relatively recent option is the "travelling grate" (or CID) kiln [16.50] (developed in Germany). It consists of a rectangular shaft preheating zone, which feeds the limestone into a calcining zone. In that zone, the limestone slowly cascades over five oscillating plates, opposite which are a series of burners. The lime passes into a rectangular cooling zone. The CID kiln can burn gaseous, liquid or powdered fuels and is reported to produce a soft-burned lime with a residual $CaCO_3$ content of less than 2.3 %. The four kilns built to date have capacities of 80 to 130 t/d of quicklime.

Another relatively new design of kiln, developed in Japan, the "top-shaped" or Chisaki kiln, accepts feedstone in the 5 to 40 mm range [16.52]. It consists of an annular preheating zone from which the limestone is displaced by "pusher rods" into a cylindrical calcining zone. Combustion gases from a central, downward facing burner, fired with oil and positioned in the centre of the preheating zone, are drawn down into the calcining zone by an ejector. The lime then passes down into a conical cooling zone. The kiln is reported to produce high quality quicklime, suitable for steel production and precipitated calcium carbonate. Kiln capacities are up to 100 t/d, and heat usages of 1,100 to 1,250 kcal/kg of lime are quoted. It is reported that, because of its relatively low height, the kiln can accept limestones with low strengths.

16.5   Selection of Lime Kilns

The selection of a lime kiln for a particular installation is a complex process [16.53]. The following paragraphs can, at best, provide an initial check list on which to build.

A starting point is the characteristics of the limestone. Its particle size, coupled with the strength of the limestone/lime and their calcination properties may limit the choice of kiln.

The next step is to establish, as clearly as possible, the quality of the quicklime required, in terms of % $CaCO_3$, sulfur, reactivity and other key parameters, as well as the "lead grade" in terms of particle size. The impact of the design of the quicklime processing plant on product quality should not be overlooked (see chapter 17).

The availability and cost of suitable fuels is an important factor, which interacts with the choice of kiln.

Other factors, such as planning constraints on height and the area available for the kiln installation (and for any future extensions) may also limit the choice of kiln.

Thereafter, the standard project considerations apply (e.g., capital/operating costs and the logistics of number/size of kilns in relation to current and forecast market demand). The extent to which a new kiln might fit into the existing infrastructure, and particularly to existing lime and limestone handling and storage equipment, can also have a marked effect on the capital cost and ease of installation.

Many lime producers have installed both shaft kilns (for the larger grades of limestone) and rotary kilns (for the smaller sizes). This makes good use of the processed stone and produces a range of grades and qualities which can meet a variety of market demands. A recent trend, however, has been to install shaft kilns designed to accept smaller sizes of stone.

Where the lime kilns are part of an integrated plant, involving lime grinding and/or hydration, the requirements of those plants may also have a significant bearing on the choice of kiln.

16.6 Kiln Control

16.6.1 General

The detailed approach to kiln control depends on the design of kiln, the market being supplied and the culture of the organisation. The following general comments are offered as basis for consideration in a particular context.

Before developing a control philosophy, it is important to answer two questions.

a) How quickly can significant changes be detected with confidence?
b) How rapidly does the kiln respond to significant changes?

It should be noted that the response time of lime reactivity, which partially reflects conditions in the calcining zone, may be less than the residence time of the burden in the kiln, whereas that of % $CaCO_3$, which reflects the time-temperature history throughout the kiln, may be longer than the residence time.

In the context of the principles of calcination of limestone (chapter 15), the objective of kiln control is to produce a time-temperature profile such that the calcination and sintering of the various sizes of particles produce the required product quality in terms of % $CaCO_3$, reactivity and, in some cases, % sulfur. This is achieved by varying the fuel to stone/lime ratio, the air to fuel ratio and the residence time.

Section 15.4 and Table 15.3 illustrate the challenge of achieving an adequate degree of calcination of the larger particles of limestone fed to the kiln, without over-burning the smaller particles.

As kiln control involves a number of operators working on shifts, it is preferable that, as far as possible, they all work to the same set of procedures. This has two principal advantages.

a) it reduces the amount of "knob-twiddling" that tends to occur at shift change-over, with the new operator changing the kiln settings to his preferred levels, and
b) if set up correctly, it reflects "best practice", which helps the less experienced operators and provides a sound basis on which improvements can be made and built in to the system.

As part of the above philosophy, appropriate corrective actions for specific circumstances should be developed and agreed. Where, for example, the primary product quality variables are % $CaCO_3$ and reactivity, either or both can be low, within the target range, or high. This results in a 3×3 matrix, with 9 "boxes", in only one of which both variables are on target. For the remaining 8 situations, an advisory corrective action procedure (with suitable checks) would assist all operators to respond to each situation in a similar and appropriate manner.

16.6.2   Shaft Kilns

The staring point for any control procedure is obtaining valid test results at an appropriate frequency.

Where practicable, the sample should be of run-of-kiln lime. However, shaft kilns are usually fed with lump limestone and it may be impractical to take representative samples and to install sample preparation equipment large enough to handle the product.

In such cases, samples should be taken of smaller fractions of lime produced by crushing and/or screening. Test results from such samples can be correlated with those for run-of-kiln lime if required.

When designing lime handling systems, due consideration should be given to the provision of a suitable sampling point(s).

Sampling for kiln control purposes should be done at a frequency within the shortest response time of the kiln. For example, if the response time to reactivity were 12 hours, routine samples should be taken every 8 hours and more frequently during periods of disturbed operation.

Increasing the fuel to stone/lime ratio results in a reduction in % $CaCO_3$ and a lowering of reactivity. Lowering the air to fuel ratio, reduces temperatures in the calcining zone and tends to increase the volume of the zone, resulting in increased reactivity for a given % $CaCO_3$. It may also have an effect on the levels of CO and O_2 in the exhaust gases and reduce thermal efficiency.

A danger with shaft kiln control is that changes are made too frequently. In principle, the kiln should be allowed to respond fully to a change — whether it be a shut-down or an adjustment to the settings — before a further change is made. It is recognised that, in practice, such an approach may not always be feasible. In addition, the effects of routine events, such as shut-downs, may be countered by a prescribed response such as an increase in heat input per tonne of stone for a specified period.

The control procedure should differentiate clearly between the actions which an operator on a particular shift may take and those which the manager/supervisor may take on a daily basis. Many lime works refer to the shift operator as a plant attendant, which emphasises the nature of his role.

16.6.3   Rotary Kilns

As rotary kilns normally burn relatively small stone, sampling run-of-kiln lime (or, indeed, lime as it is discharged from the kiln into the cooler) does not present a problem. Routine samples should be taken at suitable intervals and tested immediately, preferably by the operator, so that corrective action can be taken as quickly as possible. Additional samples should be taken as and when the operator requires them.

Where kilns are fired with a fuel containing significant amounts of sulfur, and low sulfur lime is required, the air to fuel ratio and/or kiln temperatures need to be controlled to ensure that calcium sulfate on the surface of the particles is decomposed and that sulfur dioxide is eliminated from the kiln in the exhaust gases.

Because of the relatively short response times, identification of cause and effect is more straightforward than with shaft kilns. More detailed advice on the control of rotary kilns is given in [16.1].

16.7 Mass and Heat Balances

16.7.1 General

For most lime producers, the production cost is dominated by the cost of fuel per tonne of lime. As a result, efforts to reduce heat usage frequently results in worthwhile savings.

A prerequisite for efficient control is instrumentation, of which measuring the input of limestone/the output of lime and the quantity of fuel used are the most important (see section 16.8). Optimising thermal efficiency, however, requires detailed knowledge of how the heat is used/wasted, and where savings might be made while maintaining output and product quality at acceptable levels.

16.7.2 Mass Balances

Mass balances are a way of ensuring that the measurements of inputs correspond with outputs. This serves as a valuable check that the information is substantially correct, particularly when balances are made for each chemical element. The prime purpose of determining a mass balance is to provide a basis for calculating the heat balance. Table 16.2 lists the input, intermediate and output flows for an annular shaft kiln.

Table 16.2. Inputs, intermediate flows and outputs for an annular shaft kiln

Inputs	Intermediate flows	Outputs
Limestone	Kiln gas recycle +	Lime
Fuel	cooling air	
Cooling air	Kiln gas to recuperator	Exhaust gases + dust
Primary air	Kiln gas ex recuperator	
Mantle cooling air	Primary air ex recuperator	

For each flow, the mass and its elemental composition need to be determined so that, for example, for the fuel, the input rates of carbon and hydrogen can be determined.

Inevitably, it will be found that the calculated total mass of an element entering the kiln differs from that leaving it. If the difference is small, it may reflect experimental error and an intermediate value should be taken. If, however, the difference is larger than expected, it may reflect a measurement error, which should be investigated before proceeding with calculating the heat balance. Often, such an error is associated with measurement of the quantities of limestone used and lime produced.

16.7.3   Heat Balances

Having established an acceptable mass balance, the heat balance may be calculated. A theoretical treatment of the subject is given in [16.54].

The main heat input is the mass of fuel multiplied by its calorific value (either the gross or net value may be used – see section 14.3.6 — the net value is preferred as it simplifies the calculation slightly and provides a better basis of comparison for different fuels). Heat inputs in the limestone and air flows should be calculated relative to a convenient reference temperature (0 °C is often used).

Heat outputs, based on the mass, temperature difference (temperature minus the reference temperature) and integrated specific heat (between the reference and actual temperature – see Fig. 16.18 for $CaCO_3$ and CaO [16.55]), should then be calculated for each component of the output stream. The calorific value of any combustibles should not be overlooked.

(N.B., the values in [16.55] for CO_2, H_2O, N_2 and O_2 are not given in Fig. 16.18, as they are understood to differ slightly from currently accepted values.)

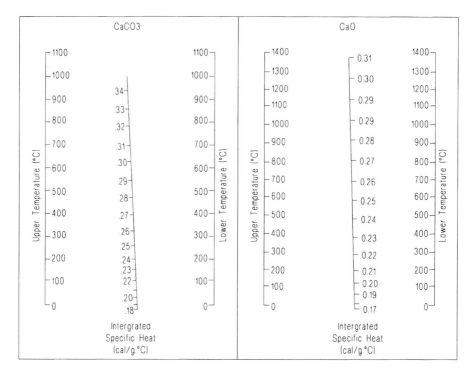

Figure 16.18. Integrated specific heats for $CaCO_3$ and CaO

The overall heat balance should than be calculated. It may be appropriate to determine subsidiary heat balances for the preheating/cooling zones and for any recuperator in the system. The difference between heat input and output is generally ascribed to radiation and convection losses. If a difference is unexpectedly large, additional work may be required to cross-check measurements or to measure losses directly.

Having obtained a heat balance, it is possible to identify and quantify where reductions in unit heat usage may be made. Areas meriting consideration include:

a) maximising output (particularly on less efficient kilns) to reduce the radiation and convection losses per tonne,
b) improving insulation,
c) optimising product quality — avoiding over-burning and unnecessarily low % $CaCO_3$ levels,
d) optimising the balance between excess air and CO,
e) optimising heat recovery in recuperators.

16.7.4 High-grade and Low-grade Heat

"High-grade" heat may be defined simply as that heat which is available for calcination. The residue is "low-grade" heat. The concept may appear obvious, but it is not universally accepted. There are, however, significant ramifications arising from the concept, which merit further consideration. A detailed mathematical model is given in [16.56].

The following crude example may help to illustrate the concept.

a) The theoretical flame temperature for a particular fuel, burning with air at ambient temperature, may be 2300 °C. To a first approximation, the percentage of high grade-heat available for calcination at 900 °C is $\{(2300-900) \times (100 \div 2300)\} = 61$ %.
b) If the combustion air were pre-heated to 700 °C, the theoretical flame temperature would rise to 2700°C and the percentage of high-grade heat would rise to $\{(2700-900) \times (100 \div 2700)\} = 67$ %.
c) In practice, however, the kiln operates with (say) 20 % excess air, which does not recover any additional heat from the lime (see N.B. below). The excess air cools the combustion products and reduces the theoretical flame temperature from 2,300 to 1800 °C, giving a percentage of high-grade heat of $\{(1800-900) \times (100 \div 1800)\} = 50$ %.

(N.B. For the purposes of this model, additional excess air can be regarded as entering the calcining zone at ambient temperature as the heat in the lime entering the calcining zone is already insufficient to pre-heat the stoichiometric amount of air required for combustion to 900 °C.)

The amount of high-grade heat increases directly with increases in the amount of heat recovered by the air and stone entering the calcining zone. It also increases when oxygen-enriched air is used for combustion as this reduces the amount of inerts (nitrogen) passing through the calcining zone and removing heat as low-grade heat.

Conversely, fuels with low calorific values, e.g. producer gas and gassified coal, can reduce the percentage of high-grade heat dramatically. It also follows that radiation and convection losses from the cooling and calcining zones count as high-grade heat, whereas losses from the pre-heating zone do not (assuming that the stone remains fully pre-heated).

Thus, an increase in heat recovery by the combustion air of 25 kcal/kg lime, increases the amount of high grade heat by 25 kcal/kg. If the theoretical flame temperature is 1800 °C, 50 % of the heat is high grade, and the fuel input can be reduced by 50 kcal/kg to re-establish the original input of high-grade heat. Thus, the reduction in heat input is double the *additional* amount recovered.

16.8   Instrumentation

16.8.1   General

Instrumentation can help the operator to operate the kiln:

a) safely,
b) at maximum output,
c) at minimum heat usage,
d) with a low down-time,
e) while producing a consistent, high quality quicklime.

The instrumentation requirements vary from one kiln design to another and also depend on the fuel used. There are, however, a number of generalisations which can be made for shaft kilns as a group and also for rotary kilns.

The advent of computers has transformed kiln control (e.g. [16.57, 16.58]), but the following sections concentrate on the instruments which enable individual parameters to be measured with the required precision. It should be noted that, in most cases, consistency (i.e., precision) is more important than accuracy.

16.8.2   Shaft Kilns

Most shaft kilns have residence times in the range of 16 to 60 hours, and response times that may be 2 to 3 times longer. Instrumentation should, therefore, be such as to assist the operator to maintain steady conditions as far as possible.

Weighing of the stone and/or the lime is, obviously, an important parameter. Belt weighers have generally proved to give poor precision in the dirty/dusty environments involved. Increasingly, batch weighing (or, to be more precise, measurement of weight loss), particularly of the stone, is being used to give high levels of precision. As most shaft kilns are charged with batches of stone, this technique fits in well with kiln operation.

Measurement of the quantity of fuel and its calorific value (CV) is of equal importance, a realistic objective being to maintain a target heat input with a precision of $\pm 0.5\%$.

For solid fuels, batch weighing is increasingly being favoured over belt weighers. Their calorific values tend to be variable owing to changes in moisture and ash content. It may be necessary to determine CV using composite samples representing periods less than the kiln residence time.

Liquid fuels have traditionally been measured using reliable positive displacement meters. As their CV is generally consistent, controlling the heat input should be relatively straightforward, given a consistent fuel temperature.

The measurement of the volume of gaseous fuels using turbine meters, or modern equivalents, linked to devices which correct for the temperature and pressure of the gas has proved to be fully satisfactory. Surprisingly, the CV of natural gas can be variable, owing to supply from two or more gas fields. A number of producers have found it necessary to install CV meters on site.

Depending on the design of kiln, many operators monitor carbon monoxide or combustibles in the exhaust gases, using meters to assist in maintaining the optimum air to fuel ratio. Some have also installed oxygen meters to monitor combustion efficiency. A combustibles meter can also serve a useful safety role during start-ups, as it detects any un-burned gaseous fuels.

Cooling air and combustion air flows are measured to ensure that the required air-fuel ratio is maintained. Changing that ratio has implications for product quality (reactivity–$CaCO_3$ relationship) as well as for the exhaust gas composition (CO–O_2 relationship).

The exhaust gas and lime discharge temperatures are generally monitored. The former has a safety role for the exhaust fan (if used) and also is an indication of any problems with heat transfer in the calcining and pre-heating zones. The lime discharge temperature should be monitored to protect rubber belt conveyors handling the lime: it is also an indication of the efficiency of heat recovery in the cooling zone.

Prolonged shut-downs have an adverse effect on kiln output, owing to radiation losses and to migration of heat up the kiln, especially if the seal at the top of the kiln is not effective. Various corrective measures are used including increasing heat input per unit of lime after the shut-down and pressurising the kiln top with air to counter the chimney effect.

16.8.3   Rotary Kilns

The residence time of rotary kilns is generally in the range 3 to 6 hours and response times are within the normal operator's working shift. Instrumentation should, therefore, assist the operator to detect changes at an early stage and to take appropriate corrective action.

Calibrated weigh feeders coupled with belt weighers have generally proved to provide the consistency required for controlling the feed rate of stone and solid fuel. Measurement of the volume of liquid and gaseous fuels is generally done using the techniques described in section 16.8.2. Changes in calorific value are much less critical than for shaft kilns and are generally compensated for by normal kiln control practice.

Monitoring temperatures is more critical. The temperature of the lime as it discharges from the kiln into the cooler is an indication of conditions in the burning zone. The "back-end" temperature of the gases leaving the kiln is also an important control parameter.

Combustion conditions are generally monitored by a carbon monoxide (or combustibles) meter and by an oxygen meter. Control of the CO level is important for thermal efficiency, for sulfur elimination and for safety reasons when an electrostatic dust collector is fitted.

16.9   Production of Calcined Dolomite

Four qualities of calcined dolomite are produced — half-, light-, hard-, and dead-burned.

Half-burned dolomite ($CaCO_3 \cdot MgO$) is produced in relatively small quantities in Germany by the controlled calcination of dolomite at 800 to 900 °C. Little appears to have been published about the process. A small scale process was developed pre-1940 using a rotary kiln in which steam was injected into the flame to moderate the temperatures and to reduce the CO_2 partial pressure [16.6].

Light-burned dolomite (or dolime) is produced in either rotary or shaft kilns. The principles are similar to those of making high calcium quicklime. Heat usages are presumably somewhat lower, owing to the lower heat of calcination of dolomite and its lower dissociation temperature.

Hard-burned dolomite is generally produced in mixed-feed shaft kilns, operating under reducing conditions. Hard-burned dolomite is generally produced in mixed-feed shaft kilns, operating under reducing conditions.

Dead-burned dolomite is produced in two grades. The high purity grade, made for the manufacture of refractories, may be produced by the direct calcination of dolomite at temperatures of up to 1,800 °C in either rotary or shaft kilns. The sintered product has an apparent density of about 3.2 g/cc. Alternatively, it may be produced by sintering pelletised, clacined dolomite at about 1,800 °C in either a rotary or a shaft kiln. The "fettling" grade is produced by the calcination of dolomite with iron oxide in a suitable kiln (usually a rotary) at 1,400 to 1,600 °C [16.6, 16.59]. The exhaust gases from these processes are cooled to below 400 °C using heat exchangers, dilution with air, or injection of atomised water to enable mild steel fan impellers to be used.

16.10 Production of Hydraulic Limes

16.10.1 Natural Hydraulic Limes

Natural hydraulic limes, as defined in [16.60] (see section 26.9), are produced from siliceous or argillaceous limestones containing more or less silica, alumina and iron [16.61]. Typical levels in the limestone are [16.6]:

SiO_2: 4 to 16 %,
Al_2O_3: 1 to 8 %
Fe_2O_3: 0.3 to 6 %.

The calcium plus magnesium carbonate content can range from 78 to 92 %.

The limestone is generally calcined in shaft kilns [16.59], which must be controlled closely to ensure that as much of the silica and alumina as possible reacts, without sintering the free lime. Typical calcining temperatures are 950 to 1250 °C: the required temperature rises as the cementation index (see section 26.9.2) increases (i.e. from feebly to eminently hydraulic limes).

The calcined lime is hydrated with sufficient water to convert the free CaO into $Ca(OH)_2$. If the free CaO content is greater than 10 to 15 %, the hard sintered lumps disintegrate into a powder. Otherwise, the lime must be ground before hydration. It may also be necessary to grind the hydrated product to achieve the required degree of fineness and setting rate.

16.10.2 "Special" Natural Hydraulic Limes

"Special" natural hydraulic limes, as defined in [16.60] are produced by intimately blending powdered natural hydraulic limes with powdered pozzolanic or hydraulic materials.

16.10.3 Artificial Hydraulic Limes

These are also defined in [16.60] and are produced by intimately blending powdered hydrated limes with pulverised pozzolanic or hydraulic materials.

16.11 References

[16.1] F. Schwarzkopf, "Lime Burning Technology — a Manual for Lime Plant Operators", 3rd ed., Svedala Industries, Kennedy Van Saun, 1994.
[16.2] A.B. Searle, "Limestone and its Products", Ernest Benn, 1935.
[16.3] M. Wingate, " Small-scale Lime-burning", Intermediate Technology Publications, 1985, ISBN 0-946688-01-X.
[16.4] T.-P. Hou, "The manufacture of soda, with special reference to the ammonia process", Hafner Publishing Co., 1969.

[16.5]   G. Chenxiang, "The structural characteristics and rationality of China new type ordinary lime shaft kiln", Proc. International Lime Congress, Berlin, 1994.

[16.6]   R.S. Boynton, "Chemistry and Technology of Lime and Limestone", John Wiley & Sons, 1980.

[16.7]   C.L. Verma et al., "Performance estimation vis-à-vis design of mixed-feed lime shaft kilns", Zement Kalk Gips **9**, 1988, 471–477.

[16.8]   C.L. Verma, "Simulation of lime shaft kilns using mathematical modelling", Zement Kalk Gips **12**, 1990, 576–582.

[16.9]   P. Mullinger, B. Jenkins, R. Manning, "The potential for up-grading vertical shaft lime kilns", World Cement, July, 1996.

[16.10]   A. Terruzzi, "Vertical lime kilns with U.C.C.-FX burner beams and two-way pressure system", World Cement, May 1990, 215–218.

[16.11]   A. Terruzzi, "Converting existing vertical lime kilns", World Cement, August 1990, 340.

[16.12]   A. Terruzzi, "Vertical lime kilns with U.C.C.-FX burner beams and two-way pressure system", Zement Kalk Gips **8**, 1990, 369–371.

[16.13]   A.M.D. Jorge and D. Nogueira, "The new Itau de Minas lime plant in Brazil", Zement Kalk Gips, 11/1994, 627–630.

[16.14]   A. Terruzzi, "Lime shaft kilns using the "two-way pressure system" — a new generation of vertical lime kilns", Zement Kalk Gips **6**, 1994, 322–326.

[16.15]   "The `two-way pressure system' in the Fercalx lime shaft kiln", Zement Kalk Gips **10**, 1996, A31–A32.

[16.16]   "Malaysian Lime order Fercalx kiln", World Cement, Oct. 1996, 38.

[16.17]   P. Accinelli, "A new (central burner) lime kiln", World Cement, Feb. 1997, 46–47.

[16.18]   W. Höltje, "Improvements in process technology in a double inclined lime shaft kiln", Zement Kalk Gips **9**, 1991, 467–471 (English: **11**, 1991, 246–249).

[16.19]   P. Zeisel, "First operating results obtained with the new multi-chamber shaft kiln", Zement Kalk Gips **5**, 1986, 243–250 (English: **7**, 1986, 204–205).

[16.20]   D. Lewerenz, P.Zeisel, "Design construction and commissioning of the new multi-chamber shaft kiln, with hot air extraction at a West German lime plant", Zement Kalk Gips **9**, 1987, 463–468.

[16.21]   P. Zeisel, "Operational capabilities of the advanced multiple chamber shaft kiln", Zement Kalk Gips **1**, 1989, 46–51.

[16.22]   P. Zeisel, "Operating results from a multi-chamber shaft kiln", Zement Kalk Gips, **6**, 1992, 278.

[16.23]   H.W. Marti, A. Schmidt, "Optimisation of a shaft kiln by modification and use of a regenerator", Zement Kalk Gips **9**, 1991, 472–477 (English: **11**, 1991).

[16.24]   K. Bechenbach, "Lime burning in the annular shaft kiln", Zement Kalk Gips **5**, 1970, 206–209.

[16.25]   C.E. Dandois, "The annular shaft kiln", Pit and Quarry, May 1973, 79–82.

[16.26]   K.W. Plessmann, J. Wetzlar, "An approach to computer-controlled automation of an annular shaft kiln", Zement Kalk Gips **8**, 1990, 372 (English: **10**, 1990, 223–226).

[16.27]   P. Zeisel, "Annular shaft kilns with the TREIVO system", Zement Kalk Gips **9**, 1996, 530–539.

[16.28]   W. Arnold, "Injector air preheating for annular shaft kilns", World Cement, Febr. 1997, 28–33.

[16.29]   G. Schaefer, "Heat consumption limits of parallel-flow/counterflow at combustion of lime regenerative kilns", Zement Kalk Gips **1**, 1981, 36–41.

[16.30]   X. Liang et al., "The first parallel-flow regenerative lime shaft kiln in the Peoples' Republic of China", Zement Kalk Gips **7**, 1988, 328–333 (English: **9**, 1988, 208–211).

[16.31]   N. Rauber et al., "Designing lime kiln plants with the aid of a process simulation model for a VALEC twin-shaft kiln", Zement Kalk Gips **9**, 1988, 428–433 (English: **11**, 1988, 253–256).

[16.32]   J. Liebl, "Construction and operation of the first pulverised coal-fired 300 t/d parallel flow/counter-flow regenerative kiln in the Federal Republic of Germany", Zement Kalk Gips **9**, 1988, 417–422 (English: **11**, 1988, 246–249).

[16.33]   F. Mereu, "Maerz vertical lime shaft kiln for the calcination of small size stone", Proc. International Lime Congress, Berlin, June 1994.

[16.34]   G. Gnecchi, "Maerz vertical double shaft parallel-flow kiln for the production of fine-size lime", Proc. International Lime Congress, Berlin, June 1994.

[16.35]   G. Leidner, "Operating experiences with a Maerz kiln for small limestone at the 'Kaltes Tal' lime works", Zement Kalk Gips **6**, 1995, 312–322.

[16.36]   G.M. Cella, T.L. Christiansen, "Commissioning of a Cim-reversy kiln for small-size limestone calcining at Buccino, Italy", Zement Kalk Gips **9**, 1990, 365–368.

[16.37]   G.M. Cella, "Cimprogetti's `TWIN-D' lime kilns, the new shape of a proven technology", World Cement, Febr. 1995.

[16.38]   G.M. Cella, "The TWIN-D lime shaft kiln — a new generation", Zement Kalk Gips **12**, 1995.

[16.39]   D.J. Kramm, "Selection and use of the rotary lime kiln and its auxiliaries — Parts 1 and 2, Paper Trade Journal, July/Aug. 1972.

[16.40]   D. Opitz, "Heat consumption of rotary lime kilns", Zement Kalk Gips **1**, 1981, 42–46 (German).

[16.41]   R.A. Schafer, "Heat exchanging internal mechanisms in rotary kilns", Rock Products, Oct. 1981.

[16.42]   W.J. Smithwick, "Rotary lime kilns — factors in process control", Pit and Quarry, Nov. 1981, 91–97.

[16.43]   F. Kaup, "Technology and experience in the use of solid fuels for the firing of various kiln systems — rotary kilns", Zement Kalk Gips **1**, 1985, 11–16 (English: **3**, 1985, 8–12).

[16.44]   L.M. Ludera, "Rotary kilns for lime burning", Zement Kalk Gips **5**, 1986, 235–242 (English: **7**, 1986, 201–204).

[16.45]   "Brazilian plant produces high-quality lime with soft stone", Rock Products, March 1988, 40A–40B.

[16.46]   P.J. Mullinger, "Flame control in rotary lime kilns", World Cement, June 1993.

[16.47]   T. Bisgaard, "Rotary lime kiln modifications", Proc. International Lime Congress, Berlin, June 1994.

[16.48]   J. Baeynens et al., "The development, design and operation of a fluidised bed limestone calciner", Zement Kalk Gips **12**, 1989, 620–627.

[16.49]   J. Baeynens et al., "Design of a limestone preheater and lime cooler", Zement Kalk Gips **8**, 1990, 389–394.

[16.50]   F. Hebbler, "Operating experience with the CID lime kiln", Zement Kalk Gips **11**, 1994, 631–634 (English: **1**, 1995, 44–46).

[16.51]   J. Dutton, "Lime Wars Down Under", Gypsum, Lime and Building Products, April 1997.

[16.52]   "Environmentally Conscious Lime Kilns from Japan", World Cement **28**, No 8, Aug. 1997, 26.

[16.53]   D. Sauers, N. Biege, D. Smith, "Comparing lime kilns", Rock Products, 1993, 1. Introduction: March, 2; Feed size: April, 3; Quarry utilisation: May, 4; Product Quality and fuel types: June, 5; Capital and operating costs: July.

[16.54]   H. Ruch, "The theoretical limits of heat consumption in lime burning considered on the basis of the physico-chemical relationships", Zement Kalk Gips **1**, 1981, 20–26.

[16.55]   J.A. Murray, "Specific Heat Data for Evaluation of Lime Kiln Performance", Rock Products, Aug. 1947.

[16.56]   H.-P. Thomas, "Simplified mathematical model for the heat economy of counter-current lime kilns", Zement Kalk Gips **9**, 1992, 446–450 (English: **11**, 1986, 282–286).

[16.57]   "ABB Linkmann Systems Ltd", National Lime Association's Lime-Lites **63**, No 1, 1996, 9.

[16.58]   G. Bittante, J. Kay, "Automation of a multiple lime plant", World Cement, Feb. 1997, 12.

[16.59]   M. O'Driscoll, "Burnt Lime/Dolime — Seeking Markets Green", Industrial Minerals, May 1988, 23–51.

[16.60]   ENV 459-1: "Building Lime: Definitions, specifications and conformity criteria", 1995.

[16.61]   J. Ashurst, "The technology and use of hydraulic lime", The Building Conservation Directory, 1997, 128–131.

[16.62]   J. Salisbury et al., "Mathematical Modelling of a Rotary Lime Kiln for the Use of Innovative Oxygen Technology to Improve Thermal Performance", Proc. 8th TOTem Meeting on High Temperature Combustion, Loughborough, 1993.

17 Processing Storage and Transport of Quicklime

17.1 Processing

17.1.1 General

The objective of processing run-of-kiln (ROK) quicklime is to produce a number of grades with the particle sizes and qualities required by the various market segments. A number of unit processes are used, including screening, crushing, pulverising, grinding, air-classifying and conveying.

A well-designed lime processing plant achieves a number of objectives, namely:

a) maximising the yield of "lead products",
b) minimising the yield of surplus grades (generally fines),
c) improving the quality of certain products, and
d) providing flexibility to alter the yields of products in response to changes in market demand.

The processing plant should include adequate storage, both for the products and intermediates, to provide a buffer between the kiln, which is best operated on a continuous basis, and despatches, which tend to be at low levels overnight and at weekends.

17.1.2 Outline Flowsheet

Figure 17.1 gives an outline flowsheet of a basic lime processing plant. ROK lime is often screened (typically at about 5 mm) to remove an impure "primary" fines fraction.

In traditional lime processing plants serving shaft kilns, the screened ROK lime then passed along a "picking" conveyor belt, where partially calcined lumps (sometimes called "bullheads") are removed, together with selected grades of lime. However, economic and social pressures have resulted in the phasing out of this activity in many countries.

If the ROK lime has a top size in excess of (say) 45 mm, it is reduced in size with the minimum production of fines. Jaw and roll crushers are widely used for this duty.

The crushed ROK lime is then fed to a multi-deck screen, which produces a secondary fines fraction (e.g. less than 5 mm), granular, and "pebble", lime fractions (e.g. 5 to 15 mm and 15 to 45 mm). Oversize lumps (e.g., greater than

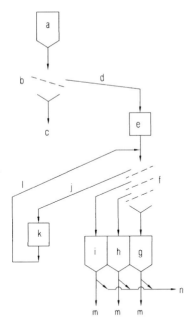

Figure 17.1. Outline flowsheet of a basic lime processing plant
(a) shaft kiln; (b) fines screen; (c) primary fines; (d) picking belt; (e) primary crusher; (f) multi-deck screen; (g) secondary fines bunker; (h) granular lime bunker; (i) pebble lime bunker; (j) oversize lime; (k) secondary crusher; (l) recycle of crushed oversize; (m) loading-out points; (n) internal feed to ground lime and/or hydrating plants

45 mm) may be crushed in a secondary crusher and recycled to the multi-deck screen.

The products are stored in bunkers, from which they can be either loaded out, or transferred to another plant for grinding (section 17.2) or hydrating (section 20.3).

17.1.3   Run-of-kiln Lime Characteristics

17.1.3.1   Shaft Kilns

The primary fines fraction of ROK lime from shaft kilns typically amounts to 5 to 15 % of the total and generally contains higher levels of :

a) silica, iron and alumina (i.e. clay materials),
b) particles of fuel ash (where solid fuel is used),
c) sulfur,
d) trace elements such as lead, and
e) over-burned particles (originating from fragments of limestone, produced by breakage, which have passed through the kiln.

ROK lime from most shaft kilns is too coarse for most customers and requires to be crushed and screened as described in section 17.1.2. The secondary fines (typically 5 to 10 % of ROK lime), have a similar chemical analysis and reactivity to the coarser fractions. Some producers use them to "sweeten" the primary fines.

Others keep the two products separate, with the primary fines being sent to less demanding customers (including in-house ground lime and hydrating plants).

As explained in chapter 15, the coarser fractions of ROK lime tend to be higher in $CaCO_3$ content than the other fractions. Crushing tends to accentuate this effect, so that the oversize produced by crushing and screening may have significantly more $CaCO_3$ than the other grades. This effect can be exploited by feeding the higher $CaCO_3$ fraction to more tolerant customers (including ground lime and hydrating plants) and reserving other fractions for customers requiring lower $CaCO_3$ levels.

17.1.3.2   Rotary Kilns

The fines fraction in ROK lime from rotary kilns also contains enhanced levels of impurities. It does not, however, contain a significant amount of over-burned lime, as the fines tend to concentrate at the centre of the rotating bed and are insulated by the larger particles. This enables the % $CaCO_3$ in the coarser fractions to be reduced to low levels without over-burning.

The feedstone to rotary kilns is generally selected to produce ROK lime that does not need crushing.

17.1.4   Product Size and Quality

Before the 1960s, the premium products were lump limes, with particle sizes up to 250 mm. With the advent of the Basic Oxygen Steelmaking process (section 27.4), however, the lead product for many producers became a 10 to 40 mm fraction with a low sulfur content (the so-called "BOS-quality" lime).

More recently, the increased use of pneumatic conveying systems and automated handling systems have raised demand for finer products, such as the 5 to 20 mm range and the various ground lime products.

Other significant trends in recent years have been:

a) demand for reduced levels of impurities,
b) movement from low/medium reactivity products to high reactivities (encouraging the replacement of traditional shaft kilns with modern designs and rotary kilns), and
c) increased consistency of product quality (placing greater emphasis on kiln control and blending of product).

17.1.5   Fine Lime

As explained above, fine lime may amount to 10 to 25 % of the output. It is, therefore, essential to find economic outlets for that product.

Traditionally, the primary fines were sold as agricultural lime, but, with the increased availability of ground limestone, the price of agricultural lime ceased to be attractive. Some primary fines are used in less demanding applications such as effluent neutralisation.

Generally, the best outlets for both primary and secondary fines are as a feed to ground lime and hydrating plants. They have also been used as a source of calcium oxide in dry process cement plants, where their use resulted in a significant reduction in heat requirements [17.1].

An alternative approach, particularly for secondary fines, which has been used by a number of lime producers, involves briquetting. The resulting pillow-shaped or almond-shaped briquettes are mixed with suitably sized screened lime. However, the briquettes are much weaker than granular quicklime and considerable breakage occurs during handling. Moreover, the capital and operating costs of presses are relatively high.

17.2 Production of Ground Quicklime

The demand for various grades and qualities of ground quicklime has grown rapidly ever since the 1950s. Particle size requirements vary from relatively coarse products used for soil stabilisation (e.g. 99 % less than 2 mm with 50 % less than 75 μm) to very finely divided products for specialist applications (e.g. more than 99 % less than 50 μm).

The coarser products may be produced relatively cheaply in a single pass through a beater mill fitted with an integral basket, which acts as a screen. Finer products are generally produced in tube mills and vertical roller mills [17.2]. In the latter case, a variable speed classifier is fitted above the mill to control the grading of the product, and recycle over-sized particles.

In the late 1980s, the high pressure roll mill was developed for the cement industry [17.3] and is increasingly being used for quicklime. The product is passed through the grinding rolls, which effectively produce a flake. It is then fed to a dis-agglomerator and an air classifier which removes particles of the required fineness and recycles the coarse fraction [17.2]. The power requirements of this system are less than half those of ball mills and less than 60 % of the ring-roll mills (requirements range from 4 to 10 kWh/t for heater mills to 10 to 40 kWh/t for the other types).

All grinding systems use air to convey the ground quicklime and, although part of the air is recycled, the remainder is purged via a bag filter to remove fines and heat. This inevitably leads to a limited amount of air slaking in addition to that which already occurs in the cooling zone of the lime kiln. Under typical conditions, the measured combined water may rise from 0.5 % in the feed to 0.7 % in the ground quicklime.

While the combined water content does not cause problems for many users of ground lime, variations caused by changes in temperature and humidity can become an issue. For example, the saturated vapour pressure of water rises rapidly from 4.6 mm of mercury at 0 °C to 31.8 mm at 30 °C, which has been observed to cause a corresponding increase in the combined water content.

17.3   Storage, Handling and Transport

17.3.1   General

Quicklime should be stored in dry conditions, free from draughts to limit air slaking. Because of the high temperatures which can be produced when quicklime comes into contact with water, it should not be stored in, or adjacent to, flammable materials.

The methods of storage, handling and transport used depend on the particle size of the product and the scale of use. A comprehensive description of current handling, application and storage methods is given in [17.4]. To simplify the presentation, the following sections describe the equipment that might be installed at a customer's works. They are equally relevant to the equipment at a lime works.

17.3.2   Transport

Sheeted tipper waggons or covered rail hoppers are generally the most economical form of transport for lump and granular quicklimes which have been screened to remove the fines (e.g. <5 mm particles). Air pressure discharge vehicles are used for fines and ground lime products and may also be used on product with a top size of up to 20 mm. Intermediate Bulk Containers are also used.

It should be noted that quicklime is classed as an "irritant" material (see section 34.1.2), but that no special preparations are required for its transport by road or rail (see section 34.2.2 regarding hazard information and packaging requirements). Water transport is used to a limited extent using inland waterways and coastal waters. Great care is exercised to ensure that water is excluded from the cargo, as hydration liberates heat and causes expansion, both of which could be dangerous. For air transport, however, quicklime is included in the list of corrosive materials (Class 8, Packing Group 3, Corrosive Material, Code UN 1910) [17.5]. In general, air transport of quicklime is only used for samples.

17.3.3   Reception Hopper

The reception hopper into which road and rail waggons deliver should be sited under cover (e.g. within a structure with three walls, a roof and a curtain on the open side to assist in containing dust during discharge). It should be constructed from mild steel or concrete and, for safety reasons, should be covered with a grid. Its capacity should be at least equal to that of the largest delivery vehicle.

The discharge mechanism from the hopper should be fast enough to ensure that the hopper can be emptied before the next delivery. Where the lime is to be reduced in size, a crusher should be installed immediately after the reception hopper so that subsequent handling equipment can be kept as small as possible. A "tramp-metal" detector should be fitted before the crusher to prevent damage.

17.3.4   Storage

The storage bunker should be totally enclosed and constructed of mild steel or concrete. Cylindrical section bunkers with discharge cone angles of not less than 60° to the horizontal, or rectangular section bunkers with valley angles not less than 55° to the horizontal are satisfactory. The bunker should be fitted with the usual access and safety equipment. A filtered vent may be appropriate in some situations.

Where the consumption of lime is insufficient to justify reception hoppers and storage bunkers, the product should be stored on a concrete base, preferably in a separate bay within a building to prevent excessive air slaking.

17.3.5   Air Pressure Discharge Vehicle Deliveries

Air pressure discharge vehicles are able to blow directly into the storage bunker, which should be as described in section 17.3.4 and fitted with a filter to remove dust from the conveying air. The filter should be weatherproof and watertight and should discharge the collected dust into the bunker. A pressure relief device should be fitted as well as the usual accesses and safety features.

17.3.6   Handling Equipment

17.3.6.1   Discharge from a Hopper, Bunker or Silo

Storage containers handling granular or lump materials are often fitted with feeder tables to control the discharge rate. As a general rule, the discharge aperture should be not less than 6 times the top size of the material being handled.

As with most powders, fine quicklime may compact, leading to "arching". Bunkers should, therefore, be fitted with aeration pads, vibrators or mechanical devices to break any hold-ups. Conversely, aerated fine lime flows readily and equipment to prevent "flooding", such as rotary valves, should be fitted to silo discharges.

All storage containers should be fitted with devices which can positively seal the base of the bunker to enable maintenance work to be done on the discharge mechanism.

17.3.6.2   Transfer of Quicklime

Many types of equipment are suitable for transferring the product and new ones are continually being developed. The following items have been used successfully, but may not be suitable in all applications.

- *Skip hoists* can be used for all granular and lump grades but are more suitable for particles greater than 100 mm. *Elevators* — both belt-and-bucket and chain-and-bucket types — have been used for all grades of quicklime. The system should minimise spillage within the equipment and ensure that wear on linkages caused by the quicklime is minimised.

- *Drag-link* conveyors are suitable for granular and fine quicklime. They are generally used for horizontal or inclined transfer. *Conveyor belts* are widely used for transferring lump and granular grades horizontally and on an upwards slope, providing the angle of inclination does not exceed 17°. Specialised types of conveyor belt are designed to operate at steeper angles. Screw conveyors are widely used for fine quicklime. *Vibrating trough* conveyors have been used for particle sizes up to 40 mm. They operate more successfully when there is a slight downward slope from the feed to the discharge point.

- *Pneumatic* conveying can be used for products with a maximum size up to 20 mm and is often lower in capital cost, but higher in operating costs than alternatives. The product is fed into a rotary blowing seal connected to a Rootes type of blower. The pipeline bore, and volume/pressure of the blowing air should be designed in the context of the size of lime being conveyed, the transfer rate and the length/route of the pipeline. The receiving silo should be fitted with an air filter and a pressure relief valve.

17.3.6.3 Weighing

For many applications, volumetric measurement of feedrate is accurate enough and various proprietary feeders are available. Alternatively, some of the types of transfer equipment described in section 17.3.6.2 may give the required accuracy. Proprietary belt weighers are also available. For the highest degree of accuracy, however, batch weighing is recommended.

17.4 References

[17.1] R.S. Boynton, "Chemistry and Technology of Lime and Limestone", John Wiley & Sons, 1980.
[17.2] F.W. Plank, "Experience with a new method of grinding in the lime industry", Zement Kalk Gips **2**, 1991, 63–69 (English: **4**, 1991, 66–69).
[17.3] N. Patzel, "Development trends in grinding technology", Zement Kalk Gips **5**, 1989, 264–268 (English: **7**, 1989, 190–192).
[17.4] "Lime handling, application and storage", National Lime Association, Bulletin 213, 5th ed., 1988.
[17.5] "The Transportation of Dangerous and Hazardous Goods Regulation" (49 CFR 172.101; 59 FR 67309, Dec. 29, 1994 and 60 FR 26796, May 18, 1995).

18   Sampling and Testing of Quicklime

18.1   Introduction

Sampling of lime products may be required for quality control, for the assessment of the suitability of a product, or for monitoring compliance with a specification. Testing may be done to assess a product's mechanical, physical or chemical properties.

While sampling and sample preparation may appear, at first sight, to be an unskilled and low technology operation, they are probably the most common causes of disagreement between laboratories, and are critical to the characterisation of product quality. Problems arise as a result of the failure to obtain representative samples. They also arise, particularly with quicklime, as a result of inadequate precautions when storing or handling the sample. It is, therefore, important that samples are taken, prepared and stored in accordance with specified procedures.

As most lime products are used within 300 km of the producer, many countries have their own standards for sampling, sample preparation and testing. This chapter refers mainly to British and European Standards to illustrate the principles and procedures involved. In other geographical areas, the reader should refer to the most relevant national or international standards.

18.2   Precautions

Section 34.2 describes the personal protection measures necessary when handling quicklime. Because quicklimes react with the atmosphere, care should be taken at all stages of sampling, sample preparation and testing to minimise exposure.

Some of the properties of quicklime, particularly its reactivity, can be significantly altered by the absorption of atmosphere moisture and carbon dioxide (see section 18.8). As it is impractical to handle the required quantities of material in an inert atmosphere, exposure times should be kept to a minimum and, where possible, mechanical sample preparation equipment should be used (e.g. crushers, sample dividers and pulverisers), taking care to avoid cross-contamination between samples.

All lime samples should be stored in such a way that absorption of water vapour and carbon dioxide is minimised. Sealed plastic bags have proved to be acceptable, with double wrapping being used for long-term storage and for lump samples that might puncture a single layer. In the laboratory, desiccators may be used, providing the desiccant has a negligible water vapour pressure (e.g. phosphorous pentoxide). Silica gel is not suitable.

18.3   Sampling

18.3.1   General

The principles involved in sampling quicklime are similar to those described in section 6.2 for limestone.

Definitions of the terms increment, sample, spot sample, composite sample and laboratory sample are given in Annex 1. Figure 6.1 illustrates how they are related.

For process control and general purposes, simplified sampling procedures may be appropriate, but they should be assessed relative to the standard methods referred to below, to ensure that they give the required accuracy and precision. Some parameters require less intensive sampling than others, e.g., reactivity and sulfur can often be determined more readily than residual $CaCO_3$, particularly in lump lime products produced by crushing.

18.3.2   Powdered Quicklime

Spot samples of powdered products (generally less than 1 mm) should be sampled in accordance with EN 196-7 [18.1], as amended by clauses 3.1 and 3.2 of EN 459-2 [18.2]. These require that a sufficient number of increments be taken to give a spot sample of (20 ± 5) kg.

Where a composite sample is required, BS 6463, Part 101, [18.3] requires that at least 10 spot samples should be taken, each containing a sufficient number of increments to produce a composite sample of at least 15 kg.

18.3.3   Granular and Lump Products

Where a spot sample of granular or lump products is required, at least 10 increments should be taken and the volume of each increment should correspond to the values given in Table 18.1. Where a composite sample is required, at least 10

Table 18.1. Increment volumes

Maximum particle size (mm)	Minimum volume of increment (l)
3.4	1
6.3	1.5
10	2
14	2.5
20	3
28	3.5
37.5	4
50	4.5
63	5

spot samples should be taken, each consisting of 1 or more increments with a volume corresponding to the appropriate value in Table 18.1 [18.3].

It should be noted that obtaining representative samples of lump products (with a top size greater than 20 mm) is more difficult than for granular and considerably more difficult than for powdered products. Particular care should, therefore, be taken to follow the prescribed procedures.

18.4 Sample Preparation

Preparing a sample for testing and analysis in the laboratory generally involves two or more of the following operations:

a) blending increments and/or spot samples,
b) reducing the quantity of the sample,
c) dividing the sample to produce two or more equivalent sub-samples, and
d) reducing the particle size by crushing and grinding.

[18.3] gives specific guidance on how these operations should be done for powdered, granular and lump quicklimes.

18.5 Packing and Marking of Samples

[18.1] gives procedures for packaging powders. Granular and lump quicklimes should be at least double-wrapped, and should preferably be sealed in a rigid outer container (e.g. a drum). [18.3] gives guidance on labelling samples.

Regulations regarding the transport of quicklime samples should be observed see [18.4].

Additional precautions are required when lime samples are to be carried in aircraft (see section 17.3.2) [18.4].

18.6 Physical Testing

Table 18.2 summarises the physical tests for quicklime products specified by various standards [18.2, 18.3, 18.5].

18.7 Chemical Analysis

Tables 18.3 and 18.4 summarise the chemical analyses for lime products specified in the various standards [18.1–18.3, 18.5].

Table 18.2. Physical tests for quicklime products

Test	Category of lime					Reference standard
	1	2	3	4	5	
Fineness	✓	–	–	–	–	EN 459-2
	–	–	✓	–	–	prEN 12485
	–	–	–	–	✓	BS 6463: Pt 103
Reactivity	✓	–	–	–	✓	EN 459-2/6463
Soundness	✓[a)]	–	–	–	–	EN 459-2
Yield	✓	–	–	–	–	EN 459-2
Bulk density	✓	–	–	–	–	EN 459-2
Residue on slaking	–	–	–	–	✓	BS 6463: Pt 103
Particle density	–	–	–	–	✓	BS 6463: Pt 103
Specific surface area	–	–	–	–	✓	BS 6463: Pt 103
Workability	–	–	–	–	✓[b)]	BS 6463: Pt 103

1: building lime — quicklime
2: high calcium quicklime — lump
3: high calcium quickline — pulverised
4: half-burnt dolomite
5: quicklime

[a)] After slaking.
[b)] On putty of standard consistence.

Trace levels of free carbon, produced by pyrolysis of organic matter in the limestone, is of interest in a few applications. The technique described in [18.6] differentiates between combined carbon in carbonates and free carbon. Acid dissolution is used to decompose carbonates and the residue is heated in oxygen to oxidise free carbon. The carbon dioxide produced is determined using its absorption in the infra-red spectrum.

18.8 Absorption of Moisture and Carbon Dioxide from the Atmosphere

Atmospheric moisture and carbon dioxide are absorbed in the cooling zone of the kiln, during handling and storage, and during sampling and sample preparation. Under normal conditions, providing the precautions described in section 18.2 are observed, their absorption does not significantly affect the properties of the lime. Table 18.5 summarises the author's experience in the U.K. Levels of moisture pick-up increase under hot, humid conditions and are, therefore, generally greater in summer than in winter.

In the context of meeting customer specifications, the most pronounced effect of absorption of water and carbon dioxide is on the reactivity of the lime. Combined water contents greater than 1.0 %, coupled with the associated, but smaller, increases in combined CO_2, can have a significant effect on reactivity. Such levels

Table 18.3. Chemical analysis of quicklime products by traditional methods.

Test	Category of lime					Reference standard
	1	2	3	4	5	
CaO + MgO	✓	–	–	–	–	EN 196-2
MgO	✓	–	–	✓	–	EN 196-2/12485
CO$_2$	✓	–	–	–	–	EN 196-2
	–	–	–	✓	✓	EN 12485/6463
Loss on ignition	–	–	–	–	✓	EN 196-2/6463
Sulfate	✓	–	–	–	–	EN 196-2
	–	–	–	✓	–	EN 12485
Total sulfur	–	–	–	–	✓	BS 6463: Pt 102
Neutralising value	–	–	–	–	✓	BS 6463: Pt 102
Available lime	–	–	–	–	✓	EN 459-2/6463
Free CaO in half-burnt dolomite	–	–	–	✓	–	EN 12485
Water sol CaO/Ca(OH)$_2$	–	✓	✓	–	✓	EN 12485/6463
Water insol constituents	–	✓	✓	–	✓	EN 12485/6463
Insol in acetic acid	–	–	–	–	✓	BS 6463: Pt 2
Insol in HCl	–	–	–	–	✓	BS 6463: Pt 102
HCl extractable fluoride	–	–	–	–	✓	BS 6463: Pt 102
Si, Fe, Al (titration)	–	–	–	–	✓	EN 196-2/6463
Ca, Mg (titration)	–	–	–	✓	✓	EN 12485/6464

Category: 1: building lime — quicklime
2: high calcium quicklime — lump
3: high calcium quickline — pulverised
4: half-burnt dolomite
5: quicklime

Table 18.4. Chemical analysis of quicklime products by atomic spectrometry

Test	Category of lime				Reference standard
	2	3	4	5	
Al, Fe, Mg, Mn, Si					
ICP-OES	✓	✓	–	–	EN 12485
Flame AAS	✓	✓	✓	✓	EN 12485 / BS 6463
Cd, Cr, Pb, Ni					
ICP-OES	✓	✓	–	–	EN 12485
Graphite tube AAS	✓	✓	✓	✓	EN 12485 / BS 6463
Cu, Ag, Sn, Zn					
Graphite tube AAS	–	–		✓	BS 6463: Pt 102
Sb, As, Se,					
Hydride & flame AAS	✓	✓	✓	✓	EN 12485 / BS 6463
Hydride & AFS	✓	✓	–	–	EN 12485
Hg cold vapour & AAS	✓	✓	✓	✓	EN 12485 / BS 6463
& AFS	✓	✓	–	–	EN 12485

Category: 2: high calcium quicklime — lump
3: high calcium quickline — pulverised
4: half-burnt dolomite
5: quicklime

Table 18.5. Moisture absorption by quicklimes

Sample point (discharge from)	Normal range of measured combined water (% m/m)
Calcining zone	0.15–0.25
Lime cooler	0.3–0.5
Lump lime bunker	0.4–0.7
Ground lime bunker	0.4–0.7

suggest that inadequate precautions have been taken during sampling, sample preparation, and/or storage.

Absorption of H_2O and CO_2 inevitably increases the levels of $Ca(OH)_2$, $CaCO_3$ and loss on ignition, but rarely causes those parameters to become out of specification.

18.9   References

[18.1]   EN 196-7: "Methods of taking and preparing samples of cement", 1992.
[18.2]   EN 459-2: "Building lime — Test methods", 1995.
[18.3]   BS 6463, "Quicklime, hydrated lime and natural calcium carbonate", Part 101: "Methods for preparing samples for testing", 1996; Part 102: "Methods for chemical analysis", in preparation; Part 103: "Methods for physical testing", in preparation.
[18.4]   "The Transportation of Dangerous and Hazardous Goods Regulation" (49 CFR 172.101:59 FR 67309, Dec. 29 1994; 60 FR 26796, May 18 1995).
[18.5]   prEN 12485: "Chemicals used for treatment of water intended for human consumption — calcium carbonate, high-calcium lime and half-burnt dolomite — Test methods", in preparation.
[18.6]   P. M. Fouché et al., "The determination of trace levels of free carbon in lime", Zement Kalk Gips **6**, 1992, 302-305.

Part 4 Production of Slaked Lime

19 Physical and Chemical Properties of Slaked Lime

The Chemical Abstracts Service (CAS) registry number for $Ca(OH)_2$ is 1305-62-0. The European Inventory of Existing Commercial Substances (EINECS) reference number is 215 137-3.

19.1 Physical Properties

- *Molecular weight* of calcium hydroxide is 74.09. That of magnesium hydroxide is 58.33.

- *Colour.* Most hydrated limes are white. High levels of impurity can result in a grey or buff colour.

- *Odour.* Hydrated lime has, like quicklime, a slight earthy odour.

- *Crystal structure.* Calcium hydroxide crystals have a hexagonal symmetry. The particles in commercial hydrated limes consist of micro-crystalline agglomerates.

- *Specific gravity.* High-calcium hydrated lime has a specific gravity of 2.24 g/cm^3 [19.1]. The values for partially and fully hydrated dolomitic limes are about 2.7 and 2.5 g/cm^3, respectively [19.2].

- *Bulk density.* Reported values for the compacted bulk density vary from 450 to 640 kg/m^3, with a typical value of 560 kg/m^3 [19.2]. In the as-poured state, it can be as low as 350 kg/m^3, owing to air-entrainment.

- *Angle of repose.* The angle of repose depends on the moisture content and on the degree of fluidisation (see section 21.1).

- *Hardness.* This is between 2 and 3 Mohs [19.2].

- *Refractive index.* Calcium hydroxide is bi-refringent, with indices of 1.574 and 1.545 [19.1].

- *Specific heat.* The specific heat of calcium hydroxide rises from 0.270 cal/g °C at 0 °C to 0.370 cal/g °C at 400 °C [19.3]. That of dolomitic hydrate is believed to be about 5 % higher [19.2].

- *Vapour pressure.* See *stability* in section 19.2 and Table 19.1.

- *Solubility of calcium hydroxide in water.* The solubility decreases from about 1.85 g $Ca(OH)_2$/l water at 0 °C to 1.28 g/l at 50 °C and to 0.71 g/l at 100 °C [19.4].

- *Solubility in water of magnesium hydroxide and calcined dolomite.* Magnesium hydroxide is only sparingly soluble in water (about 0.01 g/l) [19.5]. The magnesium hydroxide present in calcined dolomite is reported not to affect the solubility of calcium hydroxide (but see rate of solution below).

- *Solubility in solutions of inorganic compounds.* Some inorganic compounds affect the solubility [19.6]. The effect of calcium sulfate is of particular interest: a 2 g/l solution reduces the solubility to 0.06 g $Ca(OH)_2$/l.

- *Solubility in solutions of organic compounds.* Organic compounds can increase the "solubility" of calcium hydroxide. Sugar has one of the greatest effects, as a result of the formation of calcium saccharate [19.6].

- *Heat of solution.* The heat of solution of calcium hydroxide is about 2 800 cal/ mole at 18 °C [19.2]. As its solubility in water is only 1.65 g/l, or 0.022 mole/l, the heat released in forming a saturated solution only causes a temperature rise of 0.06 °C.

- *Rate of solution.* The rate of solution depends on the particle size distribution of the hydrated lime. Three test methods are described in section 20.7.1. The presence of magnesium oxide and other impurities can markedly reduce the rate of solution of the calcium hydroxide component [19.4].

- *Freezing point.* Although calcium hydroxide is a strong, dibasic alkali, its low solubility limits its effect on the freezing point of its solutions. A saturated solution is calculated to have a freezing point of approximately -0.2 °C.

- *Electrolytic conductivity.* The conductivity of a saturated solution of calcium hydroxide is in the order of 10 mS/cm.

19.2   Chemical Properties

- *Stability.* At elevated temperatures, calcium hydroxide decomposes to quicklime and water. The partial pressure of water vapour above calcium hydroxide is given in Table 19.1 [19.5]. It reaches 1 atmosphere at 547 °C. The quicklime so produced has an exceptionally high reactivity to water and acidic gases.

  Magnesium hydroxide decomposes at much lower temperatures: its vapour pressure reaches 1 atmosphere at about 190 °C.

Table 19.1. Vapour pressure of calcium hydroxide

Temperature °C	Pressure (atmospheres)
366	0.01
417	0.05
448	0.13
500	0.42
524	0.67
547	1.00

- *pH in water.* The pH of a saturated solution of calcium hydroxide at the standard temperature and pressure of 25 °C and 1 atmosphere is 12.4 (to the nearest decimal point [19.7]).

  (N.B., this value is of particular significance in the USA, where the definition of hazardous corrosive wastes specifically excludes lime sludges, by defining a liquid as hazardous when it has a pH of 12.5, or greater, at the standard reference temperature of 25 °C.)

- *Neutralisation of aqueous acids.* Hydrated lime, whether $Ca(OH)_2$, $Ca(OH)_2 \cdot MgO$, or $Ca(OH)_2 \cdot Mg(OH)_2$, reacts readily with aqueous acids. The rate of reaction depends in part on the particle size. The heats of neutralisation of calcium hydroxide with strong acids are similar to those of other strong bases. The values for the reaction of calcium hydroxide with sulfuric and hydrochloric acids are 31,140 and 27,900 cal/mole respectively. The corresponding heats of reaction of magnesium hydroxide are reported as 31,220 and 27,690 cal/mole [19.2] (see also section 28.3).

- *Acidic gases.* Powdered hydrated lime, whether in the form of $Ca(OH)_2$, $Ca(OH)_2 \cdot MgO$, or $Ca(OH)_2 \cdot Mg(OH)_2$, reacts readily with acidic gases such as the oxides of sulfur and nitrogen, and carbon dioxide. The rate of reaction depends largely on the particle size and the degree of dispersion of the hydrate (see sections 29.4 and 29.5).

- *Impurities and trace elements.* The major impurities in hydrated lime are calcium carbonate, calcium oxide and magnesium oxide. Minor impurities include silica, alumina, iron oxide and calcium sulfate. Their significance depends on the use to which the hydrated lime is put. As with quicklime (section 13.2) high levels of "toxic substances" (lead, antimony, arsenic, cadmium, chromium, nickel and selenium) can prevent the use of a hydrated lime in applications such as treatment of water for human consumption [19.8], in foodstuffs and the production of toothpaste (e.g., sections 28.1.8, 30.4, 30.8, 30.9, and parts of chapter 31).

- *Disinfection.* Water and sewage sludge can be disinfected by adding hydrated lime to raise the pH to above 11 for 1 to 2 days (see sections 28.1.7 and 28.4.4).

- *Carbon dioxide.* Calcium hydroxide reacts readily with carbon dioxide in the absence of water at all temperatures below its dissociation temperature (equation 19.1). It should be noted that the reaction of quicklime with carbon dioxide below 300 °C only proceeds in the presence of calcium hydroxide. The water released by equation (19.1) is available to hydrate more calcium oxide (19.2), and thereby permits further carbonation to occur.

$$Ca(OH)_2 + CO_2 \rightarrow CaCO_3 + H_2O \qquad (19.1)$$

$$CaO + H_2O \rightarrow Ca(OH)_2 \qquad (19.2)$$

- *Reaction with chlorine.* Dry hydrated limes readily react with chlorine forming bleaching powder — see section 31.3. Milks of lime also react with chlorine to produce "bleach" (a solution of calcium hypochlorite — equation 19.3).

$$2Ca(OH)_2 + 2Cl_2 \rightarrow Ca(OCl)_2 + CaCl_2 + 2H_2O \qquad (19.3)$$

- *Reaction with alkali carbonates (causticization).* Hydrated lime reacts with soluble metal carbonates to produce insoluble calcium carbonate and the metal hydroxide. This reaction (19.4) is used to produce caustic soda (see section 31.20).

$$Ca(OH)_2 + Na_2CO_3 \rightarrow CaCO_3 + 2NaOH \qquad (19.4)$$

- *Reaction with silica and alumina.* Hydrated lime reacts with pozzolans (materials containing reactive silica and alumina) in the presence of water to produce hydrated calcium silicates and aluminates. The reactions may take months to proceed to completion at ambient temperatures, as in mortars (section 26.6) and lime treated soil (section 26.3), but proceed within hours at elevated temperatures and water vapour pressures (e.g., in steam at 180 °C and a pressure of 10 bar — see sections 26.10, 26.11 and 26.12). This pozzolanic reaction is the basis of the strength generated by hydraulic quicklimes (section 26.9).

- *Reaction with metals.* Hydrated lime does not react with iron or steel, which are commonly used as materials of construction for hydrating plants and hydrated lime handling systems. It reacts readily with aluminium and also attacks lead and brass [19.2], which should not be used as storage containers.

- *Precipitation of metal ions.* Many metals can be removed from solution by adjusting the pH. Table 28.9 summarises the conditions under which various metal ions are precipitated [19.9, 19.10] (see sections 28.1.3, 28.3.3 and 28.4.3, as well as parts of chapters 31 and 32).

19.3 References

[19.1] CRC "Handbook of Chemistry and Physics", 77th ed., Harlow Longman, 1995.
[19.2] R. S. Boynton, "Chemistry and Technology of Lime and Limestone", John Wiley & Sons, 1980.
[19.3] "International Critical Tables", 5, McGraw Hill, New York, 1929, pp. 98–99.
[19.4] R. Haslam, G. Calingaert, C. Taylor, J. Am. Chem. Soc., **46**, 1924, 308.
[19.5] N. Knibbs, "Lime and Magnesia", E. Benn, London, 1924, pp. 39, 111.
[19.6] A. Seidell, "Solubilities of inorganic and metal organic compounds", 4th revision, American Chemical Society, 1958-65.
[19.7] H. L. Francis, National Lime Association, Circular, Dec. 1995.
[19.8] "Chemicals used for treatment of water intended for human consumption — high calcium lime", Draft prEN 12518.
[19.9] L. Hartinger, "Taschenbuch der Abwasserbehandlung", Vol. 1, Carl Hanser Verlag, Munich, 1976.
[19.10] "Acid Neutralisation with Lime", National Lime Association, Bulletin 216, 1976.

20 Production of Hydrated Lime

20.1 Introduction

The term "hydrated lime" is widely used to describe a powdered calcium hydroxide product made by reacting quicklime with a controlled excess of water. The product is essentially dry and generally contains less than 1 % of unreacted water.

The process is called "hydration" and should be differentiated from "slaking" which involves the production of a dispersion of calcium hydroxide in water (i.e. a milk of lime or a lime putty, see chapter 22). However, the expression "slaked lime" is used as a generic term for hydrated lime, milk of lime and lime putty.

An estimated 10 to 15 % of the quicklime produced in developed countries is converted into hydrated lime and the percentage may be higher in countries which do not have a large steel industry. Because hydrating plants are relatively complex and can be fed with surplus grades of quicklime, there are relatively few of them and they are normally located at a lime works.

While the chemical reactions involved in the formation of hydrated lime are simple, the physical chemistry (i.e. the kinetics, crystallisation and agglomeration) is complex. Some of the information in the literature is apparently contradictory, presumably owing to unquantified variations in process conditions and the quality of the reactants. The following sections aim to present a coherent account of the topic. References are given for those wishing to delve more deeply into the subject.

Because of the above complexities, before deciding to invest in a hydrating plant, the hydration characteristics of the available grades of quicklime should be evaluated. The quality of the water supply should also be assessed, bearing in mind any seasonal variations.

20.2 Physico-chemical Aspects of Hydration

20.2.1 The Chemical Reactions

At temperatures below 350 °C, the calcium oxide component of high-calcium quicklimes generally reacts readily with water, liberating heat (276 kcal/kg CaO — equation 20.1) [20.1]. At higher temperatures, the reverse reaction occurs (see section 19.2).

$$CaO + H_2O \rightleftarrows Ca(OH)_2 + heat \qquad (20.1)$$
$$56.1 \quad 18 \qquad 74.1$$

The MgO component of high-calcium quicklimes and of dolomitic limes is relatively unreactive to water. Generally less than 25 % of it reacts under normal hydration conditions (equation 20.2) [20.2].

$$CaO \cdot MgO + H_2O \rightleftarrows Ca(OH)_2 \cdot MgO + heat \qquad (20.2)$$
$$96.4 \qquad\quad 18 \qquad\qquad\quad 114.4$$

Highly hydrated dolomitic lime is generally produced by reacting the quicklime with water under hydrothermal conditions (i.e. under steam pressure at temperatures in excess of 100 °C in an autoclave) for the requisite time. The heat of hydration is 211 kcal/kg CaO.

$$CaO \cdot MgO + 2H_2O \rightleftarrows Ca(OH)_2 \cdot Mg(OH)_2 + heat \qquad (20.3)$$
$$96.4 \qquad\quad 36 \qquad\qquad\quad 132.4$$

20.2.2   Physical Processes

High-calcium and partially hydrated dolomitic limes are hydrated at temperatures approaching 100 °C. This is achieved by adding a sufficient excess of water to moderate the temperature by boiling. In practice, with high-calcium quicklime, approximately double the stoichiometric quantity of water is added, with most of the excess being vented to atmosphere as steam, after absorbing much of the heat of reaction.

The addition of water to the lime should be done under well agitated conditions. This helps to avoid localised overheating.

The process is believed to proceed via the migration of water into the pores of the lime particles. Hydration then occurs, associated with both expansion (see section 32.23) and the liberation of heat. This causes the particles to split, exposing fresh surfaces into which more water can migrate. The raw hydrate so produced consists largely of "fluffy" agglomerates of very fine crystals.

"Water-burning" leads to the production of "gritty hydrate", produced by overheating moist particles. It is believed to occur for two reasons. Firstly, particles of highly reactive lime can hydrate so rapidly that water does not penetrate sufficiently far into the particle to cause it to split. As a result, the particle (which may be up to 3 mm in diameter) may have a core of unhydrated quicklime surrounded by a hard shell of a baked putty of hydrated lime. Secondly, inadequate agitation and/or uneven addition of water can lead to localised hot spots where particles of hydrated lime can agglomerate and "bake" together.

As described in section 20.3.1, other causes of grit are calcium carbonate and hard-burned lime. The production of grit is minimised by pulverising the lime and many processes reduce the particle size to below 10 mm and, in some cases, to below 5 mm. Dolomitic lime is generally ground more finely (see section 20.7).

20.3   Raw Materials

20.3.1   Quicklime

Factors which affect the hydration of quicklime include:

a) the reactivity,
b) the mean apparent density,
c) the distribution of particle density,
d) the particle size distribution,
e) % carbonate,
f) % sulfur,
g) % magnesium oxide,
h) other impurities, particularly if on the surface, and
i) whether the quicklime is crushed.

The *reactivity* influences both the rate of reaction and the particle size distribution. Good quality hydrated limes can be made from quicklimes with a wide range of reactivities, providing the plant is correctly designed.

Thus, while "highly reactive" quicklimes (see section 13.2) hydrate quickly and produce particles with a small median particle size, they can also produce coarse particles as a result of "water burning". Quicklimes of "moderate" reactivity generally hydrate well. Those of "low" reactivity hydrate slowly, produce particles with a large median particle size and are likely to contain un-hydrated grit in the raw hydrate.

The *mean apparent density* depends on both the porosity of the parent limestone and the degree of burning of the lime. Thus, a quicklime produced from a porous limestone may have a similar reactivity to one made from a dense limestone, but different hydration characteristics.

The *distribution of particle density* can also be a significant factor, particularly where a proportion of over-burned, low reactivity lime is present. In such a case, precautions may need to be taken to ensure that low reactivity particles are retained in the hydrator until they react. This helps to prevent excessive quantities of unhydrated lime passing into the product and causing expansion and unsoundness.

The *size distribution* of the feed lime and the level of impurities can be important factors. Where the impurities are acceptably low, the feed lime can be pulverised and most of the impurities allowed to pass into the product.

Often, however, it is desirable to enhance the purity of the hydrate. This may be achieved by *crushing* to a limited extent (e.g. to a top size of 10 mm, or even larger) and by using the hydrator and the processing system to remove coarser particles which tend to be rich in *carbonate*. The situation can become complicated when significant quantities of water-burned lime and/or of unreacted lime are present as grit in the raw hydrate.

The presence of *sulfur* in the quicklime affects its reactivity, but does not generally adversely affect the hydration process.

Magnesium oxide in quicklime tends to become over-burned and to be slow to hydrate. It can cause expansion of the hydrate. Magnesian and dolomitic quick-limes generally require pressure hydration to produce a non-expansive product (see section 20.8).

Other impurities in the quicklime cause contamination of the hydrate. Where impurities form a surface layer on the quicklime particles, they may inhibit the hydration process. Such an effect can be overcome by *crushing* the quicklime. Where the hydrate is used in applications which are particularly sensitive to impurities (e.g. production of drinking water [20.3]), the levels of trace elements in the feed lime may need to be limited.

20.3.2   Water

Although impurities in water affect the rates of reaction between quicklime and water when producing milks of lime (sections 22.2 and 22.8.2), relatively little has been reported on their effect on the production of hydrated lime. This could be because:

a) most applications are not sensitive to their effects, or
b) the process can be adjusted to offset the effects.

On a pragmatic basis, the acceptability of a proposed water supply may be assessed by using a representative sample in a standard reactivity test and com-paring the result with that obtained using distilled water. If the difference is within the normal variability of the quicklime reactivity, the water supply should be acceptable. If otherwise, hydration tests using the water should be carried out to ensure that the required quality of hydrated lime can be produced consistently.

20.4   Design of Hydrating Plants

20.4.1   General

Quicklime was traditionally hydrated on the small scale by sprinkling water on to a pile of lime until no more water was absorbed. The lime was left for some time to "mature" and was then passed over fine sieves to remove grit. However, such a product was of poor and variable quality and required a high labour input.

Virtually all hydrated lime is now produced in purpose-built plants. Figure 20.1 illustrates a basic plant, which consists of four stages:

a) quicklime handling and crushing,
b) hydration,
c) classification, and
d) storage and despatch (see chapter 21).

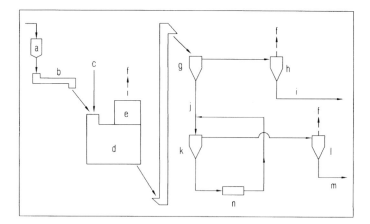

Figure 20.1. Diagram of a basic hydrating plant
(a) quicklime bunker; (b) weigh feeder; (c) water; (d) hydrator; (e) dust collector; (f) discharges
to atmosphere; (g) primary classifier; (h) cyclone; (i) primary fines; (j) primary tailings;
(k) secondary classifier; (l) cyclone; (m) secondary fines; (n) mill

Some plants mature the raw hydrate for 12 to 24 hours in bunkers before classification to reduce the level of free lime. Others may not be required to achieve
high levels of hydration/classification and can be of a very simple design [20.4]

20.4.2   Quicklime Handling and Crushing

Some buffer storage of quicklime is usually provided between the lime kilns and
the hydrating plant to provide operating flexibility. Where the removal of impurities in the hydrating plant is not important, the quicklime is generally reduced in
size using impact breakers. Otherwise, rolls and jaw crushers, or cone mills may
be used (see section 5.2).

20.4.3   The Hydrator

A variety of designs of hydrator are used (e.g., Fig. 20.2). A typical design can be
regarded as consisting of three sections – prehydrator, hydrator and finishing
stage. Each consists of a horizontal trough which is agitated by paddles mounted
on one or two horizontal shafts (Fig. 20.3). The paddles are angled to convey the
material towards the discharge end of the trough. Weirs placed at the discharge
help to provide the necessary residence time.

The prehydrator often has a smaller volume than the later stages and is highly
agitated to mix the water and quicklime quickly and efficiently.

The hydrator typically provides an average residence time of 10 to 15 min. The
heat of reaction causes most of the surplus water to boil off. The steam produced
fluidises finely divided hydrated lime which flows over the weir to the finishing stage.

Figure 20.2. A modern 3-stage hydrator with integral dust collector
(by courtesy of Cimprogetti s.p.a.)

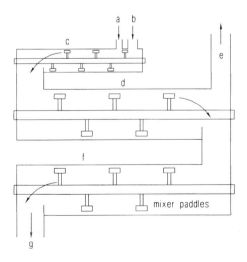

Figure 20.3. Diagram of a 3-stage hydrator
(a) water; (b) quicklime; (c) pre-hydrator;
(d) hydrator; (e) steam and air to dust
collector; (f) finishing stage; (g) raw hydrate

Coarse grit, which consists of calcium carbonate, any unslaked quicklime and water-burned hydrate is retained by the weir. It may be removed at intervals (e.g. at the end of shifts) by opening a "purge hole" in the weir or by lowering the weir. In such cases, the grit is generally rejected to tip. Alternatively, the grit may be removed continuously from the hydrator by a "purge paddle" which lifts it over the weir. In this case, the grit is removed from the "raw hydrate" by air classification or screening. Such "hydrator rejects" are either tipped, blended in a controlled manner into selected products, or sold as agricultural lime.

In the finishing stage, residence time is provided to:

a) reduce the level of free lime, and
b) allow surplus moisture to evaporate.

The plant is kept under slight suction to prevent dust emission and to maintain a low flow of air through the finishing stage to facilitate drying. The moisture content of the "raw hydrate" on discharge from the finishing stage is generally up to 1 %.

20.4.4   The Classification Stage

A large number of variations of classification plant have been used successfully. The choice depends largely on the quality of the feed lime(s) and the required quality of the product(s). In the lay-out shown in Fig. 20.1, the raw hydrate is fed to the primary air classifier, which can be adjusted to give a cut size meeting the required particle size specification.

Oversized particles rejected by the classifier (the primary tailings) consist of varying proportions of unburnt limestone, hydrated quicklime and gritty hydrate. They are generally processed in one of three ways.

a) The simplest approach is to mill the primary tailings, which are then fed to the secondary air classifier. The resulting fine fraction (secondary fines) are either blended with the primary fines or used to produce a less pure grade (e.g. for building applications [20.5]). The coarse fraction from the secondary classifier, (secondary tailings), are recycled to the mill.
b) If the secondary fines contain unacceptable amounts of free CaO, they may be recycled to the hydrator.
c) Where it is required to use the classification plant to remove some of the unburnt limestone, various options exist for isolating a $CaCO_3$-rich fraction. One option is to pass the primary tailings through a beater mill, which is designed to preferentially reduce the particle size of the relatively soft quicklime and gritty hydrate, leaving the size of the harder limestone substantially unchanged. Air classification of the milled product produces a lime-enriched secondary fines fraction and a carbonate-enriched secondary tailings fraction. The latter may be tipped, or sold as agricultural lime. The secondary fines can be transferred to a product bunker, or, if they contain excessive free lime, recycled to the hydrator.

20.5   Control of Hydrating Plant

The primary controls are the feed rates of the water and quicklime to the plant. These should be designed to give high levels of consistency (e.g., to better than ±0.5 %), so that the required level of excess water in the hydrate can be maintained to within the required range. The preferred approach is to feed a constant weight of lime and to adjust the water flow rate to achieve the required temperatures and moisture contents [20.5]. Variations in the quality of the feed lime (e.g. in the reactivity or % $CaCO_3$) can change the water requirement significantly. Precautions should be taken to minimise such variations, particularly when several grades of lime are fed to the plant.

For optimum control, the temperatures of the raw hydrate should be monitored as it passes through the hydrator and finishing stage. The temperature profile depends on the level of excess water and on the reactivity of the lime. For a given feed lime, a decrease in temperature towards the discharge of the hydrator and in the finishing stage is indicative of an increase in moisture content, and vice versa. The current drawn by the motors driving the paddles can also be indicative of moisture content.

It is particularly important to limit the level of excess moisture in the plant, as the handling characteristics of hydrated lime deteriorate rapidly when the moisture content rises above 1.5 or 2 %. Excessive levels of moisture result in blockages within the plant and hold-ups in bunkers.

Various techniques are used for determining the moisture content of the raw hydrate. They include visual inspection, a rapid "carbide" test method, infra-red moisture meters and conductivity probes. The latter can be built into the wall of the finishing stage vessel [20.5].

The grading of the fines streams may be determined by air-jet sieving and adjusted by varying either the speed or the settings of the air classifiers. Some customers require other properties to be controlled such as the specific surface area (measured by air permeability [20.6]) and the settling rate of milks of lime produced from the hydrate.

20.6   Production of High Surface Area Hydrated Limes

20.6.1   General

In the 1980's the opportunity for using hydrated lime to remove acid gases from processes such as incinerators and small boilers was recognised. However, trials showed that relatively large excess of commercially available hydrate had to be added to reduce the acid gas concentrations to the required levels. It was postulated that increasing the *effective* surface area would increase absorption efficiencies. The challenge was how to produce a hydrate with significantly improved properties.

It has been recognised for many years that the particles of calcium hydroxide in "milks of lime" have higher surface areas (up to $30 \, m^2/g$), than normal hydrated lime (typically 15 to $20 \, m^2/g$). However, the cost of drying a milk to produce a high surface area hydrate would be excessive. While milling hydrated lime improves its plasticity [20.7], it does not increase its specific surface area significantly, presumably owing to the production of relatively dense agglomerates.

Research has shown that hydrates with high specific surface areas could be produced by slaking with water-alcohol mixtures, and with aqueous solutions of sugar, lignosulfonate, or amine derivatives [20.8–20.10]. This effect was regarded as little more than a curiosity until the requirement for a high surface area hydrate was identified. Two approaches have resulted in products which have significantly larger effective surface areas. They are described below.

It should be noted that these developments are on-going. Questions are still being asked as to how the effective surface area correlates with:

a) the specific surface area, as measured by the BET (Berkland Eyde and Teller) technique,
b) the total volume of the pores,
c) the size distribution of the pores,
d) the particle size, and
e) the degree of agglomeration.

20.6.2   Hydration with a Methanol-Water Mixture

The laboratory work leading to this development is described in [20.11]. Ground quicklime is used to avoid the need to classify the product. It is mixed with a water-methanol solution, the composition and volume of which provides sufficient water for the hydration reaction and sufficient methanol to remove, by evaporation, the heat of hydration (the boiling point of methanol is $65 \, °C$)

When most of the lime has hydrated, most of the water has reacted and most of the methanol has evaporated, so that the mixture passes through a solid phase. Subsequent hydration and evolution of methanol vapour and steam causes the mixture to break down into a finely divided powder.

The process can be regarded as consisting of six stages [20.12, 20.13].

a) In the mixing vessel, ground quicklime is mixed with the methanol-water solution (typically containing 60 % of methanol) at below $45 \, °C$. The presence of the alcohol and the low temperature inhibit the hydration reaction.
b) The suspension is then heated to between 50 and $70 \, °C$ and transferred into the main reaction vessel, which is agitated by plough-type mixers.
c) In the vessel most of the water reacts with the quicklime and most of the methanol evaporates.
d) The partially hydrated solid is then transferred into the second reaction vessel.
e) In the second reaction vessel, which is also agitated, hydration proceeds to completion and the temperature rises to 95 to $110 \, °C$. The resulting evolution of methanol vapour and steam causes the granules to break up into a fine powder.

f) The fully hydrated product then passes into a de-gasifier, which removes most of the residual methanol and water, using either a nitrogen purge or a partial vacuum. It is then discharged into a cooler from which it is conveyed to a storage silo.

The finished product has a specific surface area of 35 to 43 m^2/g. It is reported to have good handling properties, to disperse well, and to be significantly more effective at removing acid gases than normal hydrated limes [20.12].

Inevitably, the capital and operating costs of the plant are relatively high. Apart from the de-gasser, the equipment is kept at above atmospheric pressure by a supply of nitrogen to prevent the ingress of air. Methanol vapour, together with some steam, from the reaction vessels and the de-gasifier, is condensed and recycled. In view of the explosion hazard, the equipment has to be well engineered and all electrical equipment within the plant is of flame-proof quality.

20.6.3 Hydration with Aqueous Solutions

Details of this process are still largely confidential. An early patent [20.14] described the hydration of reactive, fine lime (e.g., less than 6 mm) with water in the presence of an amine/glycol additive. Specified compounds were mono-, di- and tri-ethyleneglycols and mono-, di- and tri-ethanolamines, and mixtures thereof. Hydrates with surface areas of 46 m^2/g were cited.

It is understood that the process has since been developed and that further patents are pending [20.17].

20.7 Performance Criteria for Hydrated Limes

A variety of test methods are available to characterise hydrated lime. As they all measure different parameters, the choice of the most appropriate method(s) depends on the application.

20.7.1 Chemical

The chemical activity of hydrated lime can be measured in at least three ways, namely by:

a) neutralising value ($CaO + Ca(OH)_2 + MgO + Mg(OH)_2 + CaCO_3$ + part of the CaO in silicates/aluminates),
b) available lime ($CaO + Ca(OH)_2$), and
c) water soluble lime ($CaO + Ca(OH)_2$ + part of $Mg(OH)_2$ and part of the CaO in silicates/aluminates).

High-calcium and dolomitic hydrated limes for use in building are classified in terms of their CaO and MgO contents [20.19].

The "chemical reactivity" of a hydrated lime nay be measured by preparing a milk of lime, and determining its rate of:

a) causticisation of sodium carbonate solution [20.2],
b) neutralisation of acids [20.21], and
c) solution in water, as measured by conductivity (see section 22.8.4) [20.20].

20.7.2   Physical

Hydrated lime should be substantially free of grit and have an acceptable fineness. The amount of free water should be limited to less than 2 %, and preferably to below 1 %, so that they can be handled without difficulty. For particular applications (e.g. masonry mortar), it should be sound and produce mortars with acceptable properties. Some of the requirements of EN 459 for hydrates used in building are summarised in Table 26.2.

Some applications require the specific surface area, as measured by air permeability [20.6], to be controlled within a specified range. Other applications require the determination of the BET surface area (most commercial hydrates have BET areas in the range 10 to 20 m^2/g).

Most commercial hydrated limes have median particle sizes, as measured by laser granulometry, of about 5 to 10 μm. Hydrated lime quality may also be assessed by testing either a milk or a lime putty prepared from the hydrate under specified conditions (see section 22.7.2).

20.8   Dolomitic Hydrated Limes

20.8.1   Hydration at Atmospheric Pressure

Some very light-burned dolomitic limes are hydrated at atmospheric pressure in a similar way to high-calcium limes. They use maturing bunkers to provide 12 to 24 hours of residence at 80 to 90 °C in the presence of excess water.

However, even under such conditions, only part of the MgO component hydrates. Such partially hydrated products (designated Type N in ASTM C 206 and 207 [20.15, 20.16] are unsound and unsuitable for use in building and in mortars.

20.8.2   Hydration at Elevated Temperatures and Pressures

Highly hydrated dolomitic lime is produced at steam pressures in the range 1.7 to 7 atmospheres, corresponding to temperatures of 115 to 165 °C. The optimum pressure and hydration time depends on the degree of burning of the quicklime.

A typical pressure hydration plant is illustrated in Fig. 20.4. The quicklime is generally ground or pulverised. In some plants the lime and water are blended in

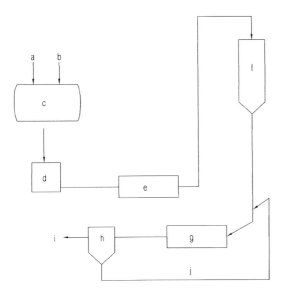

Figure 20.4. Diagram of a pressure-hydration process
(a) quicklime; (b) water; (c) pressure hydrator; (d) collection pit; (e) drier; (f) silo; (g) ball mill;
(h) air classifier; (i) finished product; (j) recycle of over-size

a pre-mixer to largely pre-hydrate the CaO component before charging into the autoclave. In most, however, the lime and water are fed separately into the auto-clave. A considerable excess of water over the stoichiometric amount is used.

When the hydration is completed, the autoclave is discharged into a collection pit or vessel and much of the excess water flashes off. The damp hydrate is dried and milled. The milled product is then air classified, with oversize being recycled to the mill.

20.8.3   Properties of Type S Hydrated Lime

The requirements for highly hydrated dolomitic lime, (also called di-hydrated or double hydrated lime) are specified in [20.15, 20.16]. It is required to contain less than 8 % of free lime, to ensure that it is sound.

The product has a high and uniform plasticity that is developed rapidly follow-ing mixing with water (within 30 min.). It also gives high mortar strengths [20.2].

20.9   Hydraulic Limes

Natural hydraulic limes are hydrated with sufficient water to convert free CaO into $Ca(OH)_2$, without hydrating significant amounts of the calcium silicates and aluminates (see section 16.10.1).

20.10   Carbide Lime

Calcium carbide is produced by reacting quicklime and coke at 1800 to 2100 °C (see section 31.4). The carbide is then reacted with water to form acetylene and calcium hydroxide as a by-product. A few acetylene producers use a dry generation process (otherwise known as the "water to carbide" process), which results in a powdered hydrated lime, commonly called "carbide lime". A high level of control is exercised over the water addition and the calcium carbide/hydrate mixture is agitated continuously to prevent localised over-heating and the formation of undesirable polymerised by-products.

The dry hydrate is more convenient to handle than the milk of lime produced by the "carbide to water" wet generation processes (see section 22.9). While it may be less pure than many commercial hydrates, it is used for the production of masonry mortar, acid neutralisation, water purification and water treatment. It has also been calcined to regenerate quicklime for recycling to the carbide furnace [20.2, 20.18].

20.11   Air Slaked Lime

Air slaking is an undesirable process resulting from the exposure of quicklime to the atmosphere. The quicklime reacts with water vapour and atmospheric carbon dioxide to produce a mixture of CaO, $Ca(OH)_2$ and $CaCO_3$. Air slaked lime has a low reactivity, even when only a few percent of the CaO has reacted, and is of little use in most applications.

20.12   References

[20.1]   "Selected Values of chemical thermodynamic properties", Circular 500 of the National Bureau of Standards, Feb. 1952 (reprinted in "Handbook of physics and chemistry", 57th ed., Chemical Rubber Corporation Press).

[20.2]   R. S. Boynton, "Chemistry and Technology of Lime and Limestone", John Wiley & Sons, 1980.

[20.3]   EN 12518: "Chemicals used for treatment of water intended for human consumption — High-calcium lime", in preparation.

[20.4]   L. Bier, H. Sauer, W. Bennewart, "Construction and operation of a hydrated lime plant for flue gas desulfurisation if the Siersdorf power station", Zement Kalk Gips **6**, 1992, 285–288 (English: **8**, 1992 E207–E209).

[20.5]   O. Collarini, T. Christiansen, "New lime hydration plant in northern Italy", Zement Kalk Gips **9**, 1996, 540–544.

[20.6]   BS 6463, Part 4: "Methods of test for physical properties of hydrated lime and lime putty", 1987.

[20.7]   American Marietta Corp., U.S. Patent 2,894,820 (M. Rickard).

[20.8]   H. Staley, S. Greenfield, Proc. ASTM 47, 1947.

[20.9]   J. Murray, "Summary of fundamental research on lime", Research Report, National Lime Association, 1956.

[20.10] T. Miller, "A study of the reaction between CaO and water", National Lime Association, 1961.
[20.11] F. Schmitz, H.-P. Henneke, H. Bestek, A. Reeder, "Dry hydrated lime for the binding of acid exit gas constituents, Part 1: Production in the laboratory and scope for use in flue gas cleaning", Zement Kalk Gips **10**, 1984, 530–533 (English: **12**, 1984 290–292).
[20.12] H.-P. Henneke et al., "Dry-hydrated lime with a large surface area — an effective reagent for waste gas cleaning, Part 2: Set-up and operation of a pilot plant, results of experiments relating to the dry sorption of pollutants from different waste gases", Zement Kalk Gips **5**, 1986, 251–258 (English: **7**, 1986, 209–214).
[20.13] Rheinische Kalksteinwerke, U.S. Patent 4,636,379 (H. Bestek et al.) 1987.
[20.14] Lhoist Recherche et Developpement, U.S. Patent 5,173,279 (P. A. Dumont, R. Goffin) 1992.
[20.15] ASTM C 206-84: "Standard specification for finishing hydrated lime".
[20.16] ASTM C 207-91: "Standard specification for hydrated lime for masonry purposes".
[20.17] E. Béchoux, private communication, July, 1997.
[20.18] "Carbidkalk, Hinweise für seine Verwendung", Fachbuchreihe Schweißtechnik, 9th ed., Deutscher Verlag für Schweißtechnik (D.V.S.), Düsseldorf, 1968, p. 16.
[20.19] ENV 459-1: "Building lime — Definitions, specifications and conformity criteria", 1994.
[20.20] D. J. Wiersma, P. Hubert, J. N. Bolle, "Chemical reactivity and other relevant properties of milks of lime as applied to water treatment", Proc. International Lime Congress, Berlin, June 1994.
[20.21] H. Becker, H. von Zander, "The rate of neutralisation of wet or dry slaked hydrated limes as a function of the conditions of manufacture", Zement Kalk Gips **8**, 1976, 381–387.

21 Handling and Storage of Hydrated Lime

21.1 General

Hydrated lime is generally supplied to small users in paper sacks or intermediate bulk containers (IBCs or "big bags") of 0.5 or 1 t capacity. Where larger amounts are used, the product is delivered in air pressure discharge vehicles (APDVs).

The flow characteristics of hydrated lime can vary widely, with the angle of repose ranging from almost 0° when it is aerated, to 90° when it is compacted and/or damp. Its handling properties, however, do not generally cause problems, providing its moisture content is less than 2 % and normal powder handling techniques are used.

Hydrated lime absorbs carbon dioxide from the atmosphere, forming calcium carbonate and water. It should, therefore, be stored in dry, draught-free conditions.

21.2 Bagged Hydrate

Hydrate is generally bagged in 2 or 3-ply paper sacks. They may be delivered loose, or on pallets, and are handled by bag truck or by fork lift truck. Their effective bulk density is about 500 kg/m^3.

To avoid deterioration of the bags by moisture and of the hydrated lime by re-carbonation, bagged hydrate should be stored under cover. The ideal store is a brick or concrete building constructed to minimise draughts. It should have a low roof to minimise internal air space and be unheated. Under such conditions, bagged hydrate can be stored for a year without significant deterioration [21.1].

If hydrated lime is kept in a general store, care should be taken to ensure that it does not come into contact with other chemicals with which it might react. The sacks should be covered by an impervious sheet.

Pallets of bagged hydrate have been stored successfully out-of-doors. The pallet is covered by a plastic sheet, the bags placed on the sheet and the pack shrink-wrapped. Such pallets should be moved into a covered store before unloading.

Handling arrangements for bagged hydrate are usually quite basic, involving manual loading into a hopper with a screw or vibrating feeder arrangement to transfer the hydrate into the process.

Intermediate bulk containers should be stored under cover. They are generally transported to the plant using a fork lift truck (or equivalent) and are discharged into a hopper, from which the hydrate is conveyed into the process.

21.3 Bulk Hydrate

Bulk hydrate is discharged pneumatically from a vehicle into a reception silo, via a filling pipe (typically 100 mm nominal bore), which may be up to 40 m long. Bends should have a radius of at least 1 m.

The silo may be constructed of steel, concrete, glass fibre, or wood and must be completely weatherproof. It should be vented via a bag filter. For calculating silo capacity, a bulk density of 480 kg/m^3 is generally appropriate although the value may range from below 400 kg/m^3 when the hydrate is aerated to 560 kg/m^3 when it is fully compacted.

The bag filter should be weatherproof and be capable of handling the delivered air flow (typically 10 to 20 m^3/min.). Where the filter is fitted on top of the silo, the collected dust is discharged back into the silo. The silo top should be fitted with an inspection manhole and dust-tight cover, and a pressure relief valve. A high level indicator, or alarm should also be fitted to prevent over-filling.

The base of the silo should be at an angle of at least 60° to the horizontal. The discharge aperture should be not less than 200 mm and a positive cut-off valve should be fitted to the outlet to permit equipment beneath the silo to be maintained.

Because hydrated lime is prone to "arching", suitable arch-breaking devices should be fitted. These include aeration pads, vibrators and mechanical devices. The choice of device depends on the pattern of use and the required accuracy of measurement of the delivery rate to the process. Conversely, precautions need to be taken to prevent "flooding" of aerated powder (see section 21.4).

21.4 Conveying and Dosing

All conveying systems should be fully enclosed and dust tight. A detailed description of suitable equipment is given in [21.1]. Where there is a risk of "flooding" of aerated material, e.g. at the discharge from a silo, a rotary valve should be fitted.

Screw conveyors are widely used. Where some control over the flow rate is required, tubular screw conveyors are often installed. Air slides are a convenient and economical way of transferring hydrate down a slight slope. They should be enclosed and fitted with a positive cut-off valve at the discharge end.

Pneumatic conveying is frequently a cost effective option, using a Rootes-type blower and a rotary blowing seal. Alternative proprietary units are available. Pipelines should normally have a nominal bore of at least 50 mm and bends should have a radius of at least 1 m. The reception hopper should be fitted with an air-pressure relief valve. Powder pumps are also used successfully. They tend to use less air than pneumatic conveying systems.

For many applications, a sufficiently accurate dosing rate can be obtained using volumetric feeding (e.g. with powder pumps, tubular screw conveyors and rotary valves). Where more accuracy is required, belt weighers and screw-feeder weighers are available. For the highest degree of accuracy, batch weighing is generally recommended.

21.5   Production of Milk of Lime

A high proportion of hydrated lime is slurried in water and used as a milk or putty. For more details, see chapter 23. Particular attention is drawn to the need for precautions to be taken to control the formation of calcium carbonate scale when pumping milk of lime (sections 23.1.3 and 23.1.4).

21.6   Reference

[21.1] "Lime handling, application and storage", National Lime Association, Bulletin 213, 5th ed., 1988.

22 Production of Milk of Lime and Lime Putty

22.1 Introduction

The term "milk of lime" is used to describe a fluid suspension of slaked lime in water. Milks of lime may contain up to 40 % by weight of solids. Milk of lime with a high solids content is sometimes called "lime slurry".

"Lime putty" is a thick dispersion of slaked lime in water. Putties typically contain 55 to 70 % by weight of solids. A semi-fluid putty is sometimes described as a "lime paste".

The production and properties of milk of lime and lime putty are important for two reasons.

a) Some 40 to 60 % of quicklime, other than that used for steelmaking, is slaked and most hydrated lime is converted into a milk before use.
b) The vast majority of customers use lime products in the form of either a milk or a putty.

Although most of the parameters affecting the production of milks of lime and lime putties have been known for many years, much of the evidence was anecdotal and contradictory. It was based on microscopy and measurements of the bulk properties of the products.

In recent years, the availability of laser beam granulometry has enabled a more fundamental approach to be adopted. This has lead to a better understanding of the parameters which affect the physical and chemical properties of milks of lime and lime putties and which, in turn, affect the suitability of the products for their various applications. As a result, products can now be tailored more precisely to the requirements of the customers' processes.

22.2 The Principles of Slaking

The chemical reactions involved in slaking are identical to those of hydration and are described in section 20.2.1.

The main variables which affect the quality of slaked lime are:

a) reactivity (to water) of the quicklime,
b) particle size distribution of the quicklime,
c) amount and quality of water used,
d) temperature of the water,
e) the pattern of addition of the lime and water, and
f) agitation during slaking.

The *initial* conditions have been shown to be very important as they determine how rapidly the calcium oxide dissolves [22.1]. When quicklime is mixed with water, the most highly reactive particles dissolve producing a very high level of super-saturation with respect to calcium hydroxide. This results in heavy "primary" nucleation (i.e., the formation of a very large number of calcium hydroxide nuclei). Less reactive particles dissolve more slowly and produce a lower degree of supersaturation, which largely results in crystal growth on the primary nuclei.

As described in section 15.4, commercial quicklimes consist of particles with a distribution of apparent densities. The rate of solution of a given particle depends on its apparent density (or reactivity) and its particle size. Thus finely divided particles/those of high reactivity dissolve and hydrate first, producing primary nuclei. Coarser particles/those of lower reactivity dissolve and hydrate more slowly and contribute to crystal growth.

Rates of solution of calcium oxide increase rapidly with increasing temperature. Thus raising the temperature of the water used for slaking a given quicklime, results in increased primary nucleation. Moreover, as the reaction is highly exothermic, the initial hydration raises the temperature and accelerates the hydration of the coarser/less reactive particles. Reducing the water to lime ratio reduces the thermal capacity of the mix, increases the rate of temperature rise and accelerates the rate of slaking.

As a converse of the above, the addition of an excessive volume of relatively cold water to a low reactivity quicklime can result in "drowning" of the lime. Under such conditions, the surface of the quicklime particles hydrate, but the particles do not disintegrate effectively and relatively little primary nucleation occurs. Indeed, some of the quicklime may fail to slake fully, resulting in unsoundness and grit.

Agitation of the slaker can have two effects. Relatively low shear rates can help to ensure an adequate dispersion of the lime in the water. This reduces the risk of local over-heating, which can lead to localised boiling and "water-burning" of the lime. Water-burning arises from the generation of very high temperatures which can "bake" calcium hydroxide putty, resulting in gritty particles of slaked lime and possibly in unsoundness caused by unreacted calcium oxide. High shear rates also accelerate solution rates, thereby increasing primary nucleation. They reduce agglomeration and cause crystals to break up, producing "secondary" nucleation.

The degree of agglomeration can also be reduced by slaking at high water to lime ratios [22.2]. The addition of dispersing agents such as alcohols (e.g. methanol, ethanol and butanol) and lignosulfonates to the slaking water also reduces agglomeration and increases the effective fineness of the calcium hydroxide. Sugar, which greatly increases the "solubility" of calcium hydroxide, has a similar effect, although the mechanism is probably different.

Conversely, the presence of carbonate hardness, sulfites and sulfates in the water (or quicklime) tend to inhibit the slaking process, resulting in less primary nucleation and increased crystal growth [22.1] contrasts slaking with a municipal drinking water, which gave a median particle size of $15\,\mu m$, and slaking with distilled water under identical conditions, which gave a median particle size of $3\,\mu m$. Chloride ions increase slaking rates slightly [22.3], but do not have a pronounced effect on particle size. Brackish waters can, therefore, be used for slaking, providing they do not contain excessive levels of the inhibiting ions mentioned above.

22.3 Slaking Practices

Slaking practices vary with the quality of the quicklime and the water and with the required properties of the milk of lime or lime putty. Generally, it is advisable to adopt the procedures recommended by the suppliers of the lime and the slaker, as they are familiar with the slaking characteristics of the quicklime and the performance of the slaker. Typically, the conditions are controlled to give a slaking time of about 10 min., using a water to lime ratio in the range of 3:1 to 4:1 (see section 22.4).

For some applications, it may be necessary to optimise standard practices to match the properties of the slaked lime to the requirements of the process. The following paragraphs describe some of the principles involved.

The properties of slaked lime (see section 22.6) depend, in part, on the reactivity of the quicklime, but can be altered considerably by adjusting the other variables. Most applications require the slaked lime to be fairly finely divided: some require relatively coarse particles, while others require the particles to be as finely divided as possible.

As described in section 22.2, for a given set of slaking conditions, reactive quicklimes produce more primary nucleation than less reactive products, and result in a lower median particle size and a higher proportion of particles less than $1\,\mu m$ [22.1].

Reactive quicklimes produce slaked limes with acceptably fine particles if added to a limited volume of cold water, with stirring. When the temperature of the suspension rises to 70 to 90 °C, further additions of quicklime and cold water are made to maintain the temperature in that range.

If quicklimes of moderate to low reactivity are slaked with cold water, they produce relatively coarse particles of slaked lime. However, they can produce finely divided slaked lime if action is taken to increase the initial slaking rate. This can be done by crushing the quicklime finely (e.g., to 90 % less than 2 mm), and by using a controlled amount of hot water to start the slaking process.

Certain impurities in the water (e.g. temporary hardness, sulfite and sulfate ions) reduce the initial reaction rate and increase the median particle size.

The most finely divided milks of lime produced by traditional slaking practices are obtained by adding finely crushed, high reactivity quicklime (e.g. 5 mm to dust) to hot water. This produces a high level of primary nucleation. Subsequently, cold water may be added, providing the temperature of the mix is kept above 80 °C. Secondary nucleation may be increased by using high shear rates. Dispersing agents (see section 22.2) may also be used, if acceptable to the customer.

The above practices relate to batch slaking. In continuous slakers, primary nuclei are always being recycled and reduce the levels of super-saturation. As a result, for the same conditions, continuous slakers tend to produce coarser particle size distributions than batch slakers. If required, this effect can generally be offset by adjusting the slaking conditions.

Regardless of the practices adopted, slaking of most quicklimes results in the formation of grit. This may arise from uncalcined particles of limestone, from discrete impurities (e.g. flint), from water-burning of reactive lime and from drowning of low reactivity lime.

Grit in milks of lime can cause a variety of problems. It results in abrasion of pumps and pipelines. It settles rapidly and may not re-disperse, necessitating shut-downs to remove it. Grit can also cause problems in the process using the milk. It is, therefore, normal practice to remove grit from milks of lime immediately after slaking.

There are three ways of avoiding the production of grit, namely:

a) slaking in a mill which grinds any grit produced into fine particles,
b) slaking finely ground quicklime, and
c) producing a milk of lime using hydrated lime.

Milks of lime and lime putties "mature" on storage. The effects of maturing (e.g. slower settling, increased volume yield and plasticity) are beneficial for most applications. Most of the benefits generally occur within 12 to 24 hours. However, storage at temperatures approaching 100 °C encourages unwanted crystal growth, so many slakers produce a relatively thick milk (or paste), which is diluted and cooled using cold water. The quality of water used for dilution is not critical, thus water containing significant levels of the inhibiting ions mentioned above may be used, without appreciably affecting the properties of the milk.

22.4   Slaker Design

22.4.1   Batch Slaking

The traditional method of hand slaking uses a wooden or metal trough. Water is added to cover the base and sufficient quicklime is added to bring the water to the boil. Long handled rakes or hoes are used to agitate the contents. Further additions of cold water and lime (in the ratio of about 3 to 1 by weight) are made to keep the mixture at or near boiling point. When the trough has been filled, agitation should continue for at least 5 min. after boiling has ceased. The milk of lime is discharged through a stop-cock or slide valve and should be passed through screens to remove lumps of uncalcined limestone and grit. For more details, see [22.4].

Mechanical slakers need to be robustly built to handle lumps of unburned limestone, refractory etc. without jamming or causing damage. Various designs are used. The traditional pan slaker (e.g., Fig. 22.1) is cylindrical and incorporates a cruciform stirrer from which heavy chains are suspended to agitate the lime particles that settle on the base of the slaker. Water and lime are added progressively, keeping the temperature in the range 85 to 95 °C. The tank is fitted with a discharge cock and a drain cock, through which grit can be removed.

A modern design of batch slaker is described in [22.2]. It consists of a horizontal cylinder agitated along its length with paddles. Quicklime (up to 20 mm in size) is blown at a controlled rate into the cylinder and water is added to maintain the temperature at 85 to 95 °C. The unit is able to slake a truckload of 20 t in 1 to $1^1/_2$ hours. It can be fitted with a wet scrubber for dust control and with a de-gritting system.

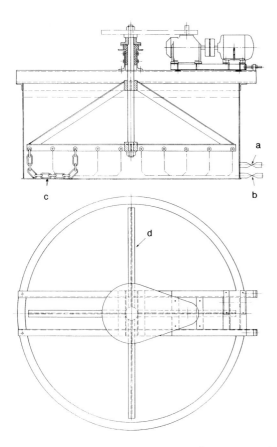

Figure 22.1. Diagram of a batch slaker (by courtesy of Buxton Lime Industries Ltd.)
(a) discharge; (b) drain; (c) stirrer chains; (d) stirrer arms

22.4.2   Continuous Slaking

Many designs of continuous slaker have been developed (see [22.2] for descriptions of some of the units available in North America). Indeed, several major lime
manufacturers offer their own design of slaker.

Most slakers are of the "detention" type, which essentially consist of a tank
agitated with an impeller (e.g., Fig. 22.2). They produce a fluid milk of lime, and
generally use between 3.5 and 4 parts by weight of water to 1 part of high-calcium
quicklime. For dolomitic quicklimes, the water to lime ratios are lower.

The quicklime is fed to the slaker at a controlled rate (usually volumetric) and
water is added to maintain the required temperature (generally 80 to 85 °C). The
tank typically gives an average residence time for the water of 10 min. Slow slaking particles of lime are detained until they react, or are removed as grit.

The paste-type of slaker is widely used (Fig. 22.3). It operates on a higher water
to lime ratio of 2 or 3 to 1 by weight. The slaking compartment consists of a horizontal trough with two sets of counter-rotating paddles, which act as a pug mill.

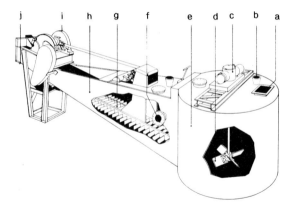

Figure 22.2. Diagram of a continuous detention-type of slaker (by courtesy of Dorr-Oliver Inc.)
(a) quicklime; (b) slaking water; (c) stirrer drive; (d) impeller; (e) slaking compartment;
(f) milk of lime discharge; (g) reciprocating rakes; (h) classification compartment; (i) rake drive;
(j) washed grit discharge

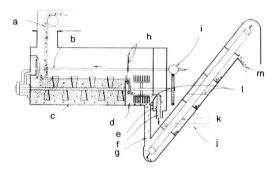

Figure 22.3. Diagram of a paste slaker (by courtesy of Wallace and Tiernan Ltd.)
(a) quicklime; (b) slaking water; (c) slaking compartment; (d) grit removal section; (e) slurry discharge compartment; (f) milk of lime discharge; (g) classifier; (h) water jet; (i) torque controlled
water valve; (j) water for grit washing; (k) grit conveyer; (l) liquid level; (m) grit discharge

The consistency of the mixture is controlled by sensing the torque on the paddle
drive and adjusting the water addition rate. The residence time in the trough is
typically 5 min. The paste flows over a weir and is diluted with a water spray.

Most of the continuous slakers may be fitted with grit removal systems.

The need to remove grit can be avoided by using a ball mill as a slaker. The
high levels of abrasion ensure rapid slaking, even with low reactivity limes. Relatively poor quality water may be used, as any surface coating of insoluble salts is
rapidly broken up and, by classifying the milk of lime, any residual grit in the discharge from the mill can be recycled.

The attrition slaker uses twin propellers with opposed pitches to slake the lime and break down the weaker particles of grit. The remaining grit may be removed in a spiral classifier and fed to a ball mill.

Dust emission from slakers is rarely a problem. Slakers should be kept under slight suction to prevent emissions into the plant, with the displaced air being vented to atmosphere via a suitable dust collector (see section 33.3.6).

22.5   Dispersion of Hydrated Lime

Hydrated lime disperses readily in water. In small-scale applications, batches of milk of lime are produced in a stirred vessel by manually charging hydrate from bags into a known volume of water. The resulting milk of lime is then fed via a solution/slurry feeder into the process [22.2].

On the larger scale, bulk hydrate is discharged continuously via a powder feeder into either a batch or continuous mixer. The resulting milk of lime is then fed via a solution/slurry feeder to the process. Details of various types of feeders are given in [22.2].

22.6   Lime Putties

Lime putties are generally produced by allowing a milk of lime to settle and drawing off the layer of clear lime water. The settled solids are then mixed [22.4]. The properties of the putty improve with ageing and it is generally recommended that several hours be allowed between slaking and use [22.5].

Alternatively, a milk of lime may be produced using hydrated lime and an appropriate amount of cold water. Such putties initially have inferior plasticity and water retention and should generally be matured for at least 24 hours [22.4, 22.6].

It should be noted that putties should not be produced by direct slaking of quicklime with a limited amount of water. The evolution of heat would cause the mix to boil vigorously and to eject drops of hot milk of lime. Moreover, slaking with such a restricted amount of water would lead to water-burning, grit, poor physical properties and possibly unsoundness.

The properties of lime putties improve progressively with time. They may be stored indefinitely under moist conditions. If left uncovered, a calcium carbonate crust forms that protects the putty from further carbonation.

22.7   Performance Criteria for Milks of Lime and Lime Putties

22.7.1   Chemical

The chemical activity and reactivity of milks of lime and lime putties are determined as described for hydrated lime in section 20.7.1.

22.7.2   Physical

As a general rule, all milks of lime and lime putties should be de-gritted before use. EN 459 [22.6] permits up to 2 % of particles greater than 200 μm.

The solids content of both milks of lime and lime putties is an important parameter, which affects their handling characteristics.

Many of the properties of slaked lime are related to the particle size distribution, which may be measured using laser beam granulometry. Other factors, such as particle shape and degree of agglomeration affect the handling characteristics and performance.

The settling rate and viscosity for a given % solids are used to monitor the "quality" of milks of lime. For a given solids content, slower settling rates and higher viscosities are indicative of increasing fineness.

The quality of building limes is specified in [22.6]. It requires minimum values for standard mortars produced from lime putties — namely penetration and air content — and lists water demand and water retention as parameters that customers may request. It also specifies a minimum volume yield of putty per kg of quicklime. Building limes must also be sound (i.e., pass the popping, pitting and expansion tests). A workability test, based on the spread of a standard mortar on a jolting table, was specified in the 1972 edition of BS 890 [22.11], but is no longer included.

It should be noted that the physical properties of all lime putties improve on maturing. This is particularly marked when using hydrated lime, which should be matured as specified (generally for 24 hours). Milling milks of lime (e.g. in a colloid mill) also improves physical properties and accelerates the maturing process.

22.8   Ultra-fine Milks of Lime

22.8.1   General

While slaking under optimum conditions produces milks of lime containing fine particles, there are benefits from producing milks with still finer particles, namely:

a) a very high chemical reactivity,
b) improved handling characteristics.

Two techniques are used commercially to produce ultra-fine milks of lime, namely slaking at high shear rates and milling conventional milks.

22.8.2   Slaking at High Shear Rates

As mentioned in section 22.2, slaking at high shear rates and at relatively high dilutions favours the production of well dispersed and finely divided particles of calcium hydroxide.

A patented process [22.7, 22.8] produces a milk of lime with a median particle size of less than 1 μm, and a high chemical reactivity. To a large extent it can be handled as a liquid rather than as a suspension. The process uses highly reactive quicklime with low $CaCO_3$ and S contents and with a maximum particle size of 5 mm. The water should be relatively soft (less than 0.2 g $CaCO_3$/l) and have a low sulfate content. A slaking temperature of 70 to 80 °C is used.

Figure 22.4 illustrates the process. Quicklime is fed into a mixing chamber into which milk of lime is recycled through tangential feed pipes. The resulting vortex ensures that the lime is rapidly dispersed in the milk. The milk is then drawn through the recycling pump and is forced through a static in-line mixer. Most of the hydration occurs under high shear rates in the pump and in-line mixer.

The milk then enters a cyclone vessel tangentially. Coarser particles remain close to the wall of the vessel and are drawn back into the mixing chamber by a tangential off-take. The finely dispersed particles are drawn from the centre of the cyclone vessel into a stirred vessel. In that vessel, the milk is subjected to further high shear rates and is diluted with cold water to the required concentration.

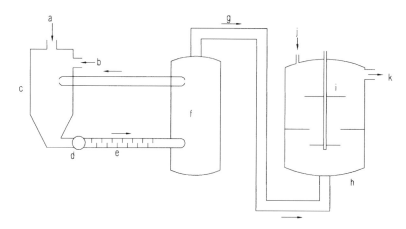

Figure 22.4. A high-shear slaker plant
(a) quicklime; (b) slaking water; (c) mixing vessel; (d) pump; (e) in-line mixer; (f) cyclone vessel; (g) milk of lime; (h) stirred vessel; (i) stirrer; (j) dilution water; (k) diluted milk of lime

22.8.3   Use of Bead Mills

The bead mill (also known as the agitated, or stirred, ball mill [22.9]) has become widely used in wet grinding of ultra-fine products such as pigments. It consists of a horizontal or vertical cylinder which is largely filled with glass beads. A rotating shaft along the axis of the cylinder is fitted with pins, or other protrusions, which cause the beads to tumble. The beads are retained in the cylinder by a screen at one or both ends.

Ultra-fine milks of lime may be produced by pumping a conventional milk of lime through the mill at a suitable rate. The milk may be recycled if a single pass does not give the required size reduction.

The properties of the milled milk depends on the grinding media and the throughput rate. A median particle size of 2 μm and below can be achieved. The chemical reactivity and handling characteristics of such milks tend to be intermediate between those of conventional milks of lime and of milks of lime produced as described in section 22.8.2.

22.8.4   Measurement of Chemical Reactivity

The chemical reactivities of ultra-fine milks of lime can be measured using a conductimetric method [22.1]. A quantity of milk sufficient to give a concentration equivalent to (say) half of the saturation concentration is added, with stirring, to a standard volume of de-ionised water. The rate of increase in the conductivity of the resulting mix is a measure of the rate of solution of the milk of lime.

Ultra-fine milks of lime can have solution rates, which are measured in seconds, and which are more than 10 times greater than those of standard milks of lime.

22.8.5   Handling Characteristics

For a given solids content, the viscosity of milks of lime increases with decreasing particle size. It is, therefore, necessary to maintain the concentration of solids below a critical level, which, for milks with a median particle size just below 1 μm, is about 20 % by weight of slaked lime.

Because the settling rate decreases with increasing fineness, ultra-fine milks can, to a large extent, be treated as conventional liquids rather than as suspensions. This enables them to be used in place of soluble alkalis such as caustic soda.

22.9   Carbide Lime

Most acetylene generation from calcium carbide uses one of the wet processes and produces a milk of lime containing 10 to 20 % by weight of calcium hydroxide. The milk is thickened in settling pits to produce either a thick milk of lime or a lime putty (generally called "carbide lime dough"). The product is marketed as an inexpensive slaked lime and is used in water treatment, acid neutralisation, soil stabilisation, the production of sand-lime bricks, and as agricultural lime [22.10] (section 20.10 describes the production of dry "carbide lime").

22.10   References

[22.1]   D.J. Wiersma, P. Hubert, J.N. Bolle, "Chemical reactivity and other relevant properties of milks of lime as applied in water treatment", Proc. International Lime Congress, Berlin, June 1994.

[22.2]   "Lime: handling, application and storage", National Lime Association, Bulletin 213, 5th ed., 1988.

[22.3]   G. Frank, G. Achenbach, "Effect of chloride ions on the slaking of lime", Zement Kalk Gips 9, 1987, 479–482 (English: **11**, 1987, 277–278).

[22.4]   M. Wingate, "Small-scale Lime-burning", Intermediate Technology Publications, 1985.

[22.5]   R.S. Boynton, "Chemistry and Technology of Lime and Limestone", John Wiley & Sons, 1980.

[22.6]   EN 459: "Building Lime", Part 1: "Definitions, specifications and conformity criteria"; Part 2: "Test methods".

[22.7]   Johann Schaefer Kalkwerke, German Patent DE 2714858C3 (H. von Zander).

[22.8]   H. Becker, "Highly reactive calcium hydroxide suspensions for industrial chemical processes — production and properties", Chem. Ing. Tech. **59**, No.3, 1987, 228–235.

[22.9]   S. Bernotat, K. Schönert, "Size Reduction", Ullmann's Encyclopedia of Industrial Chemistry, 5th ed. on CD-ROM, 1997.

[22.10] F.-W. Kampmann et al., "Calcium Carbide", Ullmann's Encyclopedia of Industrial Chemistry, 5th ed. on CD-ROM, 1997.

[22.11] BS 890: "Specification for Building Limes", 1972.

23   Handling and Storage of Milk of Lime and Lime Putty

23.1   Milk of Lime

23.1.1   General

Many customers requiring to add slaked lime to their process have found that milk of lime is the most convenient form in which to store, handle and apply it. Providing certain precautions are taken, milk of lime can be handled as a liquid. Moreover, it disperses well in aqueous systems and has a high chemical reactivity.

23.1.2   De-gritting, Dilution and Dosing

For most applications, milks of lime should be de-gritted (if necessary) before storage and use (see section 22.3).

Many slakers include a dilution stage. Where this is not the case, the milk should be pumped to a holding tank and diluted to the required concentration. The choice of the strength of the milk of lime depends on several factors. Lower concentrations result in reduced viscosity and may improve dispersion of the milk in the process stream, but involve handling larger volumes and may have undesirable effects on the process. Generally, a concentration in the range 5 to 20 % of hydrated solids is favoured. The concentration can readily be measured by the specific gravity (Table 23.1) [23.1].

Table 23.1. Specific gravity vs. concentration of milk of lime

Specific gravity at 15 °C g/ml	Concentration g $Ca(OH)_2$/l [a]
1.010	15.5
1.050	82.6
1.100	166
1.150	251
1.200	337
1.250	424

[a] For a typical high-calcium slaked lime.

Milks of lime are generally "dosed" into the process. This involves preparing a milk of a nominal strength and adding a sufficient quantity to produce the required effect. pH is frequently used as the control parameter. An alternative approach is to prepare a standard strength and add a metered volume to the process.

23.1.3   Deposition of Scale

All storage and handling systems should pay due regard to the fact that any carbonate hardness in the water used to make the milk, or used to dilute it, will be deposited as calcium carbonate. Unless appropriate action is taken, this will result in scaling on the walls of pipes, storage tanks and on the impellers and casings of pumps.

Two approaches to the scaling problem can be adopted. Either the system can be designed to cope with scale formation, or action can be taken to prevent or minimise scaling [23.1].

Ways of coping with scale formation include:

a) the use of troughs (open, or with removable lids) to feed the milk by gravity into the process: they can readily be de-scaled when required,
b) using flexible rubber or plastic pipes that can be flexed to break the scale, which can then be flushed out,
c) the use of "pigs", which have abrasives embedded in the surface and which are driven down the pipe by water pressure to remove the scale, and
d) periodically using acid (inhibited as appropriate) to dissolve the scale.

Scale formation can be avoided/minimised by:

a) using water with little or no carbonate hardness to dilute the milk of lime (where milk is settled to produce putty, the clear lime water can be recycled),
b) treating the slaking and dilution water with sodium hexametaphosphate, which sequesters calcium ions and prevents scale build-up. A dosage rate of 8 to 12 mg/l is generally sufficient [23.1],
c) recycling calcium carbonate sludge (where available) to provide seed crystals (its volume should amount to 10 to 20 % of that of the milk of lime), and
d) using a polyelectrolyte formulated to minimise lime scaling.

Build-up of solids can occur at the discharge from the pipe into the dosing vessel. This can be avoided by discharging through an air gap.

23.1.4   Conveying by Pipeline

Having resolved the question of scaling, pumping milk of lime by pipeline does not present any significant problems, providing:

a) the milk has been adequately de-gritted,
b) the velocity is kept at above 0.8 m/sec,

c) when milk is not required, the pipeline is either flushed out or the milk is recirculated,

d) the number and length of "dead legs" are kept to a minimum and provision is made for rodding-out.

The viscosity of milks of lime depends on the solids content, the particle size distribution of the slaked lime, the temperature of the milk and the degree of maturing. It may be determined, for the purposes of selecting a pump, by using an instrument such as the Brookfield viscometer. The friction loss when pumping milks of lime is generally allowed for by increasing the horsepower of the pump by about 10 % [23.1].

23.1.5 Storage

In milk of lime systems, it is important to prevent settling, as the resulting putty can be difficult to re-disperse. Storage tanks should, therefore, be agitated. The degree of agitation can be low and certainly should avoid forming a vortex, which will increase absorption of carbon dioxide from the atmosphere.

The discharge pipe from a storage vessel inevitably constitutes a "dead-leg" and provision should be made for back-flushing with water to remove any blockages.

23.2 Lime Putty

23.2.1 General

Lime putty is widely used as a component of mortar for masonry, rendering or plastering. It may be produced from either slaked or hydrated lime and made on site [23.2], or in a factory. In many countries, the latter option is becoming increasingly popular. In either case, the handling and storage requirements are similar.

23.2.2 Production from Slaked Lime

Traditionally quicklime was slaked on building sites in lime pits, dug into the soil. The resulting milk of lime was allowed to settle and mature for several days. The layer of lime water was removed and the putty was scooped out of the pit using a bucket.

In factory installations, de-gritted milk of lime is fed to a settler. A convenient design of settler is a rectangular tank with a V-shaped base fitted with a screw discharge mechanism. The milk is allowed to settle and consolidate and the layer of lime water is removed. The discharge gate at the base of the settler is then opened and the screw operated to discharge the putty into buckets or drums, which are then fitted with lids. The putty may be stored for a minimum period (e.g. for a month) to ensure that it is fully matured and completely sound.

23.2.3 Production from Hydrated Lime

Putty may also be produced by mixing hydrated lime with water. The ratio is 1 of hydrate to 0.7 to 1 of water (by weight), depending on the required consistency.

With normal hydrates (i.e. "Type N", produced at atmospheric pressure) the putty should be matured for at least 24 hours to develop plasticity. Alternatively, a colloid mill may be used to accelerate the maturing process. "Type S" hydrates (produced at elevated pressure) develop plasticity within 1 hour of mixing.

23.3 References

[23.1] "Lime — handling, application and storage", National Lime Association, Bulletin 213, 5th ed., 1988.
[23.2] M. Wingate, "Small-scale lime-burning", Intermediate Technology Publications, 1985.

24 Sampling and Testing of Slaked Lime

24.1 Introduction

Samples of slaked lime products may be required for process control, for the assessment of their suitability for a particular application, or for monitoring compliance with a specification. Testing may be done to assess the product's physical or chemical properties.

Because slaked lime products are finely divided, obtaining a representative sample is much less difficult than with granular and lump products. Nevertheless, it is important that the specified sampling and sample preparation procedures are followed.

In the following sections, reference is made to the relevant British and European Standards: in other geographical regions, the reader should also refer to the relevant national or international standards.

24.2 Precautions

Hydrated lime, lime putty and milk of lime absorb carbon dioxide from the atmosphere. Their exposure during sampling, sample preparation and storage should, therefore, be kept to a minimum.

Appropriate personal protective measures should be taken, as described in section 34.2.

24.3 Sampling

24.3.1 General

The principles of sampling slaked lime products are similar to those described in section 6.2 for limestone and section 18.3 for quicklime products. Definitions of the terms increment, sample, spot sample, composite sample and laboratory sample are given in the Glossary and Fig. 6.1 illustrates how they are related.

For process control and general purposes, simplified sampling procedures may be appropriate, but they should be assessed relative to the standard methods referred to below to ensure that they give the required accuracy and precision.

24.3.2   Powdered Hydrated Lime

Spot samples of hydrated lime should be sampled as described in EN 196-7 [24.1], as amended by clauses 3.1 and 3.2 of EN 459-2 [24.2]. These require that a sufficient number of increments be taken to give a spot sample of (20 ± 5) kg.

Where a composite sample of hydrated lime is required, at least 10 spot samples should be taken, each containing a sufficient number of increments to produce a composite sample of at least 15 kg [24.3]. Each spot sample should be kept sealed and only blended with the others when all of the spot samples have been taken.

24.3.3   Suspensions of Slaked Lime in Water

Spot samples of milks of lime and lime putties should be sampled as described in EN 459-2 [24.2]. This requires a sufficient number of increments to be taken to give a spot sample of (10 ± 5) l.

Where a composite sample is required, at least 10 spot samples should be taken, each containing a sufficient number of increments to produce a composite sample of at least 5 l [24.3].

Particular care should be taken when taking increments of milk of lime to ensure that the suspension is thoroughly dispersed and mixed before sampling.

24.4   Sample Preparation

Preparing slaked lime samples for testing and analysis in the laboratory may involve one or more of the following operations:

a) blending increments and/or spot samples,
b) reducing the quantity of the sample,
c) dividing the sample to produce two or more equivalent sub-samples.

Reference [24.3] gives specific guidance on how these operations should be done for hydrated lime, lime putty and milk of lime.

24.5   Packing and Marking of Samples

Reference [24.1] gives guidance on packaging powders. Lime putties and milks of lime should be put in a rigid but non-breakable container with a well-sealed lid (e.g. a plastic container with a screw lid or a tin with a well-fitting lid) which is enclosed in water-tight outer container such as a sealed plastics bag. [24.3] gives guidance on labelling of samples.

Regulations regarding the transport and labelling of slaked lime samples should be observed [24.4] (see also section 18.5): those regarding transport in aircraft are particularly stringent [24.5].

24.6 Physical Testing

Table 24.1 summarises the main physical tests for slaked lime products, specified in EN 459-2 [24.2], prEN 12485 [24.6] and BS 6463 [24.7].

Table 24.1. Physical tests for slaked lime products.

Test	Category of lime						Reference standard
	1	2	3	4	5	6	
Fineness	✓	–	✓	–	–	–	EN 459-2
	–	–	–	✓	–	–	prEN 12485
	–	–	–	–	–	✓	BS 6463, Pt 103
Compressive strength	–	–	✓	–	–	–	EN 459-2
Setting times	–	–	✓	–	–	–	EN 196-3
Penetration/water dem'd	✓	✓	✓	–	–	–	EN 459-2
Soundness	✓	✓	✓	–	–	–	EN 459-2
Water retention	✓	✓	✓	–	–	–	EN 459-2
Air content	✓	✓	✓	–	–	–	EN 459-2
Free water	✓	✓	✓	–	–	–	EN 459-2
	–	–	–	–	–	✓	BS 6463, Pt 103
Bulk density	✓	✓	✓	–	–	–	EN 459-2
Particle density	–	–	–	–	–	✓	BS 6463, Pt 103
Specific surface area	–	–	–	–	–	✓	BS 6463, Pt 103
Volume yield	–	–	–	–	–	✓[a]	EN 459-2

Category: 1: building lime – hydrated
2: building lime – putty
3: hydraulic building lime
4: high calcium quicklime – hydrated lime
5: high calcium quicklime – milk of lime
6: hydrated lime

24.7 Chemical Analysis

Tables 24.2 and 24.3 summarise the main chemical analyses for slaked lime products, specified in EN 459-2 [24.2], prEN 12485 [24.6] and BS 6463 [24.8].

A parameter which is not included in the above, but which is of relevance to applications where delayed hydration can cause expansion is the "free CaO" content of hydrated limes. A method of calculation is given in Annex 3, Appendix B.

Table 24.2. Chemical analysis of slaked lime products by traditional methods.

Test	Category of lime						Reference standard
	1	2	3	4	5	6	
$CaO + MgO$	✓	✓	–	–	–	–	EN 196-2
MgO	✓	✓	–	–	–	–	EN 196-2
CO_2	✓	✓	–	–	–	–	EN 196-2
	✓	✓	–	–	–	–	EN 459-2
	–	–	–	–	–	✓	EN 12485/6463
Loss on ignition	–	–	–	–	–	✓	EN 196-2/6463
Free water	✓	✓	✓	–	–	✓	EN 459-2/6463
	–	–	–	–	–	✓	BS 6463, Pt 102
Sulfate	✓	✓	✓	–	–	–	EN 196-2
Total sulfur	–	–	–	–	–	✓	BS 6463, Pt 102
Neutralising value	–	–	–	–	–	✓	BS 6463, Pt 102
Available lime	–	–	✓	–	–	✓	EN 459-2/6463
Water sol $CaO/Ca(OH)_2$	–	–	–	✓	✓	✓	EN 12485/6463
Water insol constituents	–	–	–	✓	✓	✓	EN 12485/6463
Insol in acetic acid	–	–	–	–	–	✓	BS 6463, Pt 102
Insol in HCl	–	–	–	–	–	✓	BS 6463, Pt 102
HCl extractable fluoride	–	–	–	–	–	✓	BS 6463, Pt 102
Si, Fe, Al (titration)	–	–	–	–	–	✓	EN 196-2/6463
Ca, Mg (titration)	–	–	–	–	–	✓	EN 12485/6464

Category: 1: building lime — hydrated
2: building lime — putty
3: hydraulic building lime
4: high calcium quicklime — hydrated lime
5: high calcium quicklime — milk of lime
6: hydrated lime

Table 24.3. Chemical analysis of slaked lime products by atomic spectrometry.

Test		Category of lime			Reference standard
		1	2	3	
Al, Fe, Mg, Mn, Si	ICP-OES	✓	✓	–	EN 12485
	Flame AAS	✓	✓	✓	EN 12485 / 6463
Cd, Cr, Pb, Ni	ICP-OES	✓	✓	–	EN 12485
	Graphite tube AAS	✓	✓	✓	EN 12485 / 6463
Cu, Ag, Sn, Zn	Graphite tube AAS	–	–	✓	BS 6463: Pt 102
Sb, As, Se	Hydride & flame AAS	✓	✓	✓	EN 12485 / 6463
	Hydride & AFS	✓	✓	–	EN 12485
Hg	Cold vapour & AAS	✓	✓	✓	EN 12485 / 6463
	Cold vapour & AFS	✓	✓	–	EN 12485

Category: 1: high calcium quicklime — hydrated lime
2: high calcium quicklime — milk of lime
3: slaked lime

24.8 References

[24.1]   EN 196-7: "Methods of taking and preparing samples of cement", 1992.
[24.2]   EN 459-2: "Building lime — Test methods", 1995.
[24.3]   BS 6463, "Quicklime, hydrated lime and natural calcium carbonate", Part 101: "Methods for preparing samples for testing", 1996.
[24.4]   "The transportation of dangerous and hazardous goods regulation" (49 CFR 172.101:59 FR 67309, Dec. 29, 1994; 60 FR 26796, May 18, 1995).
[24.5]   The transportation of dangerous and hazardous goods regulation (49 CFR 172.101: 59 FR 67309, Dec. 29, 1994; 60 FR 26796, May 18, 1995).
[24.6]   prEN 12485: "Chemicals used for treatment of water intended for human consumption — calcium carbonate, high-calcium lime and half-burnt dolomite — Test methods", in preparation.
[24.7]   BS 6463, "Quicklime, hydrated lime and natural calcium carbonate", Part 103: "Methods for physical testing", in preparation.
[24.8]   BS 6463, "Quicklime, hydrated lime and natural calcium carbonate", Part 102: "Methods for chemical analysis", in preparation.

Part 5   Uses and Specifications of Lime Products

25 Overview and Economic Aspects of the Lime Market

25.1 General

The size and relative importance of the various market segments for quicklime and slaked lime (including products based on calcined dolomite) vary widely from one country to another. They depend on many factors, including the degree of industrialisation in general and on the specific industries which have become established, the quality/availability of limestone, and traditional building methods.

The economic factors which shape the market for lime products are common to most countries. However, specific factors such as labour costs, availability of capital, the size of the market in both geographical and tonnage terms, market penetration by competitive products and the importance of imports/exports can markedly affect the structure of the market.

This chapter aims to describe the main features to be considered when assessing a particular market. It uses information from individual countries to illustrate some of these features. It also provides a context for the following chapters on the uses of lime products.

25.2 Market Overview

Published estimates of the global production of quicklime [25.1, 25.2] suggest that the total is approximately 120 million tonnes per year (tpa), of which just under half is produced by the former USSR, China, and the USA (Table 25.1). The published figures do not include the multitude of small producers in the developing countries, nor many of the captive producers, whose lime products are not sold in the open market. If their contribution were included, the total would probably approach 300 million tpa. In most industrialised countries, the annual consumption per head of population is in the range 20 to 80 kg per year.

Lime products are used in a wide range of applications. The more significant ones are described in chapters 26 to 32 and are listed in Table 25.2.

The distribution of lime use between the main market segments varies widely from one country to another. Details for nine countries are given in Table 25.3 [25.2–25.4]. In seven of those countries, quicklime for steelmaking is the largest market segment.

The quantities of lime used for building and construction varies with both the building techniques employed and with the building materials that are readily

Table 25.1. Estimated lime production in various countries (1994)

Country	Total production[a]
Australia	1 250
Austria	700
Belgium	1 750
Brazil	5 700
China [b]	19 500
Czech Republic	1 200
Denmark	130
Finland	300
France	3 100
Germany	8 500
Greece	500
Ireland	100
Italy	3 500
Japan	7 700
Mexico	6 500
New Zealand	110
Norway	80
Poland	2 500
Portugal	200
Romania	1 600
Sweden	470
Slovak Republic	730
Spain	1 000
South Africa	1 900
Sweden	500
Turkey	4 200
USA	17 400
UK	2 500
Former USSR	16 000
Others (by difference)	8 380
Estimated world total	118 000

[a] The above estimates are based on published values for open-market
    production [25.1, 25.3].

[b] Estimates of over 100×10^6 t for China have been reported [25.5],
    but have not been substantiated.

available. In Italy and Turkey, large quantities of lime (much of it with hydraulic
properties) are used, whereas in France, where considerable amounts of gypsum
blocks are used, relatively little lime is employed. Germany uses large quantities
of lime products in sand-lime bricks and aircrete, as well as in mortar, rendering
and plaster.

Table 25.2. Uses of lime products

Section	Application	Section	Application
Construction and building		*Water & effluent treatment*	
26.3	Lime treatment of soils	28.1	Drinking water
26.4	Hydraulic road binders	28.2	Boiler feed water
26.5	Hot-mix asphalt	28.3	Waste water
26.6	Masonry mortars	28.4	Sewage treatment
26.7	External rendering		
26.8	Internal plastering	*Gaseous effluents*	
26.10	Sandlime bricks	29.2	Wet scrubbing
26.11	Aircrete	29.3	Semi-dry scrubbing
26.12	Calcium silicate – board	29.4	High-temperature dry injection
	– concrete	29.5	Low-temperature dry injection
		29.6	Reduction of dioxins/furans
Iron and steelmaking		29.7	Reduction of heavy metals
27.2	Sinter production		
27.3	Iron treatment		
27.4	Basic oxygen steelmaking	30	*Agriculture, food etc:* see Table 30.1
27.5	Electric arc steelmaking	31	*Chemicals production:* see Table 31.1
27.6	Secondary steelmaking	32	*"Other uses":* see Table 32.1

Table 25.3. Use of lime products in the main market segments (1994)

Country	Iron and steel	Non-ferrous metals	Chemical industry	Other industry	Building materials[a]	Building trade[b]	Environ-mental protec-tion[c]	Agricul-ture and food
	%	%	%	%	%	%	%	%
Belgium	70	–	7	1	6	7	8	1
Germany	36	–	7	2	27	12	14	2
France	58	3	4	–	–	–	20	15
Italy	27	–	2	–	–	65	5	1
Japan	54	1	23	3	2	5	9	3
S. Africa	42	30	16	4	–	7	1	–
Turkey	25	1	12	7	1	53	1	–
USA	31	7	5	24	–	8	25	–
UK	55	–	5	5	8	7	19	1

[a] Mainly aircrete, sandlime bricks, other calcium silicate products and refractories.
[b] Mainly mortar, render, plaster, and drying/improvement/stabilisation of soils.
[c] Mainly potable water, sewage, liquid effluent and gaseous effluent treatment.

A large proportion of the lime used in South Africa is for the production of non-ferrous metals.

The use for various aspects of environmental protection (treatment of water, sewage, liquid and gaseous effluents) is a significant proportion of the total in the USA, France, the UK and Germany and will no doubt increase elsewhere.

- *Iron and steelmaking.* As mentioned above, in many industrialised countries, the major use of quicklime, including calcined dolomite, is in steelmaking. While most of the lime is in granular form, modern developments require ground quicklime. Calcined dolomite is also used for the refractory linings of the vessels.

- *Chemical industry.* Another major use of slaked lime is in the production of sodium carbonate from salt using the ammonia-soda process. Both the lime and carbon dioxide from the kilns are used in the process. Lime products are also used in the production of a large number of smaller tonnage chemicals, including dyes, medicines, a wide range of organic chemicals, calcium carbide, bleaches, caustic soda, soda-lime and many calcium salts.

- *Other industrial applications* include: wire drawing, stabilisation of industrial wastes, production of non-ferrous metals, oil well muds, oil/petrochemical additives and refractories.

- *Construction and building* (building materials and trade). The use of quicklime and hydrated lime in soil treatment for the construction of road foundations and large areas, such as airfields, car parks, etc, is well established in the USA and France and is growing elsewhere. Lime products have been used for many thousands of years to produce mortars which have stood the test of time. Mortar for masonry, render and plaster remains an important outlet for lime. Other outlets include the production of various calcium silicate products, aircrete, and limewash, and the use as an additive for hot-mix asphalt.

- *Environmental protection.* Lime products, including half-burned dolomite, are used to adjust the pH, to clarify and to soften water. Certain sewage treatment processes use slaked lime to assist settling and compaction. Lime can also be used to disinfect sewage prior to spreading on the land. Liquid effluents are treated with slaked lime to neutralise acidity, precipitate heavy metals and to assist settling of suspended solids. Slaked lime can be used to remove acid gases from effluents, including sulfur dioxide. It is also used as a carrier for activated carbon/lignite coke in the removal of dioxins, furans and heavy metals from flue gases.

- *Agriculture and food.* Some lime products are used to adjust the pH of agricultural land. In many countries, a major use of lime is in the refining of sugar beet, where the co-produced carbon dioxide from on-site kilns is used in the process. Slaked lime is also used in the refining of cane sugar, and the preparation of a range of products, including leather, glue and gelatin, citric acid, calcium phosphates (for toothpaste and foodstuffs), insecticides and fungicides.

25.3 Economic Aspects

Quicklime is generally the least expensive and most cost effective alkali. Slaked lime is a little less cost effective, but this is offset by its greater convenience in use and lower capital costs.

In some markets (eg the neutralisation of gaseous and aqueous acids) lime products compete with the still more expensive soda ash and caustic soda. In flue gas desulfurisation and soil stabilisation, lime products compete with limestone and other products. In building and construction the main competitive products are cement, fillers and air-entraining agents.

There are, therefore, many competitive pressures on the lime market. In addition, in most countries there is fierce competition between lime producers. This often stems from over-capacity.

The *production cost* of quicklime is generally dominated by the charges for limestone and energy per tonne of product. As approximately 1.8 t of limestone are required per tonne of quicklime produced, it is generally economic to minimise transport costs by building the kilns adjacent to, or within the limestone quarry, rather than near the principal customers.

For many producers, the cost of energy is dominated by three factors, namely:

a) the energy usage of the kilns — the trend is to favour efficient kilns with low specific energy usages,
b) the costs of fuels — because these are becoming more variable, increasing effort is being directed to enable kilns to be switched between gas, oil and solid fuel (or a combination of these) to enable the cheapest acceptable option to be used,
c) the ability to select a kiln-fuel combination capable of producing quicklime for modern steelmaking processes with no more than 0.03 % of sulfur.

As with limestone, *minimising wastage* is an important factor in the economics of quicklime production. Integrated lime works often feed surplus grades to a hydrating plant or to a ground lime plant. Products which cannot be used are often blended into selected products, or sold as agricultural lime.

The production cost is also affected by charges for capital (interest and depreciation), labour and refractories. In industrialised countries, large kilns have been installed to reduce those costs per tonne of quicklime. Increasingly, shaft and rotary kilns are being installed with capacities of up to 600 and 1000 t/d respectively. In less industrialised countries, however, where labour costs and the availability of capital are lower, continuously operated shaft kilns with capacities of 1 to 100 t/d are common. Batch kilns are also widely used, where still lower production rates and/or increased flexibility is required.

The costs of *environmental control measures* are becoming increasingly more significant. Tightening environmental standards require lime kilns to be fitted with more efficient and expensive equipment to limit the emission of dust. Limitations on emissions of sulfur dioxide and oxides of nitrogen are also being imposed. Collected substances (generally dusts) may have to be disposed to land in a controlled manner. Such requirements can increase significantly capital and operating costs.

The ex-works price of lime products in Europe is generally in the range of £ 40 to £ 60 per tonne (1997). In the U.S.A. the average price was $ 59/short ton (1994) [25.1]. Specialised products, which are tailored to the needs of individual markets or customers, generally command higher prices. Ground quicklime and hydrated lime prices are typically some 25 % higher than screened quicklime, reflecting the extra operating and capital costs of the additional processing. (N.B. prices vary significantly from one country to another. While the absolute price quoted above will no doubt soon become out-of-date, their relative values should remain generally valid.)

Delivery costs often amount to 50 % of the ex-works price, depending on the distance and mode of transport. Screened products are mainly transported in bulk using tipper trucks, which are sheeted to exclude rain and to prevent dust emission. Ground quicklime and hydrated lime are transported in air pressure discharge vehicles. Such vehicles can also be used for screened products with a top size below 20 mm. Smaller quantities are packed in intermediate bulk containers of 0.5 or 1 t capacity, and in paper sacks.

The economic haulage distances for lime products are considerably greater than those for limestone. Road haulage, is often economic up to 250 to 400 km. Where appropriate, the use of rail and water transport for large customers enables that range to be extended considerably.

As a result, there is a significant amount of competition across many national boundaries. Within the European Union, for example, during 1994, some 2.2 million tonnes of lime were exported across national boundaries, of which about 200,000 t was outside the EU. Some 350,000 t were imported from other countries [25.4, 25.5].

While the delivered price is important, the physical and chemical *quality and consistency* of lime products and the *technical service* provided are becoming of increasing importance to many customers because of increasing automation of their processes.

Where both the quicklime and the co-produced carbon dioxide can be used as raw materials (e.g., in the ammonia-soda process for sodium carbonate, the sugar beet refining process and the precipitated calcium carbonate process) it is generally economically attractive to install lime kilns on site as an integral part of the process and to accept the relatively high delivery costs of the limestone.

Perhaps over 80 % of the quicklime produced, other than that used in steelmaking, is used either as a *milk of lime* or as a *lime putty*. For such applications, there are three main options, namely to:

a) purchase granular quicklime and slake it on site,
b) purchase ground quicklime and slake it on site, and
c) purchase hydrated lime and mix it with water on site.

While the cost per tonne of available lime increases progressively from option (a) to (c), the capital costs decrease and the ease of handling/automation increases. Large users generally prefer options (a) or (b), while small users tend to prefer option (c). In addition, some countries have a small, but growing market for milk of lime, generally prepared at the lime works.

25.4   References

[25.1]   "Lime", US Bureau of Mines' Mineral Industry Survey, Annual Review, 1994.
[25.2]   International Lime Association — Statistics, 1994.
[25.3]   European Minerals Yearbook, 1995.
[25.4]   Statistical Year Book, 1995, Quarry Products Association, (formerly BACMI), London.
[25.5]   G. Chenxiang, "The structural characteristics and rationality of China new type ordinary lime shaft kiln", Proc. International Lime Congress, Berlin, 1994.

26 Construction and Building

26.1 Introduction

Construction and building are treated as one industry in this chapter, to avoid the confusion that frequently arises when attempting to differentiate between them. The industry is a major user of lime products: in 1994, for example, it used 36 % of the 19 million tonnes of lime sold in the European Union, compared with 38 % used in iron and steelmaking.

26.2 Historical

Some of the early uses of lime in construction and building have already been mentioned (section 1.3.2). A major development made by the Romans was a blend of slaked lime and volcanic ash, which would harden under water, called *Roman Cement*. The volcanic ash contained reactive silica and alumina which combined with the lime in the presence of controlled amounts of water to produce a solid mass bound by calcium silicates and aluminates. Such reactive materials are called pozzolans after Pozzuoli, a city near Naples. Roman Cement was mixed with aggregate to make a lime concrete, which was used for a wide range of products and constructions.

Over the centuries, other pozzolanic substances were blended with slaked lime to produce what are now called *synthetic (or artificial) hydraulic limes*. Both naturally occurring pozzolans (such as trass, found in Germany), and synthetic pozzolans (such as ground blast furnace slag) have been, and still are used. It was also found that some impure limestones, containing silica and alumina, produced slaked limes with a range of hydraulic properties. Such *natural hydraulic limes* were widely used in construction and building for mortar and concrete.

In 1796, James Parker, of London, patented a process for calcining argillaceous limestones (or cementstone) to produce a type of cement, which was also called Roman Cement. This development was followed in 1824 by Joseph Aspdin, of Leeds, who lightly calcined a finely divided mixture of limestone and clay and ground the product to make hydraulic cement. The product was called *Portland cement*.

By the end of the 19th century, the process for the production of Portland cement had become well developed and the product was widely used in construction and building. Cement had many advantages over lime-based products, being produced from a wide range of raw materials, giving much greater strength, and being more consistent than natural hydraulic limes. As a result, Portland cement replaced lime in many aspects of construction and building.

However, in the following areas, lime still plays an important role:

a) in the drying, improvement and stabilisation of soils,
b) as a component of mortars, exterior rendering, and interior plasters,
c) as an *anti-stripping agent* in the production of asphalt and tarmac for road construction, and
d) as a *binder* in the production of a range of autoclaved calcium silicate products (including bricks, aircrete, fire-resistant board and concrete).

These uses are described in this chapter.

26.3   Lime Treatment of Soils

26.3.1   Introduction

The ability of lime to dry, modify and stabilise clay soils has been known for many years. However, its systematic use only started in the late 1940s in the USA, when the techniques of soil mechanics testing were applied to lime-soil mixtures. They demonstrated in quantitative terms that lime treated layers could, with confidence, be designed into the construction of roads, runways, car parks etc.

The benefits of lime treatment of soils, in terms of cost, performance, durability and environmental impact, have been widely promoted in the USA [26.1], where it has been used in the construction of thousands of miles of roads and in major projects such as the Dallas Fort Worth airport [26.2]. A few other countries (notably France, Germany, Sweden, South Africa, Australia and New Zealand) have recognised the benefits of lime treatment and are using it extensively as an alternative to traditional methods of construction. Typical projects which have benefited from the technique are listed in Table 26.1.

In most countries, however, progress to adopt lime treatment as a standard construction technique has been slow [26.8]. The reasons for this undoubtedly include:

Table 26.1. Typical projects using lime treatment of soils

Motorways, trunk roads & secondary roads
Runways and aprons at airports
Car parks and factory yards
Slope stabilisation
Temporary access and haul roads
Farm and forestry access tracks
Railway tracks
Reclamation of redundant docklands
Reclamation of contaminated land
Structural fill

a) conservatism of the construction industry and the designers, coupled with an absence of recognised specification standards,
b) the apparent complexity of applying the techniques to different situations, with a wide range of soil compositions under various climatic conditions and
c) the occasional failure of lime treatment projects arising from an inadequate understanding of the precautions to be observed.

However, proven ways of applying lime treatment to specific projects are now well documented (e.g., [26.1, 26.3–26.9]). In addition, work has started on the production of European Standards for lime treated materials [26.17].

26.3.2   The Effect of Lime Treatment on Soils

Lime treatment involves the use of quicklime, or slaked lime, either as powdered hydrated lime, or as a milk of lime (N.B., it does not use the product marketed as "agricultural lime", which is generally calcium carbonate and which does not have the same effect on soils). The effect of lime on soils can be divided into three stages drying, modification, and stabilisation.

Drying occurs when quicklime is used. The quicklime hydrates, absorbing water and generating heat, which in turn causes some of the water to evaporate. The total moisture loss can be up to double that required to hydrate the quicklime. The drying process occurs almost immediately with reactive quicklimes.

Modification occurs with both quick- and slaked lime which rapidly enter into physio-chemical reactions with any clay minerals present. The resulting changes, which include ion exchange, and which can be substantially complete within 6 hours, dramatically reduce the plasticity of the soil, increase its workability and improve its compaction characteristics.

The ability of lime to convert a highly plastic, sticky soil into a dry workable material, can be illustrated by reference to Fig. 26.1 [26.9]. The plastic limit [26.10] of the original clay is 25 % water. Thus at a moisture content of 35 %, the clay is 10 % above the limit and is very sticky. The addition of 2 % of lime raises the plastic limit of the clay to 40 %. By adding that amount of lime to a clay containing 35 % moisture, the clay would become 5 units of % below the plastic limit. It would, therefore, become drier and crumble easily. This effect is in addition to the drying action of quicklime mentioned above.

Stabilisation is a much slower process, which occurs progressively over several months, and involves the reaction of lime with the siliceous and aluminous components of the soil. The lime raises the pH to above 12, which results in the formation of calcium silicates and aluminates. These are believed to form initially as a gel, which coats the soil particles, and which subsequently crystallises as calcium silicate/aluminate hydrates. Those hydrates are cementitious products, similar in composition to those found in cement paste. The rate of crystallisation is temperature dependant and may take many months to reach completion. The resulting gain in strength (measured by the California Bearing Ratio Test [26.11]) is progressive, as illustrated in Fig. 26.2.

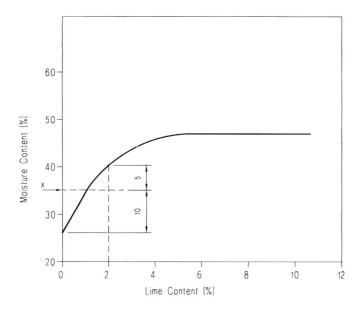

Figure 26.1. Effect of lime addition on the plastic limit of a clay (Crown copyright is reproduced with the permission of the Controller of Her Majesty's Stationery Office)

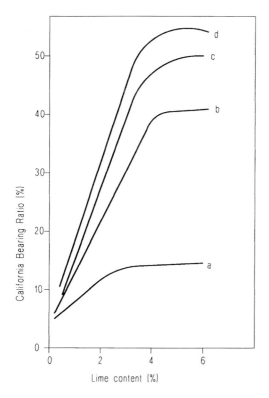

Figure 26.2. Effect of lime addition on the strength of a clay (Crown copyright is reproduced with the permission of the Controller of Her Majesty's Stationery Office)
a) immediate; b) 7 days; c) 14 days; d) 28 days

After the drying and modification stages, water is added, if required, to obtain the required moisture content for consolidation (i.e., just wet of the optimum moisture content [26.11]. The soil is then compacted to reduce the level of air voids to not more than 5 %. This ensures that the stabilisation reaction proceeds in the compacted state and results in a homogeneous, impermeable and stable layer. The stabilised layer has a low and acceptable shrink-swell potential, and improved compressive, tensile and flexural strengths. It also reduces the susceptibility of the stabilised layer to frost damage.

The timing of the compaction process is still a matter of debate. It should be delayed sufficiently to ensure that all of the quicklime is fully hydrated, to avoid expansion. If it is delayed for too long, however, the stabilisation process will have progressed to a significant degree and the final strength of the lime treated layer will be reduced. The need for a "maturing period" is discussed in section 26.3.5.

It cannot be over-emphasised that, after compaction and stabilisation, lime treated soil behaves as a completely different material from the original, both on the micro scale, in terms of soil mechanics [26.12], and on the macro scale (indeed blocks made from lime-stabilised soil are used for building in some developing countries).

26.3.3 Suitable Soil Types

Lime treatment can be effective on a wide range of soils, ranging from clayey gravels to clays. The effectiveness of lime depends both on the level of clay and on its ability to react. In all cases where lime treatment is proposed, adequate testing should be done to enable the most appropriate treatment to be determined.

Soils with a Plasticity Index [26.10] above 20 are generally suitable for stabilisation with lime, while those with an Index of less than 10 generally do not demonstrate a sufficient increase in strength [26.12]. More detailed guidance is given in [26.1] and [26.3]. Where silty soils are to be stabilised, lime may be used in conjunction with a pozzolan, such as pulverised fuel ash, or ground granulated blast furnace slag.

The amount of lime required to achieve full stabilisation can be determined using the initial lime consumption test [26.11].

Organic matter can adversely affect the development of strength. A maximum level of 2 % is generally specified. BS 1377 [26.10] specifies a test method for determining the content of organic matter.

The presence of sulfates in the soil and/or ground water, or of sulfides, which can be oxidised to sulfates, has been one of the main causes of problems with lime treatment. This is because sulfates react with the cementitious products of stabilisation to produce ettringite ($3CaO.Al_2O_3.3CaSO_4.32H_2O$), which can cause swelling and disruption of the stabilised layer.

Because of localised variations in the composition and properties of soil, a detailed survey of the soil and quality of the ground water over the proposed area for lime treatment should always be done. Both the soil and the ground water should be sampled and tested thoroughly and systematically.

The UK's Department of Transport's specification states that the upper limiting value for the total sulfate level in the soil may be equivalent to 1 % by weight (expressed as SO_3) [26.6]. It proposes an upper limit for sulfate in groundwater adjacent to lime and/or cement treated areas of 1.9 g/l. A test method for sulfate in soil is given in BS 1047 [26.14].

26.3.4 Applications of Lime Treatment

The largest applications of lime treatment are roads, airports, car and lorry parks, where lime may be used to stabilise the "subgrade" to provide a "capping layer", or to produce a stabilised "subbase". Figure 26.3 illustrates these terms.

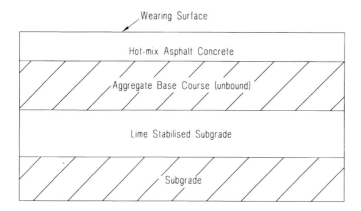

Figure 26.3. Structure of a flexible pavement with a lime-stabilised subbase

- *Subgrades and capping.* When the California Bearing Ratio (CBR) [26.11] of the subgrade (i.e. the soil over which a road is to be built) is below 15 %, the use of a suitable capping material is normally specified [26.6]. The capping is designed to provide a working platform on which placing of subbase can proceed with minimum interruption from wet weather, and to minimise the adverse effect of a weak subgrade on the compaction of the subbase.
  As an alternative to removing a layer of subgrade and replacing it with suitable material to form the capping layer, the subgrade may be stabilised using lime. Typically, such a capping may consist of two or more layers and may be up to 600 mm thick.

- *Subbase.* While the UK's Department of Transport does not recognise the use of lime stabilised subbases, they have been used successfully in private developments, where roads and parking areas are subject to lower traffic densities and where the required thickness of construction is not as great as for major roads.

The subbase is constructed between the top of the subgrade and the bottom of the roadbase. It should have a laboratory CBR of not less than 30 % [26.5]. This strength is generally produced by the use of both lime and cement. The lime is applied first to dry and modify the subgrade, which can then be scarified into a fine tilth. The cement is then dispersed, the water content is adjusted to just wet of optimum and the layer is compacted. After compaction, the cement hydrates and sets, and the lime stabilisation reaction proceeds as normal. The required levels of addition of lime and cement should be determined by testing: typical levels are 2 % of lime and 4 % of cement.

Because sub-bases are closer to the surface than subgrades and capping layers, they are more prone to damage by frost and their resistance to freeze-thaw cycles should be assessed [26.11] in the context of the conditions to which they will be subjected.

- *Bulk fill.* Lime improvement (partial lime treatment) is used to dry and/or reduce the plasticity of materials, such as silts or clays. It makes such materials more suitable for placing and compacting, whether in landfill or embankments. The Moisture Condition Value (MCV) test [26.11] and [26.13] is used to assess the suitability of a material for handling and compaction.

  The treatment can be carried out either at the excavation site or during placement of the fill. Treatment at the excavation site has the advantage of improving the handling characteristics throughout the process, but is not always feasible. Lime additions of 1 to 3 % are generally sufficient to achieve the required improvement, as measured by the MCV test.

- *Water-logged construction sites.* The addition of quicklime can often dry out wet, boggy, impassable sites, allowing construction equipment to move. The resulting lime improved layer usually prevents deterioration from rainfall, as well as providing a stable working platform. The soil and ground water should be tested to ensure that sulfate and total sulfur concentrations are within acceptable levels (see section 26.3.3).

- *Maintenance of distressed roads.* Often rural roads have a thin surface dressing and an inadequate base course. In such circumstances, when clay works into the base course, the structure becomes softened. It is often possible to provide a base for a new surface dressing by pulverising the whole of the existing road and treating it with lime (to dry and modify the clay) and cement (to raise the CBR to the required level). This reduces disruption during repair, minimises the importation of aggregates and the tipping of waste. It also conserves raw materials.

- *Temporary works.* Lime stabilisation is a viable option for the construction of temporary haul roads, site compounds and similar areas. It avoids importation of aggregate and, at the end of the project, the stabilised soil can be broken up and left in place.

- *Repair of slipped embankments.* Re-instatement of the embankment can be done by removing the slipped material to an adjacent area of flat ground, treating it with lime (typically at about 1.5 to 2.5 %) and replacing it on the embankment in layers, compacting each layer using appropriate equipment. Alternatively, embankments may be stabilised in situ by "lime piling" [26.3]. Shafts (which may be 200 mm in diameter) are drilled into the embankment, filled with quicklime and plugged. As the quicklime hydrates, it removes water from the immediate vicinity of the shaft and expands. The combination of de-watering and expansion causes consolidation around the shaft and strengthens the embankment. The process has been widely used in Japan [26.15].

- *Land reclamation.* Lime stabilisation is being used in the reclamation of a wide range of derelict sites. Not only does the lime stabilise pozzolanic material on the site, but it helps to neutralise acidic residues, immobilise heavy metals and saponify oily wastes. It can also be used to dry out a site to enable piling equipment to be brought in, or to assist with the removal of sludges.

26.3.5   Procedures

A wide range of procedures are used in lime treatment. The following is a brief summary: for more details, including equipment used, see [26.1] and [26.3].

Two principle methods are used — the static-plant mixing process, and the more widely used mix-in-place process. Mixing-in-place can be done either on site, or on "borrow" material before or after excavation.

- *Preparation.* The material to be treated should be shaped to the required level and profile.

- *Lime spreading.* The lime may be ground quicklime, powdered hydrated lime or milk of lime. It is best spread mechanically to ensure an even, controlled distribution. Hydrated lime is sometimes used when the drying action of quick-lime is not required.
  In some situations, the most convenient way of applying the lime is as a milk. For small areas, bagged quicklime, or hydrated lime can be used, in which case the bags should initially be positioned to give the required spreading rate, and the contents of each spread to give as uniform a distribution as possible. Suitable precautions should be taken to protect the operatives and to limit air-borne dust particularly with hydrated lime under windy conditions.

- *Mixing and pulverising.* It is important that the lime and soil are intimately mixed. Mixing should be done as soon as possible after spreading and certainly on the same day, although with heavy clays it is may be beneficial to carry out secondary mixing after a "maturing period" to allow the lime to partially modify the clay (see below). A UK specification [26.5] requires the pulverised soil to be at least 95 % less than 28 mm with not less than 30 % passing 5 mm.

- *Water addition.* It may be necessary to add water to the mixed soil to ensure that there is sufficient free water for the quicklime to hydrate fully and to react with the clay component.

- *Trimming and compaction.* Following each mixing pass, the material should be trimmed to shape to ensure that the processing depth is correct throughout the operation. The lime-soil mixture should be compacted (see section 26.3.2) to bring the lime into intimate contact with the clay and also to seal the surface. The latter minimises loss of water through evaporation, reduces possible damage caused by heavy rain, and limits atmospheric carbonation of the lime.

- *Maturing period.* There is a divergence of opinion on the need to mature (or "mellow") the lime treated layer before compaction. In the U.S.A., it is standard practice to allow the lime treated layer to mature overnight, or even for up to 72 hours. It is reported that this helps to obtain the most efficient use of the lime and can also help to obtain the required degree of pulverisation when treating very heavy clays. Conversely, in France, where more finely divided quicklime is usually specified, and where the clays may generally be less plastic, it is standard practice to compact the lime treated layer on the same day that the lime is applied.
  The best practice for a particular application should be established by testing. It may depend on the nature of the clay, its moisture content, the fineness and reactivity of the quicklime and on the equipment used for compaction.

- *Re-mixing and moisture adjustment.* After maturing, the treated layer is re-mixed and, if necessary, water is added to raise the % water to just above the optimum moisture content. The importance of ensuring the correct moisture content cannot be over-emphasised, as it helps to ensure that adequate compaction is achieved.

- *Final compaction and surface finishing.* Although lime treated soils can be compacted several days after mixing, the soil should not be allowed to dry out and the surface should be sealed by a light rolling. It should be recognised, however, that delayed compaction can reduce the strength of the layer and should, if possible, be avoided where full stabilisation is required.

- *Curing and protection.* For capping layers, the objective is to stabilise the material. Development of the maximum strength and durability requires that temperatures should be in excess of 7°C and that 7 days' curing are allowed after final compaction before permitting traffic on to the surface.
  Where the soil is to be modified, lower temperatures can be tolerated, although freezing of the layer should be avoided if possible. Any damage by frost or traffic can, however, be repaired by re-mixing, adjustment of the moisture content (if necessary) and re-compaction.

- *Static plant mixing.* Plant mixing of lime with the material to be stabilised may be considered if the material is being imported to site. It is most effective with more granular soils, such as clayey gravels, which are often used as a sub-base. The mixing equipment usually consists of a pug-mill type of mixer. Water is generally added to obtain the required moisture content. The mixed material should be transported in sheeted waggons to control evaporation.

26.3.6 Benefits of Lime Treatment

Summarising, lime treatment can turn unsuitable and marginal soils into useful construction materials, which can be placed and compacted to form an integral part of the project. From the viewpoint of designers producing specifications and contractors preparing quotations, the following aspects may be particularly significant.

a) Lime treatment avoids the needs to remove and tip large volumes of unsuitable material and to import similar amounts of replacement fill.
b) Lime treated soil can be used for capping, avoiding the need to quarry and import aggregates.
c) The resulting reduction in traffic helps to reduce the impact of construction work on the community.
d) Lime treatment can reduce the cost of material plus transportation.
e) It can reduce operating costs by enabling traffic to move on the subgrade shortly after final compaction and avoiding hold-ups in wet weather.
f) It can give increasing long-term strengths, thereby improving the durability of the completed project.

26.3.7 Lime Specifications

A wide variety of lime products have been used in lime treatment. In general, however, calcium limes are recommended [26.4, 26.5], which meet the requirements for building lime [26.18]. Dolomitic and hydraulic lime products are less suitable, as indicated below.

A European Standard is in preparation [26.17].

Quicklimes used for lime treatment should (provisionally) meet the requirements of Types CL 70, CL 80, or CL 90 as specified in ENV 459-1 [26.18] and summarised in Table 26.2. It requires that the MgO content in building limes used for soil stabilisation should not exceed 5 % in CL 70, and 10 % in CL 80 and CL 90. In the long-term, however, [26.17] will specify the characteristics for limes used in road construction. A "high" reactivity, as defined in section 13.2, is an advantage when quicklime is used for drying a site.

The grading of quicklimes for lime treatment is under discussion. For drying, a product with a nominal top size of 6 mm, and with at least 95 % less than 5 mm, has been found to be acceptable. For stabilisation, it is more important to obtain

Table 26.2. Specifications for Types CL 70, 80 and 90 quick- and slaked limes

Type	CaO+MgO %	MgO %	CO_2 %	SO_3 %	Soundness[a]	Yield[b] dm^3	Fineness[c] %
CL 90	≥ 90	≤ 5 [d,e]	≤ 4	≤ 2	pass	≥ 26	≤ 7
CL 80	≥ 80	≤ 5 [d]	≤ 7	≤ 2	pass	≥ 26	≤ 7
CL 70	≥ 70	≤ 5	≤ 12	≤ 2	pass	≥ 26	≤ 7

[a] In accordance with 5.3 of EN 459-2.
[b] For quicklimes only, in accordance with 5.9 of EN 459-2.
[c] For slaked limes only, % by mass retained on 0.09 mm sieve, in accordance with clause 5.2 of EN 459-2.
[d] For soil stabilisation, ≤ 10% (N.B., the revision of EN 459 will not refer to soil stabilisation – refer to [26.17]).
[e] MgO content of up to 7% is acceptable, for uses other than soil stabilisation, if the soundness test is passed.

efficient dispersion and a product with a maximum size of 3 mm, with 80 % less than 425 µm is widely used in the U.K. It is understood that, in France, a finer product is favoured, with 100 % less than 2 mm and 70 % less than 200 µm.

Dolomitic quicklimes are not generally used as:

a) the MgO component hydrates much more slowly than the CaO component and is likely to cause expansion after the final compaction,
b) greater quantities are required as the MgO content does not contribute to the drying, modification and stabilisation processes, and
c) the MgO component may slow down the solution of the CaO component and reduce the rates of reaction.

Hydraulic limes are not used because it is virtually impossible to obtain the full strength potential. Compacting the soil before the hydraulic component starts to hydrate runs the risk that the CaO component completes its hydration afterwards, resulting in expansion. Conversely, compacting the soil when the CaO component has hydrated fully, runs the risk that some of the hydraulic component has started to hydrate, resulting in a loss of strength.

When both lime and cement are used, the lime is added first and allowed to slake and modify the soil. The cement is then mixed in and the layer is compacted.

Slaked limes (including both hydrated limes and milks of lime), which meet the requirements of Types CL 70, CL 80, or CL 90 [26.18] are suitable for lime stabilisation. Those requirements are summarised in Table 26.2. (N. B., for slaked limes, the specified percentages are on the basis of water free and boundwater free products.)

Dolomitic limes which are fully hydrated (i.e Type S [26.19], or non-expansive [26.18]) can be used, providing the addition rate is increased to compensate for the reduced CaO content.

26.4 Hydraulic Road Binders

26.4.1 Introduction

A variety of binders are used for road bases and sub-bases, capping layers, soil stabilisation and soil improvement, depending on the availability of materials and local traditions. *This section refers to factory blended mixtures of binders.* In-situ treatment of materials, using lime and other binders is described in section 26.3.

26.4.2 Constituents

A European Standard for factory blended hydraulic road binders is in preparation [26.20]. Current proposals for permitted constituents are listed in Table 26.3.

Table 26.3. Draft list of permitted constituents of hydraulic road binders

Category	Permitted constituents
Major	Portland cement clinker
	Granulated blastfurnace slag
	Natural pozzolanas
	Thermally activated clays and shales
	Siliceous fly ash
	Calcareous fly ash
	Unslaked calcareous fly ash
	Burnt shale
	Limestone
	Limes
Minor	as specified, up to 5%
	Calcium sulfate (gypsum, hemihydrate, or anhydrite)
Additives	as specified, up to 1%

26.4.3 Requirements

[26.20] is expected to specify classes of binder, based on minimum compressive strength at 7 and 28 days, and to detail requirements for fineness, initial set time, soundness, sulfate content and composition.

26.4.4 Lime Specifications

Lime used in hydraulic road binders should conform to EN 459-1 [26.18].

26.5   **Hot Mix Asphalt** (see also section 8.6)

A growing use for hydrated lime, particularly in the United States, is as an additive to hot mix asphalt [26.21, 26.22]. The main benefit is that it increases the resistance of the asphalt to water stripping. Lime also acts as a mineral filler and as an antioxidant.

- *Water stripping* (also called water sensitivity). In the presence of water, many asphalt-aggregate mixtures suffer a loss of bond between the binder and the aggregate. This results in a loss in strength, which causes the mixture to fail prematurely. Various anti-stripping agents are used, including lime, cement, amines, and diamines. Experience in the United States indicates that lime is the most suitable for the widest range of aggregates and asphalts. Hydrated lime addition levels of 1.0 to 1.5 % by weight of the aggregate are usually sufficient to reduce water stripping.

- *Mineral filler.* Hydrated lime, added as a mineral filler, has been shown to increase the viscosity of the binder, and increase the stiffness, tensile strength, compressive strength and resistance to water stripping, all of which increase the durability of the mix.

- *Antioxidant.* Preliminary results (1994) from laboratory and field tests have indicated that the addition of hydrated lime also reduces the oxidation of the binder, which causes hardening and stiffening leading to premature failure of hot mix asphalts. The favoured method of addition of the hydrated lime is as a powder to a moist aggregate (containing 2 to 3 units of percent above the saturated surface-dried moisture content of the aggregate).

- *Lime specifications.* Only slaked lime should be used in hot-mix asphalt. Both powdered hydrated lime and milk of lime meeting the requirements of Types CL 70, 80, or 90 [26.18] would be fully satisfactory, as would Type S dolomitic limes [26.19].

26.6   **Masonry Mortars**

26.6.1   **Historical**

Before considering the practical and technical aspects of masonry mortars, it is appropriate to review their development, which has led to the present-day specifications for mortars and their components.

Traditional lime-sand mortars, using "fat" lime as the binder, have proved to be durable over many centuries. From the builder's viewpoint, however, they became uneconomic, as they set slowly by loss of water and hardened even more slowly by absorption of carbon dioxide, forming a calcium carbonate-bonded skin. Moreover such mortars did not set under water and were, therefore, unsuitable for use in many situations (e.g., below ground level).

Natural and artificial hydraulic limes offered faster setting and hardening rates than lime. Their hydraulic properties enabled them to set, even under water, by the formation/hydration of calcium silicates and aluminates. The disadvantages of such limes were that the raw materials were not widely available and were often variable in quality (modern hydraulic limes are generally much more consistent and are widely used in some countries).

The development of Portland Cement in the 1800s presented the opportunity to use a hydraulic binder which was very consistent and set rapidly. Its relatively rapid development of strength was a marked advantage when there was a risk of frost. However, cement-sand mortars proved to be too strong for most purposes (they are still used where it is essential to prevent the passage of water, or to resist the effects of frost/soluble salts). Cement-lime mixes were used in the late 1800s to give good "soft" properties and controlled strength. By the 1930s, the 1:1:6 mixture of cement, lime and sand had become well established.

Masonry cements were developed in the 1930s in the U.S.A. They consisted of cement blended with an inert void filler and an air-entraining agent to give the required soft properties and controlled strength.

A discovery that air-entrainment not only improved workability but also increased frost resistance led to the use of mortars with high levels of air entraining agents. While the "soft" properties of such mortars were good, very high levels of entrained air resulted in poor bond strengths, and lead to the failure of a number of walls under tension and/or flexion. Modern specifications limit the level of entrained air.

26.6.2   Requirements

26.6.2.1   General

Many factors are involved when selecting a mortar with the most appropriate combination of properties. The architect needs to consider the requirements relating to the building, e.g.,

a) the exposure of the site to wind and rain,
b) the degree of exposure of each building component,
c) the type of masonry (i.e., bricks, blocks or stone),
d) the duty (i.e., load-bearing, or not).

The architect must also strike a balance between the short term requirements of the builder, who needs to be able to lay the masonry as quickly and economically as possible, with his own and the owner's long-term requirements for a durable, attractive and trouble-free construction. Table 26.4 summarises the results of a survey made in 1990 in the USA in which architects and builders were asked to prioritise the characteristics of mortars [26.23].

Table 26.4. Prioritisation of mortar characteristics

Importance	Architects	Builders
Greater	bond strength	workability
	water leakage	water leakage
	durability	uniformity
	compressive strength[a]	durability
	drying shrinkage	bond strength
	uniformity	ease of use
	appearance	compressive strength[a]
	colour	drying shrinkage
	workability	appearance
	availability	bond life
	ease of use	availability
	cost	colour
	bond life	yield
Lower	yield	cost

[a] This relates to the *correct* compressive strength (see section 26.6.2.3).

It is significant that both architects and builders recognised the importance of minimising water leakage and having a durable mortar, while the former placed particular emphasis on bond strength, and the latter considered workability to be of greatest importance. Both groups considered the cost of mortar to be relatively unimportant, and, by implication, that the minor increase in cost associated with the addition of lime was small in relation to the benefits that it confers. The factors influencing the more important characteristics are discussed below.

26.6.2.2   "Soft" Characteristics of Mortars

The mix should have a good *workability* at the required *consistence* so that it spreads easily, and penetrates into the surface texture of the masonry.

It should be *cohesive* so that it flows from the trowel.

The *water retentivity* of the mortar should be sufficiently high to spread evenly on the masonry units. Some units, with high levels of capillary suction, may present difficulties with premature drying-out and stiffening of the mortar.

The mortar should remain plastic long enough for the masonry units to be laid and aligned without causing cracking at the mortar-masonry interface, but it should have an acceptable *setting time* and develop adequate strength to permit additional courses to be laid within an acceptable period.

Summarising, good "soft" characteristics enable the builder to lay the bricks or blocks economically (i.e., quickly and accurately), while producing good quality mortar joints.

26.6.2.3   "Hard" Characteristics of Mortars

- *Compressive strength*. Mortars hardly ever fail in compression; indeed, lime-sand mortars, which have particularly low compressive strengths, have survived intact over many centuries. Problems have, however, been experienced with high-strength mortars, as, when movement occurs after the mortar has set, the stress tends to be relieved as a small number of large cracks, which may pass through the masonry units. Such large cracks can lead to water penetration. With weaker mortars, however, such movement results in a larger number of fine cracks, which do not result in excessive water penetration. Standards for mortar, therefore, specify mixes with limited compressive strengths, which are compatible with the characteristics of the masonry units.

- *Bond strength*. Experience has shown that most failures of mortar under stress occur under tension or flexion, when the bond strength between the mortar and the masonry is not adequate. This accounts for the emphasis placed by architects on having as high a bond strength as practicable (Table 26.4). As already mentioned, bond strength can be adversely affected by increasing levels of entrained air.

- *Sulfate resistance*. Under certain conditions, and particularly with certain bricks, the mortar needs to be resistant to the attack of sulfates.

- *Frost resistance*. Air entrainment increases resistance to frost damage.

- *Efflorescence/lime bloom*. This is a complex topic. Efflorescence is often caused by soluble salts in the masonry units. It can also be caused by soluble salts in the mortar, which include $Ca(OH)_2$ released by the lime and/or by the hydration of the cement. When water containing those salts migrates to the surface of the masonry, it evaporates and deposits the salts, causing unsightly efflorescence. In the case of calcium hydroxide, carbonation results in the formation of calcium carbonate.

26.6.3   Benefits of Using Lime

26.6.3.1   "Soft" Characteristics

Lime and lime-cement mortars have good "soft" characteristics.

a) They have high workabilities.
b) Their water retentivities are very high, making them particularly suitable for use with absorptive units.
c) The set times and 7-day strengths of lime-cement-sand mortars can be controlled by the amount and type of cement.

26.6.3.2   "Hard" Characteristics

The compressive strength of lime-cement mortars can be adjusted to the required level by the selection of the mix design.

Incorporating lime in mortar is reported to improve adhesion (or bond strength) and reduce rain penetration [26.24].

There is evidence that the presence of lime can increase the resistance of mortar to attack by sulfate [26.25, 26.26]

Another major benefit of incorporating lime into mortars is that it confers "autogenous healing" of cracks. Cracking occurs for a variety of reasons both before, during and after the setting of the mortar. The formation of cracks, both at the mortar-masonry interface and within the mortar reduces the strength of the masonry unit and increases water penetration.

In mortars containing lime, carbon dioxide from the atmosphere dissolves in water in the mortar and reacts with the lime to produce insoluble calcium carbonate crystals. The crystals form in spaces such as cracks and grow, thereby sealing the cracks. This "self-healing" characteristic reduces water penetration and increases durability. The rate of carbonation is dependent on the environmental conditions. Tests have shown that some mortars are not fully carbonated after several years.

A "general use" mortar has recently been developed in the UK by the Building Research Establishment [26.27]. It is based on an intimate mixture of equal volumes of hydrated lime and cement and incorporates sufficient air entraining agent to entrain 10 to 18 % of air. Laboratory durability tests indicate that, by using an ordinary Portland cement with a low content (e.g., <9 %) of tricalcium aluminate, the mortar gains a useful degree of sulfate resistance, as well as having an adequate resistance to frost, while maintaining the high bond strength associated with lime-based mortars.

26.6.4   Specifications for Mortars and their Component Materials

Many countries have specifications and standards for mortars and the materials used in mortars, which parallel the UK and European documents quoted below.

26.6.4.1   Mortar Mixes

A draft European Standard prEN 998-2 [26.28], classifies masonry mortars into categories, based on compressive strength. It is also specifies the quality of the component materials. Table 26.5 summarises the relationship between the mix proportions of the four weaker classes, which may contain lime.

Table 26.6 summarises the designations of mortar mixes specified in BS 5628 [26.24]. It is understood that a draft revision of BS 5628 also includes the general use mixes described in section 26.6.3.2.

Mortars of designation (i), (ii), (iii) and (iv) are suitable for use in load-bearing brickwork. The strength increases from designation (v) to (i), but the ability to

Table 26.5. Relationship between mixture and minimum compressive strength

Minimum compressive strength N/mm^2	Proportions by volume		
	Cement	Air lime	Aggregates
1	0	1 to 1.5	4 to 5
1	1	1 to 2	6 to 9
2.5	1	1	6
5	1	0 to 0.5	3 to 4.5

Table 26.6. Designations of mortar mixes

Mortar designation	Type of mortar (proportions by volume)		
	Cement : lime : sand	Masonry cement : sand	Cement : sand with plasticizer
(1)	1:0 to 0.25:3	–	–
(2)	1:0.5:4 to 4.5	1:2.5 to 3.5	1:3 to 4
(3)	1:1:5 to 6	1:4 to 5	1:5 to 6
(4)	1:2:8 to 9	1:5.5 to 6.5	1:7 to 8
(5)	1:3:10 to 12	1:6.5 to 7	1:8

accommodate movements due to temperature and moisture changes increases from (1) to (5) [26.24]. The properties and applications of the designations may be summarised as follows.

a) Mortars of designation (1) and (2) are for where high strengths and durability are required.
b) Designation (3) is a general purpose mortar with good durability. It is recommended for use with all normal construction which is likely to be exposed to severe conditions.
c) Designation (4) is suitable for normal construction, which is not likely to be exposed to severe conditions.
d) Designation (5) mortars are only recommended for internal work and special uses such as the repair of historic buildings, where durability is less important than minimising the risk of damage to the masonry, as the mortar has relatively low frost resistance.

The resistance to frost attack increases in the sequence cement-lime-sand, masonry cement-sand, cement-sand with plasticiser, while adhesion and resistance to rain penetration increases in the sequence cement-sand with plasticiser, masonry cement-sand, cement-lime-sand [26.24].

Advice on the selection of the most appropriate mortar for a particular application is given in [26.24, 26.29, 26.30].

26.6.4.2   Materials for Mortar

Mortars consist of four components:

a) building (or "soft") sand, which provides the bulk of the mortar,
b) a binder, which bonds the sand particles together and the mortar to the masonry unit,
c) void-filling material, which enables the sand particles to slide over one another, and
d) water, which holds the mortar together as a cohesive mass and provides lubrication for the sand particles, the binder and the void-filling material.

The *sand* should preferably be of a rounded shape and should have a suitable particle size distribution (typically less than 4 mm and with a limited amount of fines) which gives 30 to 40 % of voids. The characteristics of the sand can markedly affect both the "soft" and "hard" properties of the mortar. Sands suitable for building are designated in the UK as either Type S (BS 1199), or Type G (BS 1200) [26.31]. As the specified sands have a voidage in the above range, it is standard practice to design mortars with three volumes of sand to one volume of binder plus void-filling material.

The *binder* can be calcium lime, dolomitic lime, a hydraulic lime, a lime-cement mix, or cement. When calcium lime, or Type S dolomitic lime is mixed with sand to produce a sand-lime mortar, its initial role is that of a void filler. It subsequently causes the mortar to harden slowly, as a result of carbonation of the calcium hydroxide (N.B. when the term lime is used in connection with mortars, it refers to fully slaked lime with a low expansion potential. This includes Type S dolomitic limes, which are widely used for mortars in the USA).

Cements for mortars are specified in ENV 197: Part 1, [26.32] (N.B., high alumina cement should never be used with lime).

Limes — see section 26.6.7.

Suitable *void-filling materials* include lime and finely divided filler such as ground limestone. Alternatively, the voids can be filled with bubbles of air, by adding air-entraining agents, (or mortar plasticisers), which should comply with BS 4887 [26.33]. To avoid excessive loss of bond strength, while maintaining adequate frost resistance, the level of air entrainment should be in the range 10 to 18 % [26.8, 26.27].

The quantity of clean water needed to give the required consistence is a function of the mix design and of the components.

26.6.5   Preparation of the Mix

Mortar mixes are generally proportioned by volume. Lime may be measured as putty or as dry hydrated lime, as the weights of calcium hydroxide per unit volume are approximately equal. Mixing should be done either by hand on a clean surface, or in a mechanical mixer.

A lime-sand mortar, known as "coarse stuff", is generally prepared first. It may be produced by:

a) mixing the specified proportions of lime putty with sand, or
b) mixing the specified proportions of hydrated lime and dry sand and then adding water.

The workability of coarse stuff prepared in this way is improved by allowing it to stand overnight before use. Alternatively, coarse stuff may be purchased ready prepared.

While coarse stuff does not need to be used immediately, it should be protected from drying out. Immediately before use, it should be gauged with the specified proportion of cement, adding sufficient water to obtain the required consistence.

An alternative approach is to prepare a dry mix of cement, hydrated lime and sand, to which water is added to obtain the required consistence. Although this method does not give as high a volume yield and workability as the others, it is often more convenient in practice.

The production of factory-prepared, dry ready-mixed mortar for masonry, rendering and plaster, has grown rapidly in recent years (e.g., [26.34–26.38, 26.54]). A major benefit of factory-mixed mortars is their high level of consistency.

26.6.6 Workmanship

The appearance, strength and durability of mortar depends on working practices, as well as on the quality of the mortar. Detailed attention should be given to the laying of masonry units (including filling of vertical joints). Bagged materials should be kept dry. Sands should be kept at a uniform moisture content and covered to prevent contamination. Some "suction" of the masonry units helps to increase bond strength; corrective action should, however, be taken if the suction is excessive and causes premature stiffening of the mortar.

Mortar should never be re-worked when it begins to stiffen. Such practice reduces workability, compressive strength and bond strength.

26.6.7 Lime Specifications

Limes for building purposes are specified in CEN Standard EN 459 [26.18]. They are divided into the 8 groups listed in Table 26.7. Calcium limes (CL, which may be in the form of quicklime, powdered hydrated lime, or putty) are sub-divided into three – CL 90, CL 80 and CL 70 — on the basis of their chemical analysis. Dolomitic limes (DL) are sub-divided into DL 85 and DL 80, also on the basis of their chemical analysis. Hydraulic limes (HL) are described in section 26.9.

Table 26.8 outlines the relatively complex physical and chemical requirements specified in EN 459-1 [26.18].

Table 26.7. Classification of building limes

Description	Form[a]	Classification
Calcium lime 90	Q, H, P	CL 90
Calcium lime 80	Q, H, P	CL 80
Calcium lime 70	Q, H, P	CL 70
Dolomitic lime 85	H	DL 85
Dolomitic lime 80	H	DL 80
Hydraulic lime 2	H	HL 2
Hydraulic lime 3.5	H	HL 3.5
Hydraulic lime 5	H	HL 5

[a] Q = quicklimes, H = hydrated limes, P = putties/milks of lime.

Table 26.8. Summary of chemical and physical tests for building limes

Requirement	Classification		
	CL	DL	Hyd[a]
CaO + MgO	✓	✓	–
MgO	✓	✓	–
CO_2	✓	✓	–
SO_3	✓	✓	✓
Available lime	–	–	✓
Strength	–	–	✓
Yield	✓	–	–
Grading	✓	✓	✓
Soundness	✓	✓	✓
Free water	✓	✓	✓
Penetration	✓	✓	✓
Air content	✓	✓	✓
Setting time	–	–	✓

[a] Hyd includes all hydraulic limes (NHL, NHL-Z and HL — see section 26.9)

26.7   External Rendering

26.7.1   General

External rendering mortars (sometimes referred to as "stucco") are used to enhance the appearance of buildings and to make the structure waterproof. Until the late 1800s, they were predominantly lime-based. It was then found that the use of cement gave faster setting and greater strengths. Subsequently, it was discovered that excessively high cement contents caused problems and that the use of cement-lime mortars gave significant improvements in both the "soft" and "hard" characteristics.

The selection of the mortar depends on the background material (i.e., the bricks, blocks or other masonry which is being rendered), the exposure conditions and the required finish.

26.7.2   Requirements

As with masonry mortars (see sections 26.6.2.2 and 26.6.2.3), rendering mortars should have good workability, adequate but not excessive strength, a good bond to the background material, durability and an attractive appearance. For rendering, the shrinkage should be low to avoid excessive cracking. The requirements are specified in prEN 998-1 [26.28].

Good workability is important to ensure that the mortar is easily applied and can be effectively worked into the joints and crevices of the background material. A high degree of cohesiveness ensures that the mortar adheres to the trowel and to the background. It is also essential when using mechanical spraying equipment.

A high bond strength to the background material is particularly important. The bond needs to withstand movement between the rendering and the backing material caused by variations in temperature and moisture content.

External rendering is generally applied in two or more layers. The strength of the backing layer(s) should be adequate for the duty, but should be lower than that of the background material, to ensure that movement is accommodated in a large number of fine cracks, rather than in a few large cracks which are more likely to permit water penetration. For the same reason, the finishing coat should have a still lower strength than the backing layer(s).

Durability results from the use of suitable materials, giving adequate long-term bond strength.

Renderings are often used to improve the appearance of the building and are treated to give various surface textures and colours. It is therefore essential that the finishing layer is compatible with the proposed finish.

26.7.3   Benefits of Using Lime

Lime-based mortars have excellent workability and plasticity, a high degree of cohesiveness and spread easily under the trowel. These properties help to increase productivity and minimise wastage by droppings. They are eminently suitable for use with mechanical spraying equipment.

Incorporating lime into the mix ensures a good bond between the rendering and the background material, particularly with porous backgrounds. As with masonry mortars, the mix design of rendering mortars should be selected to give the required strength to cope with the inevitable movement of both the background material and the mortar.

The proven excellent durability of lime-based renderings is due partly to the above properties and partly to the autogenous healing properties conferred by lime (described in section 26.6.3.2), which help to maintain a high resistance to water penetration. As with masonry mortars, the above benefits are recognised to heavily outweigh the small increase in cost associated with the addition of lime to rendering mortars.

26.7.4   Materials

The *sands* used for rendering mortars (BS 1199 [26.31]) generally contain more coarse particles in the range 1 to 5 mm than those used for building mortars. This is primarily to limit drying shrinkage. The sand should be of good quality, with low levels of loam and very fine particles. *Coarse aggregate*, with a top size of 5 to 12 mm, may be used for rough-cast, or dry-dash finishes. It should be clean and free from deleterious matter such as clay and fines.

The *cement* should comply with EN 197 [26.32] (N.B., high alumina cement should never be used with lime). *Lime*, in the form of quicklime, hydrated lime, or lime putty, should comply with EN 459 [26.18] see section 27.6.7.

26.7.5   Mix Design

The recommended mortar designation depends on the strength of the background material, the degree of exposure, whether it is to be used as an undercoat, or as a final coat and on the type of finish. Detailed guidance is given in BS 5262 [26.16].

26.7.6   Preparation of the Mix

The practices described in section 26.6.5 for masonry mortar are appropriate for external rendering mortar. As uniformity of appearance is important in rendered finishes, particular care is necessary to ensure that the materials and proportions are consistent throughout a particular job.

26.8   Internal Plastering

26.8.1   General

Internal plastering is used to cover up differences in level which are unavoidable in most internal walls and to provide a surface which is suitable for the final decorative finish. It is generally applied as one or more undercoats, followed by a finishing, or skim coat.

Until the late 1800s, most internal plasters were predominantly lime-based. Thereafter, mixes using cement and "gypsum" (actually de-hydrated gypsum) were increasingly used. These binders gave faster setting and higher crushing strengths than the traditional lime plasters. It was subsequently found that the use of lime "gauged" with cement or gypsum, to obtain the required setting time and strength, improved both the handling characteristics and the durability of the plaster.

26.8.2 Requirements

Although most of the requirements for internal plastering and those for mortar and external rendering are similar, there are several important differences [26.28]. The requirements are described in full, to avoid undue cross-referencing [26.39].

Plasters should have a good workability and be cohesive so that they can be applied easily and achieve intimate contact with the background material. They should set and dry reasonably quickly to maximise productivity.

They should have sufficient crushing strength to resist impact damage, but be weaker than the background material to minimise cracking and "shelling". A good bond to the background material is essential to withstand differential movement between the plaster and the background caused by changes in temperature and/or moisture content.

The plaster should be durable and capable of withstanding normal wear and humidity. In kitchens and bathrooms, for example, plasters gauged with cement should be used in preference to those containing gypsum.

26.8.3 The Benefits of Using Lime

The benefits may be summarised as follows.

a) Plasters containing lime have excellent workabilities, are highly cohesive and are eminently suitable for mechanical plastering techniques.
b) Setting times and crushing strengths can be controlled by selecting the most appropriate mix design.
c) The high water-retentivity of lime-based plasters, coupled with their high workability, ensures a good bond to the background material. This is particularly beneficial with porous backgrounds.
d) The ability of lime to promote autogenous healing of cracks helps to ensure its durability by reducing water penetration.
e) The high alkalinity of the plaster inhibits the growth of mould and the corrosion of iron and steel.

As with other mortars, the benefits arising from the addition of lime far outweigh the small increase in raw material costs.

26.8.4   Materials

The *sands* used for internal plasters are specified in BS 1200 [26.31]. *Gypsum building plaster* should comply with the requirements of BS 1191 [26.55]. *Cement* should comply with the requirements of ENV 197 [26.32] (N.B., high alumina cement should never be used with lime). *Lime* used for internal plastering should comply with the requirements of EN 459-1 [26.18] — see section 26.6.7.

26.8.5   Mix Design

The choice of mix for internal plastering depends on the background material, the finish required, the method of application (manual or spray) and whether the mix is being used as an undercoat, or a finishing coat [26.16].

26.8.6   Preparation of the Mix

Mixes consisting of lime, Portland cement and sand, whether for undercoats or finishing coats, should be prepared as described in section 26.6.5. When hair is to be incorporated, it should be well beaten into and mixed with the "coarse stuff" before gauging with cement.

Lime plasters gauged with gypsum should be prepared as follows. Lime putty prepared by soaking hydrated lime in water (preferably overnight), or purchased from a factory, should be thoroughly mixed with the required proportion of sand. Immediately before use, the gypsum plaster should be added in the required proportion and thoroughly mixed, together with any additional water needed to give the required consistence. With finishing coats, it is particularly desirable to soak the hydrated lime.

26.9   Hydraulic Limes

26.9.1   Use

The term "hydraulic" was adopted by Vicat [26.40] in the early 19th century to describe materials which were able to set and harden under water. In the case of hydraulic limes, all of which contain free lime, the absorption of atmospheric carbon dioxide generally contributes to the hardening process.

Hydraulic limes are used for making mortar, plaster and concrete, particularly in Italy, France and Germany (see section 16.10 for their production). In the UK, their use is largely restricted to conservation and restoration work, where their slow set times and relatively low strengths are considered to be of particular benefit when used on weathered stonework [26.41].

Guidance on the formulation of mortar mixes based on hydraulic limes is given in [26.42]. A hydraulic lime-sand mortar (consisting of 1 volume of lime to 2 to 3 volumes of sand) is considered to be equivalent to Designation (iv) (see Table 26.6) [26.43]. Hydraulic limes gain strength slowly, so that, when used in new construction, the rate of building must be restricted to allow adequate strength to develop.

26.9.2 Classifications

The term "hydraulic lime", as used in EN 459-1 [26.18], refers to three groups of products.

a) Natural hydraulic limes (designated NHL), which are limes produced by burning, at below 1250 °C, of more or less argillaceous or siliceous limestones, with reduction to powder by slaking, with or without grinding. They contain variable amounts of quicklime, hydrated lime, α and β dicalcium silicate, calcium aluminate, calcium carbonate, tricalcium silicate and α and β monocalcium silicate (in approximate order of decreasing concentration) [26.44].
b) "Special" natural hydraulic limes (designated NHL-Z), which are produced by blending natural hydraulic limes with up to 20 % of suitable pozzolanic products (e.g., pulverised fuel ash, volcanic ash and trass), or hydraulic materials (e.g., ordinary Portland cement and blast furnace slag).
c) "Artificial" hydraulic limes (designated HL) consist mainly of calcium hydroxide, calcium silicates and calcium aluminates. They are produced by blending suitable powdered materials, such as natural hydraulic limes, fully hydrated air limes and dolomitic limes, pulverised fuel ash, volcanic ash, trass, ordinary Portland cement and blast furnace slag.

Natural hydraulic limes have traditionally been classified in three categories: *feebly* hydraulic, *moderately* hydraulic, and *eminently* hydraulic.

Another classification uses the terms semi-hydraulic and "hydraulic", corresponding approximately to feebly and eminently hydraulic respectively. Eminently hydraulic lime is also called Roman lime.

Hydraulic limes are also classified in terms of the cementation index (CI) [26.45], based on the analysis of the lime. This is calculated by the following equation:

$$CI = \frac{(2.8 \times \% \ SiO_2) + (1.1 \times \% \ Al_2O_3) + (0.7 \times \% \ Fe_2O_3)}{\% \ CaO + (1.4 \times \% \ MgO)} \qquad (26.1)$$

The equation gives a high weighting to the silica content, as most of the strength of the set mortar arises from the calcium silicate component.

The properties of the categories of lime are summarised in Table 26.9 [26.18, 26.41, 26.46].

Table 26.9. Characteristics of hydraulic limes

Parameter		Type/Classification of hydraulic lime		
		2 Feebly	3.5 Moderately	5 Eminently
Compressive strength (N/mm²)	7-day	–	≥ 1.5	≥ 2
	28-day	2–7	3.5–10	5–15[a]
Active clay minerals (% m/m)		< 12	12–18	18–25
Cementation Index		0.3–0.5	0.5–0.7	0.7–1.1
Setting time in water (days)		< 20	15–20	2–4
Slaking time		slow	slow	very slow
Expansion		slight	slight	slight
Typical colours on slaking		off-white/ pale grey	pale grey/ pale buff	dark grey/ grey/brown

[a] NHL 5, with a bulk density lower than 0.90 kg/l, may have strengths up to 20 N/mm².

Up to 5.1 % of gypsum may be added to hydraulic limes to control the setting process, although it is not normally required when ordinary Portland cement is added.

The presence of added Portland cement can be detected by X-ray diffraction and studying the alite ($3CaO \cdot SiO_2$) peak at 0.176 nanometres. Portland cement is rich in alite, which is only formed at above 1250 °C, whereas natural hydraulic limes are calcined at below 1250 °C and do not contain significant amounts of the mineral [26.18].

26.9.3   Specifications and Consistency

EN 459-1 [26.18] classifies hydraulic limes into the three types described above, namely NHL, NHL-Z and HL and three strength categories with minimum 28-day compressive strengths of 2, 3.5, and 5 N/mm² respectively (Table 26.9). A summary of the specified physical and chemical tests is given in Table 26.8 (see page 278). Reference should be made to the Standard for the specific requirements.

It will be noted that the compressive strength requirements of the three classes of hydraulic limes overlap. This is to accommodate variations caused by changes in the composition of the limestone and in the calcining process (in the case of natural hydraulic limes) and by variations in the components of artificial hydraulic limes.

26.10   Sandlime Bricks

26.10.1   Introduction

Sandlime bricks, also called calcium silicate bricks, are produced by moulding, under high pressure, a moist mixture of silica sand (or crushed siliceous stone, or flint) and hydrated lime. The "green" bricks are then autoclaved using steam pressures of at least 11 atmospheres. Under these conditions, the hydrated lime reacts with the silica to form hydrated calcium silicate, which bonds the aggregate particles into a strong and durable brick [26.47]. Other shapes, such as blocks and building elements are also produced.

   Because hydrated calcium silicate is white, the autoclaved products are an attractive near-white colour. This enables pigments to be used to produce a wide range of coloured products.

   Although the calcium silicate brick process was patented in the UK in 1866, its commercial exploitation only occurred in Germany towards the end of the 19th Century. In that country, the sandlime brick industry now accounts for some 40 % of the brick production. The process is also used in many other countries, but its penetration has varied considerably, depending on traditional construction methods and on the relative costs of competitive products [26.48].

   The lime may be purchased by the brick manufacturer either as hydrated lime, or as quicklime. All of the processes can use damp sand — there is no need to pre-dry it.

26.10.2   Process Using Hydrated Lime

Sandlime brick plants using hydrated lime (see Fig. 26.4) are relatively simple to operate and low in capital cost. The operating costs, however, tend to be higher than for quicklime, owing to the higher delivered cost of hydrated lime per unit weight of available lime.

   The raw materials, sand and hydrated lime, are fed into bunkers. From there, they are discharged into a batch mixer via mechanisms, such as weigh hoppers, which enable an accurately controlled blend to be produced. Typically some 10 parts by weight of hydrated lime are used per 90 parts of sand. When coloured bricks are required, pigments are added and are thoroughly dispersed. Water is then added to produce a readily compacted mixture with a free moisture content of about 6 %.

   The damp mix is then passed to the press which moulds the material into the required shape at high pressure. The "green" bricks are stacked on trollies for transfer into autoclaves. After charging, the autoclaves are sealed and steam is blown in to raise the pressure progressively to about 12 bar (185 °C). That pressure is maintained for some 6 hours to enable the calcium hydroxide to react fully with the silica to produce hydrated calcium silicate. The pressure in the autoclave is then progressively released, with the vented steam being used to pre-heat a freshly charged autoclave. A typical cycle time, including charging and emptying,

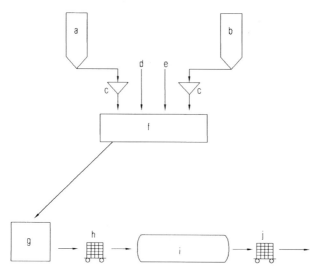

Figure 26.4. Diagram of a sandlime brick plant
(a) aggregate; (b) hydrated lime; (c) batch weighers; (d) water; (e) pigment; (f) batch mixer;
(g) brick press and stacker; (h) green brick transfer; (i) autoclave; (j) brick handling and
despatch

is 11 hours, although some plants operate on shorter cycle times by using steam
pressures of up to 14 bar.

   After autoclaving, the bricks are sampled and tested before being formed into
packs and stored in stock yards from which they are dispatched.

26.10.3   Processes using Quicklime

26.10.3.1   General

Quicklime may be purchased as either a granular or ground product. While granular quicklime has a lower ex-works price and lower delivery costs than ground
quicklime, it incurs greater capital and operating charges, because it needs to be
finely ground before use. This involves the installation of pulverising and classification equipment to produce the required particle size distribution (typically
90 % passing 75 µm).

   The ground quicklime is stored in a bunker and discharged through a similar
mechanism to that described for hydrated lime (section 26.10.2). The quicklime is
hydrated using either the drum, or the silo hydration process.

26.10.3.2   Drum Hydration Process

In this process, the appropriate proportions of ground quicklime and sand are placed in a rotating drum. When the raw materials are thoroughly mixed, sufficient

water is added to slake the quicklime and give a slight excess. The drum is then sealed. The heat of hydration may generate sufficient steam to raise the pressure within the drum to about 4 atmospheres (140 °C), although additional steam may be required with impure and/or low reactivity quicklimes. The drum is kept at pressure until the lime is fully hydrated (typically for 30 min.).

The pressure is then released and the contents discharged into a mixer. Pigment is added, if required, and the free moisture content is raised to about 6 %. The remaining stages are as for the hydrated lime process (section 26.10.2).

26.10.3.3   Silo Hydration Process

In this variant, the appropriate proportions of ground quicklime and sand are thoroughly blended in a mixer. Sufficient water is added to hydrate the quicklime and provide a controlled excess. The damp mix is then placed in a silo to allow the quicklime to hydrate completely at a temperature approaching 100 °C. This can be done by placing batches of mix into a number of insulated silos, which together provide a residence time of up to 24 hours. Alternatively, one large "mass flow" silo is used, which is continuously charged and discharged. It has been reported [26.49] that the time required for complete hydration can be reduced to as little as 3 hours, by using reactive quicklimes and carefully controlled levels of excess water.

The matured blend is then discharged from the silo(s) into a second mixer, and water is added to raise the moisture content to about 6 %. The pressing and autoclaving stages are as already described (section 26.10.2).

26.10.4   Lime Specifications

Specifications vary from manufacturer to manufacturer and from one design of plant to another. In general, most require either high-calcium quick- or hydrated lime with not more than 2 %, and preferably less than 1 % of MgO (2 % of MgO causes an expansion of 0.5 % in the autoclave). Higher levels can be accepted, where Type S hydrated lime is used, in which the magnesium is fully hydrated. The combined CO_2 should be less than 7 %. Hydraulic components, namely silica and alumina, should be less than 5 % in total.

It is important that hydrated lime is fully hydrated and non-expansive. This generally requires hydration with a higher than normal level of excess water and maturing of either the raw hydrate, or the finished product, at an elevated temperature for the necessary period of time. Non-expansive hydrate is most readily produced from "highly reactive"[*] quicklime, but can also be made from less reactive quicklimes by careful control of the maturing process. It should be noted that operation with the necessary excess of water adversely affects handling characteristics and appropriate measures need to be taken to minimise blockages in chutes and "rat-holing" in bunkers.

The hydrated lime should be air classified to ensure that at least 98 % of the product is less than 200 µm to minimise the risks of surface blemishes caused by the expansion of individual particles. In addition, at least 93 % should be less than 90 μm to ensure effective dispersion of the hydrate in the mix.

"Moderate" and even "low" reactivity quicklimes can be used in the drum hydration process, because of the pressurised hydration stage (see section 13.2, Fig. 13.3, and Annex 1 for explanations of these terms). The reactivity of quick-limes used for the silo hydration process should be "moderate" to "high", depen-ding on the silo residence time. For both of these processes, the quicklime should be finely ground (e.g., with 90 % less than 75 μm) to ensure efficient dispersion and hydration.

Consistency of lime quality is desirable for all calcium silicate processes. It helps to improve process control and to raise the quality and consistency of the autoclaved products. Thus, although the process can tolerate relatively low qual-ity lime products, in practice, high quality products, which contain consistently low levels of impurities, are often favoured.

26.11   Autoclaved Aerated Concrete

26.11.1   Introduction

Autoclaved aerated concrete (AAC, or aircrete) was first produced in Sweden in 1924. It is now made in over 200 factories in many countries throughout the world.

Aircrete is produced by mixing a source of finely divided reactive silica and water with a binder (generally lime plus cement, or cement) and then adding an aeration agent, usually in the form of aluminium powder. The ingredients are thoroughly mixed at high speed and poured into a mould. The calcium hydroxide in the binder reacts with the aluminium to produce hydrogen bubbles, which cause the mix to rise within the mould. At the same time, the binder causes a "green" set to occur. The mix is then held at an appropriate temperature until the green set is strong enough for the cake to be de-moulded. The cake is then cut to shape using wire saws and autoclaved at steam pressures of at least 11 atmos-pheres for about 8 hours.

Aircrete is used to make both blocks and reinforced products. When blocks are produced, the green cake is trimmed to the required height and cut laterally and longitudinally to produce blocks of the required dimensions.

The autoclaved material has an excellent combination of strength and density. By adjusting the density and the proportion of binders it is possible to tailor strengths and thermal/acoustic insulating properties to specific duties.

Originally only "low" reactivity quicklime was used and many producers spe-cify that quality. In the UK, however, suitable quicklimes were not generally available and many of the processes were initially modified to use cement as the sole binder. Subsequently, part of the cement has been replaced, with advantage, by "moderate" and "high" reactivity high-calcium quicklimes.

26.11.2   The Process

From the viewpoint of the lime producer, the process can be regarded as consisting of six stages — raw material handling, mixing, rise-and-set, de-moulding, cutting and autoclaving. There are many variations on the process. In some, the pulverised fuel ash (pfa)/milled sand and returns are added to the mixer as a slurry in water. However, the general principles are common. An outline flowsheet is shown in Fig. 26.5.

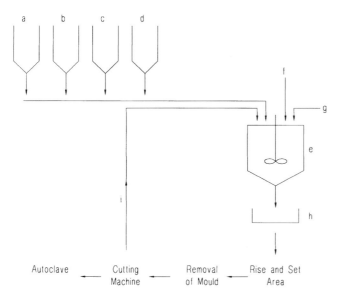

Figure 26.5. Outline flowsheet of the aircrete process
(a) pfa; (b) sand; (c) cement; (d) ground quicklime; (e) mixing vessel; (f) aluminium slurry; (g) water; (h) mould; (i) recycle of off-cuts/rejects

26.11.2.1   Raw Materials

The raw materials typically consist of reactive silica, raw silica sand, binder, and an aerating agent. Ground synthetic anhydrite ($CaSO_4$) may also be added to retard the slaking of the lime.

The *reactive silica* can take the form of pfa, or finely ground silica sand: the choice affects the rheological properties of the mix. Ground sand tends to produce a "harsher" mix, which requires closer process control than when using pfa. The *binder* consists of cement and ground quicklime. The *aerating agent* generally consists of an aqueous suspension of finely flaked aluminium powder. In addition, green *off-cuts* and returns from the cutting stage are re-cycled. Not only does this reduce wastage, but the recycled material can help to produce a better product, by stabilising the process.

The formulation depends on the raw materials used and on the required characteristics of the product. A typical formulation is given in Table 26.10.

Table 26.10. Typical formulation of an aircrete mix

Material	Proportion (% m/m)
Pulverised fuel ash (dry)	42
Ordinary Portland cement	12
Quicklime	12
Water	32
Anhydrite	2
Aluminium	0.07

26.11.2.2 Mixing

In a typical sequence, water at a controlled temperature is added to the mixing vessel, which is agitated vigorously by a high speed stirrer. The reactive silica, aggregate and recycled off-cuts are then added (as mentioned above, these materials and most of the water may be added as a premixed slurry). The temperature of the resulting slurry is typically 30 °C. Cement is added to the mixer, and dispersed. Lime is then added and the mix stirred for a further minute or so to ensure efficient dispersion. Finally, the aluminium suspension is added and dispersed for some 30 s. The mix is then discharged into a mould and immediately transported to an area where it is left undisturbed to rise and set.

Inevitably, some hydration of the lime occurs in the mixer. Typically, this results in a discharge temperature of about 40 °C and some thickening of the mix. If too much hydration occurs at this stage, the viscosity becomes excessive and problems are experienced with discharging the mix into the mould. Possible corrective action includes the addition of more water (which can adversely affect the green set and the strength of the autoclaved product), lowering the temperatures, specifying a less reactive quicklime and replacing some of the quicklime with cement.

26.11.2.3 The Rise-and-Set

In the rise-and-set stage, the reaction between dissolved calcium hydroxide and the aluminium produces bubbles of hydrogen, which cause the mix to rise. The extent of the rise determines the density of the autoclaved blocks. The primary control over the rise is the quantity of aluminium added, although secondary factors, such as the water addition, temperature and the rheology of the mix are also important.

At the same time, the ongoing slaking of the quicklime and the formation of a calcium hydroxide gel cause the mix to thicken and set.

The set time should correspond closely with the end of the reaction of aluminium to form hydrogen. Otherwise, if the set is delayed, too many of the hydrogen bubbles migrate to the surface causing the cake to subside, and, in extreme cases, to collapse. Conversely, if the set occurs too early, the ongoing evolution of hydrogen produces internal pressures which can cause cracks within the mix and a consequent loss of strength.

The required rise and set characteristics are achieved by matching the rates of reaction of the aluminium and the lime, as well as by altering temperatures and the amount of water. Various grades of aluminium powder are available, with differing reaction rates. The rate of hydration of the lime is related to its reactivity.

After the mix has set, the hydration of the less reactive fractions of the lime continues (as does the slower hydration of cement). This results in a progressive stiffening of the mix as well as a progressive rise in temperature. The latter increases the water vapour pressure within the hydrogen bubbles and if that pressure exceeds a critical level, it causes cracking within the cake, and loss of strength.

26.11.2.4   Cutting

When the green set is sufficiently advanced and the cake is of the required consistency, the mould is removed and the cake is trimmed/cut into blocks by oscillating, or vibrating, wire saws. Surplus materials (off-cuts or returns) are recycled to the mixer, as already mentioned. If, however, cutting is delayed, the mix eventually becomes too hard to be cut and must be scrapped. Thus the "cutting window" is an important factor in the routine operation of the process. The use of lime as part of the binder increases the period during which the mix can be cut.

26.11.2.5   Autoclaving

In the autoclaves, the material is heated by steam (typically at about 12 atmospheres) to approximately 185 °C for some 8 hours. Initially, any residual calcium oxide hydrates, together with the magnesium oxide present in the lime.

The hydrated lime then reacts with the reactive silica and alumina to produce hydrated calcium and magnesium silicates and aluminates. The cement also hydrates completely.

The calcium and magnesium oxides that hydrate after cutting cause expansion of the blocks in the autoclave. While a small expansion can be tolerated, excessive expansion causes the material to exceed the specified dimensions [26.50, 26.51] and can reduce strength through cracking of the mass, or micro-cracking of the walls between the voids. It is, therefore, essential to limit the amounts of overburned calcium oxide and of magnesium oxide, in the lime.

The autoclaved product (Fig. 26.6) combines a controlled low density with adequate load-bearing capacity to meet many building applications.

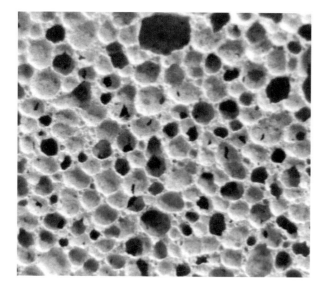

Figure 26.6. Internal structure of an aircrete block (magnified 4.5 times)
(by courtesy of Kingsway Technology Ltd.)

26.11.2.6   General

The process has many degrees of freedom including:

a) the quantities of water, active silica, aggregate, cement, lime and aluminium,
b) the reactivities of cement, lime and aluminium,
c) the times and sequences of additions, and
d) the temperatures,

each of which can affect:

a) the rheology of the mix,
b) the rise and set,
c) the generation of internal pressures,
d) the cutting window,
e) expansion after cutting,
f) the structure of the cake, its strength and density.

   Consistency of operation has always been important, but the increased use of computer control and consequent reduction in manning levels have placed even greater emphasis on the need for consistent raw materials.
   In view of the complexity of the process, it is not surprising that the know-how relating to the process is kept confidential. Nor is it surprising that, once a producer has found a successful combination of materials and operating parameters, often as a result of painstaking development work, he is generally reluctant to experiment with alternative combinations.

26.11.3   Quicklime Specifications

- *General.* An experienced aircrete producer can probably tailor his process to use quicklimes of widely differing reactivities, providing they disperse well, do not contain expansive components, and are consistent.

- *Dispersion.* With moderate and low reactivity ground quicklimes, an adequate dispersion can generally be achieved by grinding so that 90 % passes a 75 µm sieve. With highly reactive limes, however, there is a tendency for the quick lime particles to agglomerate as they enter the slurry in the mixer. Such agglomeration can be seen as white specks in the cake, which are undesirable from an aesthetic viewpoint, but can also result in reduced strengths for a given density.
  Various approaches have been developed to overcome this problem. The simplest is to add 1 to 2 % of water while grinding the quicklime. This results in a small amount of calcium hydroxide being formed on the surface of the quicklime particles. The main consequence of this water addition is an improvement in dispersion. It also reduces the temperature rise ratio and the final temperature rise (see Fig. 26.7), both of which are favourable effects.

- *Expansive components.* Expansion in the autoclave is generally caused by hard-burned calcium oxide and magnesium oxide. As described in section 15.4, during the calcination of limestone, the magnesium oxide component of lime becomes over-burned. Practical experience has shown that MgO levels up to 2 % can be tolerated. The amount of hard-burned calcium oxide generally increases as the average reactivity of the lime decreases. It also depends on the type of kiln and fuel used. In addition, malfunctions of the kiln can increase the proportion of the hard-burned fraction.
  The relative amount of hard-burned lime from a particular kiln can be monitored by study of the time-temperature curve produced by the reactivity test (Fig. 26.7). Any increase in the time at which the maximum temperature begins to fall (by 0.2°C) indicates that the amount of slow-slaking calcium oxide has increased.
  A more direct way of identifying expansive components is to make a small sandlime brick, using the coarse fraction extracted from the milk of lime produced in the reactivity test. Any expansion on autoclaving the brick at 10 Bar for 8 hours is indicative of expansive components in the quicklime.

- *Reactivity.* Various lime reactivity tests are used in the aerated concrete industry. One of the best, from the viewpoint of the information which it gives is described in Annex 3, Appendix A (see page 432). The reactivity specified by aircrete producers ranges from "low" to "high" (see below).
  Many aircrete producers specify low reactivity quicklime produced in mixed-feed kilns, fired with coke. The additional costs per tonne of lime, arising from the use of coke, and from operating at higher temperatures to produce hard-burned lime, are substantial, but are accepted as being necessary.

Other producers have found it advantageous to modify their process to accept high reactivity quicklime produced in modern kilns (see section 26.10.4). Those kilns are controlled within a narrow reactivity range, with measures being taken to prevent very high reactivity lime being produced.

As described in section 26.11.2.6, all aircrete producers require the reactivity of the quicklime to be as consistent as possible. A simple way of monitoring consistency is to record:

a) the temperature rise ratio (i.e., the temperature rise at 2 min. divided by the maximum temperature rise, which can range from a low value of 0.3 to a high value of 0.9) and

b) the maximum temperature rise, typically 65 to 70 °C (see Fig. 26.7).

a) is an indication of the shape of the reactivity curve, while b) is an indication of the reactive calcium oxide content.

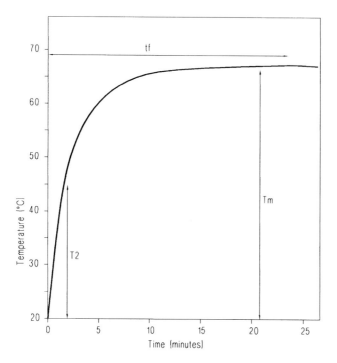

Figure 26.7. Typical quicklime slaking curve (see Appendix A for test method)
T_2: temperature rise at 2 minutes (°C); T_m: maximum temperature rise (°C); t_f: time at which temperature falls (by 0.2 °C); temperature rise ratio = $T_2 \div T_m$

It should be noted that the standard reactivity tests, using quicklime and distilled or de-ionised water, do not adequately mimic the performance of the quicklime in the process. This can be illustrated by two examples.

If a quicklime, containing little or no calcium sulfate, is divided into two samples, and calcium sulfate is added to the second, that sample would hydrate more slowly than the first in the standard reactivity test. However, in the aircrete process, the mix contains both calcium sulfate and additional calcium ions from the cement, and the two quicklimes mentioned would be expected to slake at the same rate.

In another example, two quicklimes, produced from very different limestones, but with virtually identical slaking curves in the standard test, were found to perform differently on the full scale. When a reactivity test incorporating some cement was used, the slaking time increased from about 20 to 60 min. and the difference between the quicklimes became apparent.

- *Other factors.* Other components in ground quicklime, such as calcium carbonate, silica, alumina and iron oxide can be regarded as inert. From the viewpoint of the aircrete producer, the key factor is not so much their level as their variability (as that is the main cause of variability in the CaO content). As an example, it is generally easier to control the calcium carbonate to within ± 1 % when the average level is 3 % than when it is 6 %.

26.11.4 Advantages of Using Reactive Quicklimes

There are three principle advantages to the aircrete producer of using more reactive quicklimes.

a) Reactive quicklimes help him to make the lower density blocks, with higher insulating values, which are in demand. This is because higher reactivities tend to produce a more finely divided calcium hydroxide, which enables more aluminium and water to be added without de-stabilising the wet mix.
b) The faster rate of hydration results in a quicker rise and set and a shorter cutting time. This reduces the number of cast cakes between the mixer and the cutting machine, so that, in the event of a prolonged interruption to the cutting operation, fewer cakes have to be recycled.
c) High reactivity limes are often more consistent than limes with lower reactivities. This is because most modern shaft kilns are designed to produce high reactivities and are able to be controlled more precisely than those producing lower reactivities. (N. B., although rotary kilns can be operated to produce a wide range of reactivities, the consistency tends to be low.)

26.12 Calcium Silicate Products

26.12.1 Calcium Silicate Wall-board

Calcium silicate wall-board is produced from cellulose fibre, calcium silicate-forming materials (e.g., hydrated lime, silica and cement) and fire protective fillers.

The raw materials are blended with water in a "hydro-pulper" and the slurry is transferred to the board-making machine. Excess water is removed through a filter cloth by suction. The uncured boards are then transferred to autoclaves, where they are heated under steam pressure to react the lime and silica and to hydrate the cement.

The particle size distribution of the hydrated lime should be sufficiently fine to give the slurry the required rheological properties and to produce a "green" board with the required mechanical properties. It should, however, be sufficiently coarse to enable the de-watering process to proceed at an acceptable rate.

In practice, the specific surface area, as measured by air permeability [26.52] has proved to be an acceptable control parameter. The surface area may be adjusted either by hydrating a quicklime with a different reactivity (increasing reactivity produces a more "fluffy" hydrate with a higher area for a given cut size), or by changing the settings of the air classifiers.

26.12.2   Cast Calcium Silicate Concrete Products

High density calcium silicate concrete products are made, particularly in Eastern Europe [26.46], by co-grinding silica sand and lime, adding controlled amounts of water and unground sand, and casting the mix into moulds. The products are then autoclaved at elevated temperature and steam pressure to produce hydrated calcium silicate concrete. A range of products, including slabs, beams and lintels, are produced, with and without metal reinforcement.

26.13   References

[26.1]   D.N. Little, "Stabilization of Pavement Subgrades and Base Courses with Lime", The National Lime Association, Kendall/Hunt Publishing Co, ISBN 0-8403-9632-5.

[26.2]   M. Kelley, "A Long Range Durability Study of Lime Stabilised Bases at Military Posts in the Southwest", Bulletin 328, National Lime Association, WSA, 1977.

[26.3]   "Lime Stabilisation", Proceedings of a seminar at Loughborough University, 25 Sep. 1996, ISBN 0-7277-2563-7.

[26.4]   "Lime Stabilisation Manual", British Lime Association, 1990.

[26.5]   "Specification for Highway Works", Department of Transport, HMSO, London.

[26.6]   "Design Manual for Roads and Bridges", Vol. 4, Section 1, Part 6 HA 74/95, "Design and Construction of Lime Stabilised Capping".

[26.7]   "Lime in Building", British Quarry and Slag Federation (now Quarry Products Association), 1974.

[26.8]   P.T. Sherwood, "Views of the Road Research Laboratory on Soil Stabilisation in the United Kingdom", Cement Lime and Gravel **42**, No. 9, 1967, 277–280.

[26.9]   P.T. Sherwood, "Soil stabilisation with cement and lime", HMSO, London, ISBN 0-11-551-171-7.

[26.10] BS 1377, Part 2: "Methods of test for soils for civil engineering purposes — Classification tests", 1990.

[26.11] BS 1924, Part 2: "Stabilised materials for civil engineering purposes — Methods of test for cement-stabilised and lime-stabilised materials", 1990.

[26.12]   J. Perry, R.A. Snowdon, P.E. Wilson, "Site Investigation for Lime Stabilisation of High-
          way Works", Department of Transport, 1995.

[26.13]   BS 1377, Part 4. "Methods of test for soils for civil engineering purposes — Compac-
          tion-related tests", 1990.

[26.14]   BS 1047: "Specification for air-cooled blastfurnace slag aggregate for use in construc-
          tion", 1983.

[26.15]   Lime-Lites, Vol. LXI, Jan.–June, 1995, p. 41, National Lime Association.

[26.16]   BS 5262: "Code of practice for external renderings", 1991.

[26.17]   prEN XXX: "Unbound and Hydraulically bound mixtures for roads — specification for
          lime-treated mixtures for road construction and civil engineering — definitions, compo-
          sition and laboratory mixture requirements", in preparation.

[26.18]   EN 459: "Building lime", Part 1: "Definitions, specifications and conformity criteria",
          DD ENV 459-1, 1995 (being revised); Part 2: "Test methods", BS EN 459-2, 1995 (being
          revised); Part 3: "Conformity evaluation", prEN 459-3, in preparation.

[26.19]   ASTM Specification C207-91, "Standard Specification for Hydrated Lime for Masonry
          Purposes", 1992.

[26.20]   prEN XXX: "Hydraulic Road Binders — Composition, specifications, and conformity
          criteria", in preparation.

[26.21]   J.A. Epps, "Lime in Hot Mix Asphalt: Design, Properties and Performance", Proc. of
          the International Lime Conference, Berlin, 1994.

[26.22]   "Use of Anti-stripping Additives in Asphaltic Concrete Mixtures", NCHRP Report
          No.373, ISBN 0-309-05374-9.

[26.23]   M.A. Wallace, "How mortar is chosen", Masonry Construction, Feb. 1991, 50–55.

[26.24]   BS 5628: "Code of practice for use of masonry", Part 1: "Structural use of unreinforced
          masonry", 1992; Part 2: "Structural use of reinforced and prestressed masonry", 1995;
          Part 3: "Materials and components, design and workmanship", 1985 (being revised
          1997).

[26.25]   W.H. Harrison, M.E. Gaze, "Laboratory-scale tests on building mortars for durability
          and related properties", Masonry International, Spring, 1989.

[26.26]   G.K. Bowler, W.H. Harrison, M.E. Gaze, A.D. Russell, "Mortar Durability Testing: An
          Update", Masonry International **8**, No. 3, 1995.

[26.27]   "Building Mortars", Building Research Establishment Digest No. 362.

[26.28]   prEN 998: "Specification for mortar for masonry", 1997, Part 1: "Rendering and plaste-
          ring mortar with inorganic binding agents", in preparation; Part 2: "Masonry mortar",
          in preparation.

[26.29]   "Model Specification for Load-bearing Clay Brickwork", Special Publication 56, British
          Ceramic Research Association.

[26.30]   Data Sheet No.3, Mortar Producers Association Ltd, 1972.

[26.31]   BS 1199 and 1200: "Specifications for building sands from natural sources", 1976.

[26.32]   ENV 197, Part 1: "Cement, composition, specification and conformity criteria", 1992.

[26.33]   BS 4887, Part 1: "Mortar admixtures: Specification for air-entraining (plasticizing)
          admixtures", 1986.

[26.34]   U. Dilger, "Design, construction and commissioning of a modern dry ready-mixed mor-
          tar plant at Rotterdam", Zement Kalk Gips **9**, 1988, 462–466 (English: **11**, 1988, 268–
          270).

[26.35]   P. Korf, "Heidelberger Zement's new dry ready-mixed mortar plant at Mainz-Weise-
          nau", Zement Kalk Gips **9**, 1988, 467–470 (English: **11**, 1988, 271–272).

[26.36]   U. Dilger, "Current state of dry ready-mixed mortar technology", Zement Kalk Gips **8**,
          1990, 395–398.

[26.37]   F. Fontanari, "The new Röfix dry ready-mixed mortar plant at Sennwald, Switzerland",
          Zement Kalk Gips **8**, 1990, 399–404.

[26.38]   W. Diem, "Weighing, Mixing and Controlling in Dry Mortar Plants", Zement Kalk Gips
          10, 1990, 496–505.

[26.39]   BS 5492: "Code of practice for internal plastering", 1990.

[26.40]   L.J. Vicat, "A practical and scientific treatise on calcarious mortars and cements", transl.
          by J.T. Smith, John Weale, 1837.

[26.41]   "Lime and lime mortars", DSIR Special Report No. 9, 1927.

[26.42] J. Ashurst, "The technology and use of hydraulic lime", The Building Conservation Directory, 1997, p. 128–131.

[26.43] W.H. Harrison, "Brickwork mortars — sticking together", Brick Bulletin, The Brick Development Association, Winter 1996, 25–27.

[26.44] M.H. Roberts, "The constitution of hydraulic limes", Cement and Lime Manufacture **29**, 1956, 27–36.

[26.45] A.B. Searle, "Limestone and its Products", Ernest Benn, 1935.

[26.46] R.S. Boynton "Chemistry and Technology of Lime and Limestone", John Wiley & Sons, 1980, ISBN 0-471-02771-5.

[26.47] BS 187: "Specification for calcium silicate (sandlime and flintlime) bricks", 1978.

[26.48] B. Bowley, "Calcium silicate Bricks", Structural Survey **12**, No. 6, 1993/94.

[26.49] "Sandlime (calcium silicate) bricks", Technical Service Note, No. TS/E/19, ICI Mond Division, 1980.

[26.50] BS 6073: "Precast concrete masonry units", Part 1: "Specification for precast concrete masonry units", in preparation; Part 2: "Methods for specifying precast concrete masonry units", in preparation.

[26.51] prEN 771-4: "Specification for masonry units", Part 4: "Autoclaved aerated concrete masonry units", in preparation.

[26.52] BS 6463, Part 4: "Methods of test for physical properties of hydrated lime and lime putty", 1987.

[26.53] C.T. Grimm, "Effect of Mortar Air Content on Masonry", J. Masonry Society.

[26.54] D. Neuschütz, R. Nino, "Mixing it right — the K.M.A. integrated dry mortar plant for gypsum, lime and cement-based products", Gypsum Lime and Building Products, April, 1996, 33–36.

[26.55] BS 1191: "Specification for Gypsum Building Plasters", 1973/1994.

27 The Use of Lime in Iron and Steelmaking

27.1 Introduction

As described in section 25.2, in many countries the iron and steel industry vies with building and construction as the largest market segment for lime. Most of the lime used is for fluxing impurities in the basic oxygen steelmaking (BOS) process [27.20]. Lime is also used in smaller quantities in:

a) the sinter strand process for the preparation of iron ore (in addition to lime-stone — see section 11.1.2),
b) in the desulfurisation of pig iron,
c) as a fluxing agent in other oxygen steelmaking process,
d) in the electric arc steelmaking process, and
e) in many of the secondary steelmaking processes.

The BOS process replaced the Bessemer and open hearth steelmaking processes during the 1960's and caused some major changes in both the steel and lime industries [27.1, 27.2]. The process is currently used for 70 % of the world's steel production, with most of the remainder being in electric arc furnaces (EAF). The main processes in iron and steelmaking are illustrated in Fig. 27.1.

The Bessemer process [27.2] was widely used in Europe (except the U.K.) and used "small pieces" of quicklime (typically 20 kg/t of hot metal) to produce a basic slag (CaO ÷ SiO_2 of 2.5 to 3). The preferred lime quality had a high reactivity and low silica and sulfur contents. The open hearth process [27.2, 27.3] was the principal steelmaking process in the U.S.A. and the U.K. It used limestone as the main fluxing agent, with 0 to 20 % of quicklime (up to 5 kg/t): more details are given in section 11.2.

27.2 Production of Sinter

The sinter strand process, which transforms finely divided iron oxide into sintered agglomerates, suitable for charging into the blast furnace, is described in section 11.1.2.

Some steel works add small amounts of quicklime (e.g. 1 kg/t of sinter) to bind smaller particles of ore (0.5 to 2 mm) into larger agglomerates and produce stronger "green" pellets. This helps to increase the porosity of the bed on the sinter strand and increases the potential output per unit area [27.4]. Larger amounts of quicklime (10 to 20 kg/t), are used by some producers to increase the output of the

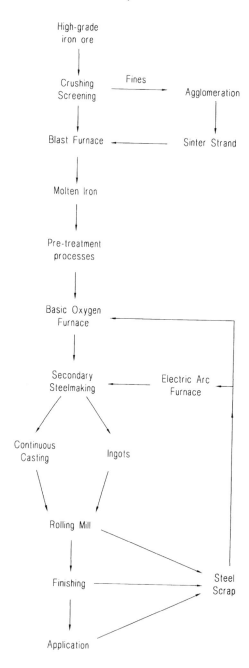

Figure 27.1. The main processes in iron and steelmaking

sinter strand, by 20 to 40 %, and reduce heat usage, by about 5 % [27.5]. The decision whether to add lime depends on site-specific economic factors.

The quicklime used in the sinter process should preferably be low in sulfur, high in reactivity, and milled – e.g., S less than 0.02 %, reactivity greater than 46 °C at 2 min. (BS 6463), and with at least 95 % passing 1 mm.

27.3   Treatment of Pig Iron

The removal of sulfur in the blast furnace from the molten iron into the slag is a partition process. As purer supplies of iron oxide have been obtained, there has been a progressive reduction in the quantity of slag produced in the blast furnace from 700 to 250 kg/t of pig iron (see section 11.1.3). This has reduced the amount of sulfur removed from the iron by the slag, and levels of 0.05 % S are typical [27.5].

This effect, coupled with the increasing need to produce low sulfur steels (less than 0.001 %) has necessitated the introduction of a desulfurisation stage between the blast furnace and the BOS process. In Germany, for example, some 80 % of pig iron is desulfurised [27.5].

Several elements are capable of removing sulfur from molten iron. In order of decreasing effectiveness, they are Ce, Ca, Sr, Ba, Mg, Na [27.4]. They may be used directly as metals, as oxides or carbonates, or indirectly via a synthetic slag.

Calcium carbide was initially used as the desulfurising agent. It was added, in a granular form, to the torpedo ladle (used to transfer the molten iron to the basic oxygen steel making process). It is now generally injected as a powdered mixture with lime (typically 60 to 80 % of calcium carbide) into the molten pig-iron in the transfer ladle. At an addition level of 7 kg/t of pig iron, such mixtures can reduce the level of sulfur to 20 % of the initial value.

Other commonly used desulfurising agents include powdered mixtures of magnesium or aluminium with lime or calcium carbide, and sodium carbonate [27.6]. The choice of reagent depends on economic factors and on the required sulfur level in the treated iron — typically in the range of 0.02 to 0.005 % [27.6].

Steels with the lowest sulfur specifications (of less than 0.001 %, or 10 mg/kg) are generally produced by a combination of a pig iron desulfurisation stage, combined with efficient desulfurisation in the BOS and secondary steelmaking processes.

A typical specification for lime for the injection process is given in Table 27.1.

Table 27.1. Ground quicklime used for injection systems

Parameter	Level
Neutralising value	> 95 %
CO_2	< 2.5 %
H_2O	< 1.5 %
S	< 0.02 %
Reactivity (BS 6463, 2 min.)	> 60 °C
Maximum size	< 1 mm
Passing 75 µm	> 95 %

In Japan, multi-stage pig iron treatments have been developed, which include de-sulfurisation, de-siliconisation (using iron oxide), and de-phosphorisation (using lime plus fluorspar, or sodium carbonate), prior to the BOS process [27.5–27.7]. Such treatments are understood to have the potential to reduce lime consumption in the BOS process by about 25 % relative to the current levels, quoted in section 27.4.

The slags produced by desulfurisation, de-siliconisation and de-phosphorisation are removed before the molten iron is charged to the steel-making furnace.

27.4   Basic Oxygen Steelmaking

27.4.1   The Processes

Several variations of the oxygen steelmaking process have been developed. They include bottom-blown converters (e.g., the OBM process, the Q-BOP process and the LSW process) and the top-blown converters (e.g., the LD process, its variant the LD-AC process, the Kaldo process and the Oberhausen process) [27.3, 27.8].

By far the most widely used is the LD (or Linz Donawitz) converter, developed in Austria. It is frequently called the BOF (basic oxygen furnace – see [27.8] and, for much greater detail, [27.9]). The process based on the BOF is widely referred to as BOS (basic oxygen steelmaking), or the BOP (basic oxygen process). Although the following description concentrates on that process, the basic chemistry and technology of the other basic oxygen steelmaking processes are similar.

A basic oxygen furnace is shown in Fig. 27.2 and a cross-section is given in Fig. 27.3. The furnace is lined with dead-burned magnesia refractory impregnated with carbon. Typically the charge consists of molten iron at about 1250 °C (80 %), scrap iron (18 %), plate iron (1 %) and iron ore (1 %). Varying the amount of

Figure 27.2. Charging a basic oxygen furnace (by courtesy of British Steel Strip Products)

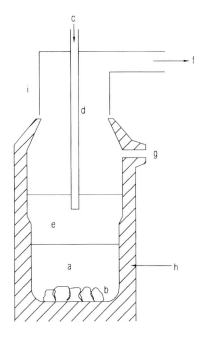

Figure 27.3. Cross-section of a basic oxygen furnace
(a) molten iron; (b) scrap; (c) oxygen; (d) lance;
(e) slag; (f) exhaust gases; (g) tap hole;
(h) refractory lining; (i) movable hood

iron ore is often used to control the final temperature of the steel. Some furnaces take a charge of 450 t of metal.

Oxygen is blown into the molten metal at supersonic speeds through a water-cooled lance for about 15 min. to oxidise some of the impurities. Two to three minutes after the start of the oxygen blow, quicklime (30 to 50 kg/t of hot metal) soft-burned dolomite (10 to 25 kg/t) [27.21] and, optionally, fluorspar (up to 2.0 kg/t) are added [27.10].

The removal of impurities from the molten metal during the course of the oxygen blow is illustrated in Fig. 27.4. Silicon is removed rapidly at the start of the blow (27.1), carbon is removed progressively throughout the blow (27.2), with phosphorous being removed towards the end of the blow (27.3). Manganese, and iron are also oxidised (27.4–27.6).

$$Si + O_2 \rightarrow SiO_2 \tag{27.1}$$

$$2C + O_2 \rightarrow 2CO\uparrow \tag{27.2}$$

$$4P + 5O_2 \rightarrow 2P_2O_5 \tag{27.3}$$

$$2Mn + O_2 \rightarrow 2MnO \tag{27.4}$$

$$FeS + CaO \rightarrow CaS + FeO \tag{27.5}$$

$$2Fe + O_2 \rightarrow 2FeO \tag{27.6}$$

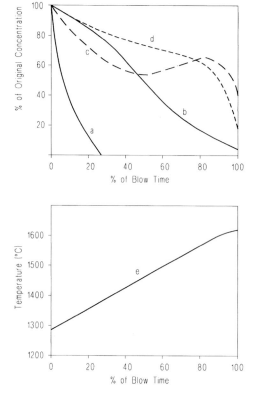

Figure 27.4. Removal of impurities in the BOS process
(a) silicon; (b) carbon; (c) manganese; (d) phosphorous; (e) temperature

The quicklime and dolomite react with the silica and iron oxide (27.7–27.9) to form a molten slag with a basicity (CaO ÷ SiO$_2$) of 2.8 to 3.5, which acts as a reaction medium for refining droplets of iron. The development of the slag and its effect on the refining processes are complex [27.8], with the chemical composition and physical properties of the slag changing throughout the blow. A typical final slag analysis is given in Table 27.2.

Table 27.2. Typical analysis of a final BOS slag

Component	Concentration (% m/m)
CaO	45
FeO	20
SiO$_2$	17
MgO	6
MnO	6
P$_2$O$_5$	1
S	0.3
Other (including Al$_2$O$_3$)	ca. 5

$$SiO_2 + 3CaO \rightarrow (CaO)_3SiO_2 \qquad (27.7)$$

$$P_2O_5 + 3FeO \rightarrow Fe_3(PO_4)_2 \qquad (27.8)$$

$$Fe_3(PO_4)_2 + 3CaO \rightarrow Ca_3(PO_4)_2 + 3FeO \qquad (27.9)$$

The magnesium component of the dolomite raises the concentration of MgO in solution in the slag to 6 to 8 %, and helps to protect the magnesite refractory lining of the furnace by reducing the rate of solution of magnesium oxide.

Fluorspar may be added to control the viscosity of the slag. The correct slag viscosity is necessary to form a foam which acts as a reaction medium. If the viscosity is too low, the foam does not form, whereas if it is too high, explosive slopping can occur. In either case, the refining efficiency is reduced.

The height of the lance is critical. If the lance is raised, the amount of FeO in the slag increases and the slag volume rises, with a consequent loss of yield of Fe. It also increases the risk of the slag slopping out of the furnace. Conversely, if the lance is lowered the reverse happens; the slag volume is reduced, and less phosphorous and sulfur are removed.

Sulfur removal is a partition process and is favoured by high basicity, a high FeO level in the slag, a large volume of slag and high temperatures. Typically 40 % of the sulfur in the hot metal is removed in the BOS vessel. Phosphorous removal is favoured by lower temperatures; otherwise the optimum conditions are similar to those for sulfur.

As silicon levels in hot metal have reduced in recent years, so have the amounts of quicklime required to combine with it. The lower slag volumes have reduced the capability of the BOS process to remove sulfur and phosphorous. This, coupled with the increasing demand for higher specification steels, has placed increased emphasis on de-sulfurisation and de-phosphorisation of pig iron (see section 27.3). At some UK plants, a siliceous stone (70 % SiO_2 and 13 to 17 % Al_2O_3) is added to increase the volume of slag.

Mixing of the hot metal and the slag helps to bring reactions towards equilibrium. In the early stages of the blow, the reaction of the oxygen with carbon dissolved in the metal produces carbon monoxide and agitates the metal. Towards the end of the blow, however, less carbon monoxide is produced and gas stirring using injected argon or nitrogen is widely used. Stirring helps to reduce the amount of FeO in the slag, thereby increasing the yield of iron.

When the oxygen blow is completed (generally after 15 to 17 min.), the temperature is measured (typically about 1,600 to 1650 °C) and a sample of the hot metal is taken for analysis of the impurities. If these are on target, the furnace is tapped: if not, it is re-blown. When the required temperature and metal analysis are achieved, the furnace is discharged into a ladle and the slag is rejected.

It is important that as much of the slag is removed as possible, as silica can displace phosphate ions, allowing phosphorous pentoxide to migrate back into the hot metal.

$$Ca_3(PO_4)_2 + SiO_2 \rightarrow (CaO)_3SiO_2 + P_2O_5 \qquad (27.10)$$

At this stage, the molten steel typically contains 99.5 % of Fe.

The exhaust gases from the BOF consist mainly of carbon monoxide, with some carbon dioxide and nitrogen. They also carry iron oxide fume and fines (mostly less than 3 mm) from the lime and other materials charged into the vessel.

27.4.2   BOS Specifications for Quicklime and Soft-burned Dolomite

The requirements of the BOS process have set the quality standards for much of the lime industry in many countries since the 1960s. They have influenced the selection of limestone, kilns, fuels, operating parameters and the design of lime handling plants (see chapters 14, and 16 and section 17.1).

Tables 27.3 and 27.4 summarise typical specifications for high-calcium lime [27.11] and calcined dolomite [27.12] used as fluxing agents in the BOS process. Both products should be:

a) low in combined sulfur (any sulfur present reduces the sulfur removing capacity of the slag),
b) consistent in chemical analysis,
c) reasonably low in $CaCO_3$ (a coolant),
d) low in $Ca(OH)_2$ (both a coolant and a source of hydrogen which can lead to embrittlement),
e) low in SiO_2 (1 % of SiO_2 combines with about three times its weight of CaO in the BOF),
f) relatively small in size, with a low fines content,
g) high in reactivity to water in the case of quicklime and lightly burned in the case of the calcined dolomite (but see section 27.4.3).

Table 27.3. Typical BOS specification for quicklime

Parameter	Criterion	Level
Neutralising value (as CaO)	average	> 95.0%
	minimum	93.0%
SiO_2	average	< 1.0%
	maximum	1.5%
MgO	average	< 1.5%
S	average	< 0.03%
	maximum	0.04%
Loss on ignition (due to CO_2)	maximum	2.5%
Reactivity (BS 6463, 2 min)	minimum	46 °C
Grading	+ 44 mm	0%
(ex works)	– 44 + 38 mm	< 10%
	– 38 + 12 mm	> 75%
	– 12 + 6 mm	< 10%
	– 6 mm	< 5%

Table 27.4. Typical BOS specification for calcined dolomite

Parameter	Level
CaO	54–63%
MgO	34–43%
SiO_2	≤ 2.0%
S	≤ 0.06%
"Reactivity"	"soft burned"
Passing 5 mm	≤ 15%

The specifications adopted by individual steelworks reflect local economic factors including the availability of suitable lime products and the qualities of steel being produced.

The chemical requirements for the injection grade quicklime used in steelmaking processes such as the LD-AC and the Q-BOP, are similar to those in Table 27.3. The lime is ground to suit the requirements of the injection system (e.g., Table 27.1).

27.4.3 Discussion of Grading and Reactivity/Density Requirements

The solution of the quicklime and calcined dolomite in steelmaking slags is a complex process which is controlled by both chemical and physical factors. It is important for the lime to react rapidly with silica and iron oxide at the start of the blow to form a molten slag and, thereafter, to dissolve rapidly in the slag.

Surprisingly, even with a rapidly reacting quicklime, only half of it is in solution when the blow is 80 % complete (i.e. after about 12 min.) [27.4]. Most of the remaining half of the lime dissolves during the final 20 % of the blow (or 3 min.), as more FeO is produced and the volume of the slag increases. The final slag contains some unreacted calcium oxide and 40 to 45 % of total calcium oxide.

Table 27.3 gives a typical BOS specification for quicklime with a moderate to high reactivity to water and with a nominal size of 6 to 38 mm.

A number of researchers have investigated how the properties of the quicklime affect its solution rate. [27.13] and [27.14] reported that the rate of solution increased with increasing reactivity to water.

Anderson and Vernon [27.1] reported that the rate of solution of lime increased as the mean apparent density decreased and as the particle size was reduced. For a given quicklime, reactivity correlated with mean apparent density (see Fig. 13.2), but reactivity was not the fundamental parameter. Moreover, reactivity could be affected by absorption of atmospheric water and carbon dioxide (see section 13.2).

A subsequent investigation into desulfurisation in the BOS process concluded that sulfur removal increased with increasing reactivity of lime and also with decreasing particle size [27.15].

Oates [27.16] reported laboratory results for a study of the solution rates, in a synthetic BOS slag, of quicklimes with a range of mean apparent densities and particle sizes. The solution curves fitted a solution rate equation, based on the external surface area per unit weight of the quicklime particles. This concept was developed to explain some anomalous results from full scale and pilot plant trials.

The external surface area concept was then used to explain the low solution rate of a commercial quicklime (below the reactivity specification in Table 27.2, but within the grading specification), which caused unsatisfactory operation in a BOF. The specific surface area of the product was increased to a "normal" value, by reducing the size of the screens on which it was produced, but the reactivity was kept below the specified value. This increased the solution rate of the quicklime and its performance in the BOF became fully acceptable.

A recent paper [27.11] indicates that at least one steelmaking company is considering the use of mean apparent density and particle size to characterise quicklime quality for the BOS process. Increasing solution rates may become increasingly important as blow times are reduced to maximise production rates.

Another aspect of reactivity and mean apparent density is that, as reported in [27.17], at the final temperature in the BOF of 1,600 °C, lightly burned quicklime sinters to "medium burned" within 5 min. and to "hard burned" within 10 min. (Table 27.5).

Table 27.5. Sintering of quicklime at 1600 °C

Time (min.)	Reactivity (ml 4N HCl)[a]	Estimated apparent density (g/ml)[b]
0	300	1.8
5	230	2.1
10	130	2.3
15	110	2.4

[a] Acid titration test result at 5 min. [27.23].
[b] Based on reactivity and on Figs. 13.1 and 13.2.

In conclusion, it appears that the fundamental quicklime parameter affecting its solution rate and performance in BOS slags is the specific external surface area. The area is related to the particle size distribution and the mean apparent density. Reactivity to water is related to mean apparent density. Other factors, such as the structure of the limestone, could be equally significant. In addition, degradation caused by handling and storage could significantly change the relative external surface areas of different quicklimes and would need to be taken into account when comparing their slag-forming characteristics.

27.5   The Electric Arc Process

27.5.1   The Process

The Electric Arc Furnace (abbreviated to EAF or ARC) is used to produce about 30 % of the world's steel. Originally, it was used to produce relatively low tonnages of special high qualities of steel, which could only be made by this process. Increasingly, however, its main role is to melt scrap iron and steel. Ultra high power (UHP) EAFs are now used to produce relatively large tonnages of steel in direct competition with those made in the BOF [27.2]. EAFs with a tapping capacity of 150 t are common. Because of its much lower capital cost than a blast furnace plus basic oxygen furnace, the EAF is economically viable in smaller units, where scrap steel is readily available and electricity costs are not excessive.

The furnace consists of a shallow cylindrical bath with a roof that can be lifted and swivelled, and through which three carbon electrodes can be inserted (Fig. 27.5). At the start of the process, the electrodes are withdrawn and the roof swung clear. The scrap is then charged into the furnace. When charging is complete, the roof is swung back into position and the electrodes are lowered to make contact with the scrap.

Figure 27.5. An electric arc furnace (by courtesy of Co-steel Sheerness plc and Slater Crosby)

A powerful electric current is then passed through the charge, an arc is struck, and the heat produced melts the scrap. Quicklime (typically 35 to 40 kg/t steel), calcined dolomite (about 10 kg/t) and fluorspar (about 7 kg/t) are added [27.2, 27.5]. A recent trend has been the use of pre-mixed blends containing 80 % of quicklime and 20 % of dolomite. The fluxing agents combine with the impurities

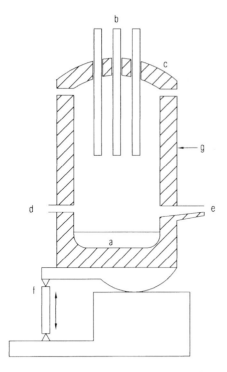

Figure 27.6. Cross-section of an electric arc furnace   (a) bath; (b) electrodes (lift and slew); (c) lid (lift and slew); (d) slag discharge; (e) steel discharge; (f) tilt mechanism; (g) refractory lining

in the metal to form a liquid slag. As the process proceeds, additional impurities migrate from the hot metal into the slag. The process removes about 40 % of the sulfur (giving about 0.02 to 0.03 % S) [27.18].

Samples of the steel are taken periodically to check its composition. When the required composition has been achieved, the temperature of the steel is measured and, if necessary adjusted to the level required for casting (about 1,600 °C).

The slag is removed by tilting the furnace so that it pours through the slag hole at the "back" of the furnace. The steel is then discharged through the tap hole at the "front" into a teeming ladle.

In some variations of the process, either iron oxide or oxygen are added to oxidise carbon and other impurities. The resulting "oxidising" slag (rich in FeO) is then removed and the steel is treated with a second reducing slag (low in FeO). This double slag technique can remove over 60 % of the sulfur in the raw steel, giving 0.015 to 0.025 % S [27.18], but is a relatively slow process.

Cycle times are generally slow relative to the BOS process. When operating in the traditional manner with an oxidising slag, followed by a reducing slag, cycle times of 4 hours are common. When the process is used primarily to melt scrap steel, cycle times of 1.5 hours are more typical [27.18].

As an indication of the flexibility of the EAF, it is used in some developing countries for the direct reduction of iron ore by carbon [27.18]. The furnace is charged with iron ore intimately mixed with ground carbon. Quicklime, calcined dolomite and fluorspar are added and an oxygen lance is used to oxidise the impurities much as in the BOS process. The slag is removed and the chemical composition of the resulting iron is adjusted by adding iron oxide or oxygen, in conjunction with a lime-based slag, to produce the required quality of steel.

27.5.2   EAF Specifications for Quicklime and Soft-burned Dolomite

Most of the quicklime and calcined dolomite supplied to the electric arc furnace is in the 10 to 30 mm size range. Because of the relatively slow refining times, the solution rate of the lime (i.e., its grading and reactivity to water) is generally not an issue. However, there is a general trend towards finer products (e.g., 6 to 20 mm) and, in the case of high-calcium lime, towards higher reactivities (e.g., > 46 °C at 2 min., using the BS 6463 method).

Where pneumatic injection systems are used, finer grades (e.g., 2 to 6 mm) are specified to suit the characteristics of the system [27.5].

Otherwise, the lime requirements are as for the BOS process (section 27.4.2).

27.6   Secondary Steelmaking Processes

27.6.1   The Processes

Secondary steelmaking is used to bring the steels produced by the BOF and the EAF into specification. In many works it acts as a physical and chemical buffer between the primary steelmaking processes and the continuous caster. It has been the main growth area in steelmaking since the early 1980s as a result of:

a) demand for improved steel quality,
b) pressures to increase productivity and reduce costs, and
c) the requirements of the continuous casting process.

Particularly as a result of (b), the roles of the primary steelmaking processes have changed, with the BOF being used mainly for the removal of carbon from pig iron, and the EAF being used mainly to melt steel scrap. This places increased emphasis on refining the hot metal either before the primary steelmaking operation (i.e., treatment of pig iron — see section 27.3), or after it (i.e., secondary steelmaking).

The processes used in secondary steelmaking are many and varied [27.2, 27.8, 27.22]. For the purposes of this section, they can be divided into five types of operations [27.5]:

a) slag removal to prevent re-absorption of impurities by the steel,
b) mixing, using gas injection (generally argon), or electromagnetic stirring,
c) vacuum treatment to remove dissolved gases and gaseous reaction products,
d) addition of alloying and fluxing agents under oxidising/reducing conditions, and
e) heating to adjust the temperature for subsequent treatments.

Many of the processes combine two or more of the above operations.

Slags play an important role in secondary steelmaking and, as the objectives of the several processes vary, so do the compositions of the slags used [27.5, 27.19].

Quicklime is added in various forms.

a) In many processes, it is injected as a powder.
b) In others, it is used in a finely divided, dust-free form, mixed with fluorspar and, depending on the application, with alumina. The mixture forms a synthetic slag containing 55 to 80 % CaO, with 15 to 30 % of fluorspar and up to 15 % of alumina. Such mixtures are often added in a bagged form. Mixtures of lime (90 %) with fluorspar, with or without ground carbon are also used to cover the molten steel to insulate it and prevent re-oxidation.
c) In the argon oxygen decarburization (AOD) process for refining stainless steel, quicklime is added in a granular form.

Addition rates of quicklime vary widely. The AOD process uses about 90 kg/t steel, whereas other processes use as little as 1 kg/t. The average total use in secondary steelmaking is in the range 5 to 15 kg/t [27.5].

Calcined dolomite is not widely used in secondary steelmaking, but is often a component of synthetic slags, which may contain up to 10 % of MgO [27.12].

27.6.2   Quicklime Specifications for Secondary Steelmaking

Chemical requirements are generally as detailed in section 27.4.2 and Table 27.3. The major exceptions are:

a) the AOD process, which requires a $CaCO_3$ content of less than 1 %, and
b) the sulfur content in limes used for synthetic slags, where levels of 0.15 % can be tolerated.

The *reactivity* does not appear to be important, presumably because most of the products are sufficiently finely divided to give the required solution rates. Indeed, some producers favour medium to low reactivity quicklimes, which are more resistant to degradation and may absorb less water vapour from the atmosphere.

As mentioned in section 26.6.1, the *grading* depends on the handling system. Processes in which the quicklime is injected in a stream of gas generally require a product with a guaranteed top size of about 1 mm. One product supplied for this application contains >95 % less than 75 μm (Table 27.1). The requirements for synthetic slag mixtures vary, with favoured sizes being 2 to 6 mm and 0.75 to 2 mm. The AOD process uses granular products in the size range 5 to 38 mm.

27.7 References

[27.1] L.C. Anderson, J. Vernon, "The Quality and Production of Lime for BOS", Proc. 73rd BISRA Steelmaking Conference, Scarborough, Oct. 1969.
[27.2] A. Jackson, "Modern Steelmaking for Steelmakers", George Newnes, 1967.
[27.3] A. Jackson, "Oxygen Steelmaking for Steelmakers", Butterworths, 1969.
[27.4] J.A.H. Oates, "The Use of Lime and Limestone in Iron and Steel Production", notes on a British Steel seminar for Buxton Lime Industries, June 1993.
[27.5] D. Springorum, "Development of the specific CaO Consumption in the Iron and Steel Industry", Proc. European Lime Association's 1st Technical Conference, Cologne, 1992.
[27.6] R. Baker, "Process Considerations and Options Available for the Production of Low Residual Steel from the Oxygen Converter", The Metallurgist and Materials Technologist, Dec. 1984, 624–627.
[27.7] R, Baker, N.J. Cavaghan, A. Herbert, N.S. Normanton, "High Quality Steelmaking", Perspectives in Metallurgical Developments, 1984.
[27.8] C. Moore, R.I. Marshall, "Modern Steelmaking Methods", The Institution of Metallurgists, Monograph 6, 1980.
[27.9] J.M. Graines (Ed.), "BOF Steelmaking", Vols. 1 & 2, Iron and Steel Society of the Americas/Institute of Mining, Metallurgical and Petroleum Industries, Warrendale, Pa 15086.
[27.10] J. Benbow. "Steel Industry Minerals", Industrial Minerals, Oct. 1989, 27–43.
[27.11] B.C. Welbourn, "Lime Quality at British Steel", P. Bromley, Proc. Réunion de Commission Aciéries de Conversion, Sollac, Dunkerque, June 1995.
[27.12] Personal correspondence from E. Parry, Redland Aggregates Ltd.
[27.13] R.O. Russell, "Lime Reactivity and Solution Rates", Journal of Metals, Aug. 1967.
[27.14] R.P. Singh, D.N. Ghosh, "Correlations of the Water Reactivity Test of Limes with their Rates of Dissolution in Steelmaking Slag", Trans Indian Institute of Metals, Vol. 38, No. 3, June 1985, pp. 207–214.
[27.15] A.A. Hejja, "The Effect of the Reactivity of Lime on Desulfurisation Efficiency in the BOF", National Inst. for Metallurgy, Johannesburg, Report No. 1425, June 1972.
[27.16] J.A.H. Oates, "Solution Characteristics of Quicklime in BOS Slags", Proc. European Lime Association's 1st Technical Conference Cologne, 1992, pp. 215–225.
[27.17] K.H. Obst et al., "Effect of Refiring on the Reactivity of Lime", Tonind. Ztg. **92**, No. 10, 1962, 389–392.
[27.18] "Desulfurisation", notes on discussions with D. Thornton, British Steel Corporation.
[27.19] W. Loscher, W. Pluschkell, R. Scheel, "Ladle Slags in Steel After-treatment". In "Slags in Metallurgy", Verlag Stahleisen, Düsseldorf, 1984, pp. 263/280.
[27.20] R. Baker, "Lime in Steelmaking", Proc. 6th International Lime Congress, London, June 1986.
[27.21] J. O'Brien, "Dolomitic Lime Flux for the BOF", Proc. International Lime Association's 3rd International Symposium, May 1974, pp. 11–18.
[27.22] W. Burgmann, "Vacuum Process Engineering and Ladle Metallurgy in the production of Steel", Steel Times International, June 1990, 11–18.
[27.23] N.E. Rogers, Cem. Lime. and Gravel, June 1970, 149–153.

28 Water and Sewage Treatment

28.1 Drinking Water

28.1.1 Introduction

Slaked lime is widely used in the production of water for human consumption. Indeed, in tonnage terms, it competes with chlorine as the major chemical used in water treatment.

Water treatment typically involves several processes, depending on which impurities need to be reduced, and includes:

a) the removal of suspended/colloidal matter (which also helps to reduce tastes, odours and coloration),
b) reducing hardness,
c) reducing the concentrations of dissolved metals,
d) disinfection,
e) pH adjustment.

A wide range of chemicals are used in water treatment. Lime is used both as an alkali and as a source of calcium ions. As an alkali, it competes in certain processes with caustic soda and soda ash. The choice between alternative chemicals is generally based on economic factors and on the ease of use in a particular installation.

Slaked high-calcium lime may be used to:

a) adjust the pH of the water to the optimum level for precipitation and subsequent treatment,
b) soften hard waters by removal of dissolved calcium and, optionally, magnesium, and
c) adjust the pH of treated water to minimise corrosion and leaching of metals.

Half-burned dolomite ($CaCO_3 \cdot MgO$) is used in some treatment plants as a closely graded granular "filter" medium to raise the pH of water without the risk of over-dosing.

The design of a water-treatment plant and the selection of chemicals is a complex operation. They depend on the levels and variabilities of the impurities in the raw water, the cost effectiveness of alternative treatment chemicals, the cost of waste disposal and the required quality of the treated water. Consideration also needs to be given to the interactions between the various treatment processes [28.1, 28.2].

Lime is generally purchased as the powdered hydrate and is dispersed in water to produce a milk of lime, which is then dosed into the process to produce the required pH. Where quicklime is purchased, it is almost always slaked before use. High purity, high-calcium lime is generally specified (section 28.1.8) to ensure that the lime does not add unacceptable impurities to the water, produce excessive quantities of sludges, or cause unacceptable turbidity.

28.1.2   A Basic Water Treatment Works

Figure 28.1 gives an example of a water treatment system for soft water, which can be divided into four stages:

a) the removal of suspended/colloidal matter and aluminium/iron at pH 5 to 7,
b) the removal of manganese at pH 8,
c) disinfection with chlorine, and
d) addition of alkali to produce the "stability" pH.

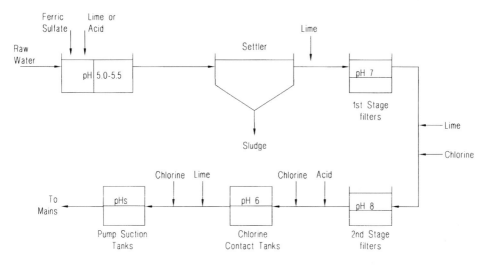

Figure 28.1. Example of full water treatment system for soft water

In the first stage, the pH is adjusted to between 5 and 5.5 by the addition of milk of lime or acid. A solution of ferric sulfate is added under turbulent conditions to coagulate both suspended solids and colloidal organic matter (see section 28.1.3). The removal of organic matter and coloration is particularly effective at relatively low pH levels (if not removed, organic matter can subsequently be chlorinated to produce undesirable compounds). The coagulated solids are then flocculated under less turbulent conditions and are removed as a sludge by settlement or flotation. The pH of the clarified water is then raised, if necessary, to 7 by the addition of milk of lime to precipitate aluminium and iron which are removed, together with any residual suspended solids, in the first stage sand filter.

In the second stage, the pH may be raised to 8 by the addition of lime and chlorine to precipitate manganese as the dioxide, which is removed in the second stage sand filter.

In the third stage, chlorine is used to disinfect the water. As the concentration of the more active agent, hypochlorous acid, is greater at reduced pH levels, the pH may be lowered by the injection of carbon dioxide. The water is then retained for the requisite period in the chlorine contact tanks.

In the final stage, the pH is adjusted by the addition of lime to produce the "stability pH" (see section 28.1.5), which minimises the corrosion potential of the treated water. If the level of residual chlorine in the treated water is below target, additional chlorine may be added: if it is above target, sulfur dioxide is added to reduce the hypochlorous acid.

28.1.3   pH Adjustment, Coagulation and Precipitation

While some slaked lime is used to neutralise acidic waters, its main use is to produce the optimum pH for subsequent treatments, which include coagulation, flocculation, and precipitation. These processes are described briefly below: for more detail, see [28.2].

Suspended particles acquire an electrostatic charge, which, in water treatment, is usually negative. The charges produce a repulsion between particles, which tends to stabilise the suspension. In colloidal suspensions, which have a maximum particle size of less than $2\,\mu m$, this repulsion effectively prevents settling.

Coagulating agents are selected to have an opposite charge to that of the suspended solids and effectively neutralise that charge (e.g., Al^{3+} and Fe^{3+}). This destabilises the suspension and allows the particles to come together. Flocculating agents form bridges between the particles and lead to the formation of large agglomerates, which can be removed by settlement or flotation.

The most commonly used coagulating agents are ferric and aluminium salts (e.g. ferric chloride, $FeCl_3$, ferric sulfate, $Fe_2(SO_4)_3$ and alum, $Al_2(SO_4)_3 \cdot 14H_2O$), which are precipitated as the hydroxides, (e.g., equations 28.1, 28.2). Both reactions generate mineral acids which may need to be neutralised by the addition of an alkali to produce the required pH.

$$Al_2(SO_4)_3 + 6H_2O \rightarrow 2Al(OH)_3\downarrow + 3H_2SO_4 \tag{28.1}$$

$$FeCl_3 + 3H_2O \rightarrow Fe(OH)_3\downarrow + 3HCl \tag{28.2}$$

Organic coagulating agents are also used, but do not remove hydrophobic organic colloids as effectively as ferric and aluminium hydroxides. In the case of waters that contain significant levels of magnesium, the addition of sufficient lime to partially soften the water in the pH range 9.5 to 10 (see section 28.1.4) precipitates magnesium hydroxide which can serve as a coagulating agent.

The removal by precipitation of iron, aluminium and manganese has already been described in section 28.1.2. Many other metals (e.g., barium, cadmium, chromium (III), lead, copper, and nickel) can also be precipitated by adjustment of the pH (see Table 28.9).

The efficiency of precipitation of the metals as hydroxides/basic carbonates depends on the detailed chemistry of the water (e.g., interferences between ions often increase residual concentrations, while removal is often enhanced by adsorption on to flocs and activated carbon). For this reason, levels of reduction reported in the literature vary significantly. It is generally recommended that proposed new water treatment installations should first be evaluated in the laboratory and pilot plant.

28.1.4   Softening of Hard Water

"Hardness" in water increases the amount of soap required to produce a lather. It is largely caused by dissolved calcium and magnesium salts, of which the most common are:

a) calcium and magnesium bicarbonates (produced by the reaction of water containing dissolved carbon dioxide with carbonate rocks such as limestone and magnesite).
b) other calcium and magnesium salts (present in some rocks, and produced by the reaction of acid rain on rocks such as limestone and magnesite).

Hardness caused by calcium and magnesium bicarbonates is referred to as "carbonate" hardness (formerly called "temporary" hardness). It can be reduced by boiling the water to drive off carbon dioxide and precipitating calcium and magnesium carbonates (28.3, 28.4). This process causes the "lime scale" frequently found in kettles and boilers.

$$Ca(HCO_3)_2 \rightarrow CaCO_3\downarrow + H_2O + CO_2\uparrow \qquad (28.3)$$

$$Mg(HCO_3)_2 \rightarrow MgCO_3\downarrow + H_2O + CO_2\uparrow \qquad (28.4)$$

Hardness caused by other calcium and magnesium salts is called "non-carbonate" hardness (formerly "permanent" hardness). It cannot be removed by boiling.

Carbonate hardness is normally reduced by treatment with slaked lime. Initially calcium carbonate is precipitated at pH levels above 8 or 9, depending on the chemistry of the water (28.5), with magnesium hydroxide being co-precipitated at pH levels greater than 9.0 (28.6).

$$Ca(HCO_3)_2 + Ca(OH)_2 \rightarrow 2CaCO_3\downarrow + 2H_2O \qquad (28.5)$$

$$Mg(HCO_3)_2 + 2Ca(OH)_2 \rightarrow Mg(OH)_2\downarrow + 2CaCO_3\downarrow + 2H_2O \qquad (28.6)$$

In practice, the softening process is more complicated than indicated by the above equations (e.g., depending on the temperature and pH, the magnesium may be precipitated as a mixture of $Mg(OH)_2$, $MgCO_3$, $MgCO_3 \cdot 3H_2O$, or $Mg_4(CO_3)_3(OH)_2 \cdot 3H_2O$, each of which has a characteristic solubility) [28.2].

Non-carbonate hardness is generally reduced by the lime-soda process, which involves treatment with slaked lime and soda ash (28.7, 28.8). The magnesium is precipitated by the lime as the hydroxide at pH 9 to 10. The co-produced calcium salt, together with other calcium salts originally present, reacts with the soda ash to form a calcium carbonate precipitate. The precipitates are generally removed as sludges by settling and/or filtration.

$$Mg^{++} + Ca(OH)_2 \rightarrow Mg(OH)_2\downarrow + Ca^{++} \tag{28.7}$$

$$Ca^{++} + Na_2CO_3 \rightarrow CaCO_3\downarrow + 2Na^+ \tag{28.8}$$

"Pellet reactors" are used in some countries for the reduction of hardness caused by calcium bicarbonate. Milk of lime is added to the hard water just before it enters the base of the reactor, causing the water to be supersaturated with respect to calcium carbonate. The water is then passed through a fluidised bed of grains of sand, which act as nuclei for the growth of calcium carbonate crystals. The grains grow into pellets, which are removed from the reactor when they have grown beyond a certain size. The pellets are relatively pure and have a low moisture content. They can more easily be sold/tipped than the traditional sludges. The ultra-fine milks of lime which have been developed in recent years (see section 22.8) are particularly suitable for use in pellet reactors on account of their high rates of solution.

Softening brings three main benefits:

a) a reduction in the amount of soap required,
b) reduced scale formation in heating systems leading to lower fuel/maintenance costs,
c) an increase in the acceptability of the water for domestic use.

However, the rising costs of softening, including the cost of sludge disposal, are causing water companies to reduce the amount of softening that they do.

28.1.5   Corrosion Control

Even at pH levels in the range 7 to 9, water can extract iron, copper and lead from piping and zinc from galvanised steel. This can result in contamination of the water as well as corrosion of the distribution system.

Contamination and corrosion are minimised by establishing the "stability pH" of the water (pHs). The stability pH is that at which the water is saturated with respect to calcium carbonate, i.e., it will neither dissolve nor precipitate $CaCO_3$. The value of pHs can vary from 7.8 to 9.2, depending on the chemistry of the water.

While the adjustment to the stability pH is often done with caustic soda, milk of lime is also suitable, providing its water insoluble content is sufficiently low (e.g., less than 2.0 %) to avoid causing an unacceptable increase in turbidity.

28.1.6   Half-burned Dolomite

Half-burned dolomite ($CaCO_3 \cdot MgO$) is used in a limited number of countries for small water treatment plants, swimming pools and waste water plants, as a closely graded granular "filter" material. Its production is described in section 16.9.

As water passes through a packed bed of half-burned dolomite, it reacts with the magnesium oxide component, which automatically raises the pH towards the stability pH (see section 28.1.5), without the cost of installing pH control equipment. While the contact time should be long enough to produce the required increase in pH, it should not be excessive when the chemistry is such that there is a risk of ferric hydroxide and/or manganese dioxide precipitating within the bed. This can lead to "blinding" of the surface of the particles and a consequent loss of reactivity. Similarly, high sulfate levels in the raw water can cause blinding with calcium sulfate.

As the particles of half-burned dolomite react, they become finer, and reduce the porosity of the bed. It then becomes necessary to back-wash the bed to remove the finest particles and to replace them with fresh material.

28.1.7   Disinfection

Chlorine, or sodium hypochlorite, is almost always used to disinfect drinking water (i.e., reduce pathogenic organisms, such as bacteria and viruses, to safe levels). The treatment has the advantage of leaving residual dissolved chlorine in the treated water, which helps to ensure that the water remains resistant to re-infection in the distribution system.

However, it should be noted that lime has been used successfully to disinfect water. The required treatment (i.e., pH and retention time) depends on the nature of the organisms and the temperature. The resulting high pH makes the water taste unpleasant, and should be reduced to about pH 9, e.g., by the injection of carbon dioxide. Further information on disinfection is given in the section on sewage treatment (see section 28.4.4).

28.1.8   Specifications

European Standards are being prepared for chemicals used for the treatment of water intended for human consumption, including high-calcium lime and half-burned dolomite [28.3–28.5]. It seems likely that they will set the norm for lime products used in many environmental applications as well as in the water industry.

High-calcium limes, which include lump and pulverised quicklime, hydrated lime and milk of lime, are categorised in [28.3] into three Types (Table 28.1), depending on the levels of the major and minor components, namely:

a) the water-soluble calcium oxide/hydroxide,
b) the main impurities (SiO_2, Al_2O_3, Fe_2O_3 and MnO_2).

Table 28.1. High-calcium lime for use in water treatment — major and minor components and grading (proposed limits)

Parameter[a]	Form of lime[b]	Type 1[c]	Type 2[c]	Type 3[c]
% CaO[d]	Ground & lump	≥ 87	≥ 84	≥ 80
% Ca(OH)$_2$[d]	Hydrate & milk	≥ 92	≥ 87	≥ 83
% SiO$_2$	All	≤ 2.0	≤ 3.0	≤ 4.0
% Al$_2$O$_3$	All	≤ 0.5	≤ 1.0	≤ 2.0
% Fe$_2$O$_3$	All	≤ 0.5	≤ 1.0	≤ 1.5
% MnO$_2$	All	≤ 0.15	≤ 0.4	≤ 0.4
% greater than 600 µm	Ground, hydrate & milk	≤ 0.1	≤ 0.1	≤ 0.1
% greater than 90 µm	Ground	≤ 7.0	≤ 7.0	≤ 7.0
% greater than 90 µm	Hydrate & milk	≤ 5.5	≤ 5.5	≤ 5.5

[a] The test methods are specified in prEN 12485 [28.5].
[b] "Ground" refers to pulverised high-calcium quicklime, "lump" refers to granular high-calcium quicklime, "hydrate" refers to hydrated high-calcium lime and "milk" refers to milks of lime
[c] % values are by mass.
[d] The % CaO and Ca(OH)$_2$ levels refer to the water soluble values.

Table 28.2. High-calcium lime for use in water treatment — toxic substances (proposed limits)

Parameter	Upper limits	
	Type A mg/kg	Type B mg/kg
Antimony	4	4
Arsenic	5	20
Cadmium	2	2
Chromium	20	20
Lead	25	50
Mercury	0.3	0.5
Nickel	20	20
Selenium	4	4

The limes are further sub-divided into Types A and B, depending on the concentrations of "toxic substances" (Table 28.2). The Standard also proposes particle size requirements (Table 28.1).

It should be noted that, in setting the proposed limits, no account was paid of the following points:

a) The amounts of each element retained by the process undoubtedly varies with the process and with the addition point. In setting the limits, a dosage rate was

assumed and the cautious assumption was made that *none* of the elements would be retained.
b) The natural variability of trace elements is very skew, based on several hundred results analyzed by the author. The upper confidence limit, as measured by the mean plus 3x the standard deviation, is over three times the mean.

(N.B., proposed revisions of the EU Directive on drinking water may lead to pressure for these limits to be reduced – particularly for lead and arsenic. However, the implications of the above points should be determined before any reductions in these limits are accepted.)

Half-burned dolomite is specified [28.3] (Table 28.3) in terms of:

a) the free MgO and $CaCO_3$ contents,
a) the levels of impurities (free $CaO/Ca(OH)_2$, SiO_2, Al_2O_3, Fe_2O_3 and SO_4^-).

It is also categorised into two Types depending on the concentrations of "toxic substances" (Table 28.4).

Requirements in the USA are specified in [28.6].

Table 28.3. Half-burned dolomite for use in water treatment — major and minor components (proposed limits)

Parameter	Requirement % (m/m)
Free MgO + Mg(OH)$_2$ (as MgO)	≥ 23
$CaCO_3$ (as $CaCO_3$)	≥ 68
Free CaO (as CaO)	≤ 2
SiO_2	≤ 2
Al_2O_3	≤ 2
Fe_2O_3	≤ 2
Sulfate (as SO_4)	≤ 1

Table 28.4. Half-burned dolomite for use in water treatment — toxic substances (proposed limits)

Parameter	Upper limits	
	Type A mg/kg	Type B mg/kg
Antimony	3	5
Arsenic	3	5
Cadmium	1	2
Chromium	10	20
Lead	15	20
Mercury	0.5	1
Nickel	10	20
Selenium	5	5

Where quicklime is purchased, its particle size and reactivity should be specified to ensure that it is compatible with the handling equipment and suitable for the type of slaker installed (see section 22.3). For pellet reactors, the solution rate of the milk of lime should be sufficient to prevent significant amounts of undissolved calcium hydroxide being carried out of the reactor by the flow of water.

28.2   Boiler Feed Water

28.2.1   Introduction

Feed water to boilers operating at moderate steam pressures should be softened and treated [28.1] to ensure that it:

a) is free from suspended solids,
b) is entirely free from non-carbonate hardness,
c) has a carbonate hardness of less than 35 mg/l,
d) is practically free from oil and dissolved oxygen, and
e) has a low dissolved silica content.

28.2.2   Softening and Coagulation

Lime-soda ash softening, followed by settling or filtration, can be used to remove suspended solids and non-carbonate hardness, and to reduce the carbonate hardness to the required level.

Colloidal silica volatilises at high temperatures and carries over in the steam. It can then re-deposit on turbine blades. The silica is effectively removed by adsorption on precipitated magnesium hydroxide. Where the feed water contains sufficient magnesium, magnesium hydroxide can be produced by softening with high-calcium lime. Where the water is deficient in magnesium, dolomitic lime is often used as the water-softening agent. The rate and efficiency of adsorption increases with temperature.

28.2.3   Specifications

Commercial hydrated limes are suitable for the treatment of boiler feed water.

28.3   Waste Water (Other than Sewage)

28.3.1   Introduction

Lime products are widely used to treat waste water to:

a) neutralise acids,
b) adjust pH prior to further treatment or discharge,
c) precipitate metals,
d) precipitate sulfate and fluoride,
e) reduce nutrients (phosphates and nitrogen), and
f) modify the characteristics of the sludges produced by the treatment.

   In these applications, lime products compete with other alkalis and with organic polyelectrolytes. The choice of process and reactants depends on the overall economics and on the acceptability of the treated water for discharge or recycle and of the sludge for disposal. The greater use of lime than of competitive alkalis reflects its higher cost-effectiveness, its efficiency in removing most heavy metals and the better physical and chemical sludge characteristics that are generally obtained.
   A comprehensive treatment of the many permutations of types of waste water, the impurities to be removed and their interactions with the form of the lime product used is beyond the scope of this work. The following sections indicate some of the more important factors to be considered. Because of the complexity of the subject, it is generally advisable to carry out detailed laboratory evaluations of the processes being considered.

28.3.2   Neutralisation of Acids and pH Adjustment

Figure 28.2 illustrates the components of a simple acid neutralisation plant.
   The alkalis commonly used for acid neutralisation are listed in Table 28.5, together with their theoretical chemical formulae and equivalent weights. Quick-

Table 28.5. Commonly used alkalis

Name	Theoretical formula	Molecular weight	Equivalent weight
Dolomitic quicklime	$CaO \cdot MgO$	48.2	24.1
High calcium quicklime	CaO	56.2	28.1
Type N dolomitic hydrate	$Ca(OH)_2 \cdot MgO$	57.2	28.6
Magnesium hydroxide	$Mg(OH)_2$	58.4	29.2
Type S dolomitic hydrate	$Ca(OH)_2 \cdot Mg(OH)_2$	66.2	33.1
High calcium hydrate	$Ca(OH)_2$	74.2	37.1
Caustic soda	$NaOH$	40.0	40.0
High calcium limestone	$CaCO_3$	100.2	50.1
Soda ash	Na_2CO_3	106.0	53.0

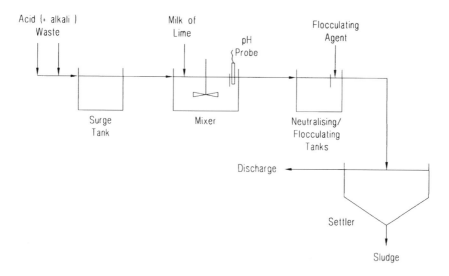

Figure 28.2. Example of a basic acid neutralisation plant

limes should be slaked to a milk of lime before addition to the waste water (see chapter 22 for the production of milks of lime).

The economics of using a particular alkali to neutralise a specified quality of acidic effluent depend on a number of factors, the more important of which include:

a) the type and concentration of the acid(s) being neutralised,
b) the cost/tonne of the alkali and its purity,
c) its equivalent weight,
d) the required pH,
e) the pH achievable with the alkali,
f) the rate of neutralisation,
g) the amount of excess alkali required for neutralisation,
h) the amount of sludge produced,
i) the settling/filtration characteristics of the sludge,
j) the acceptability of the sludge for disposal, and
k) the capital and operating costs of the plant required to use that alkali.

To complicate the situation further, some of the above factors are inter-dependent. The following paragraphs summarise the more important factors.

● *The type and concentration of the acid(s)* can have a marked effect on the neutralisation process, largely owing to the solubility of the reaction product(s). The acids most commonly found in waste water are nitric (HNO_3), hydrochloric (HCl), sulfuric (H_2SO_4), hydrofluoric (HF) and phosphoric (H_3PO_4). The solubilities of their calcium and magnesium salts are summarised in Table 28.6 [28.7] (N.B., all sodium salts are soluble).

Table 28.6. Solubilities of calcium and magnesium salts

Solubility in cold water (g/l)

$CaCl_2$	745	$MgCl_2$	553
CaF_2	0.02	MgF_2	0.08
$CaSO_4$	2.1	$MgSO_4$	449
$Ca(NO_3)_2$	1212	$Mg(NO_3)_2$	580
$Ca_3(PO_4)_2$	0.02	$Mg_3(PO_4)_2$	very low

In the case of lime products (which are themselves only sparingly soluble) the formation of an insoluble, or sparingly soluble neutralisation product can lead to a coating of that product on the surface of the lime particles, and can inhibit the reaction. The degree of inhibition rises with increasing concentrations of acid and particle size of the lime. Conversely, the neutralisation is more likely to proceed with dilute acid solutions and finely divided particles.

- *The equivalent weights* of the various alkalis (that is the weight of pure alkali, in grams, theoretically required to neutralise 36.5 g of pure HCl, or 49.0 g of H_2SO_4 etc.) is given in Table 28.5. In practice, due allowance should be made for the impurities present in each alkali, when calculating the theoretical quantities required.

- *The required pH* depends on the conditions required for subsequent treatments, or for discharge (see section 28.3.3).

- *The economically achievable pH levels* for the various alkalis vary considerably and can differ significantly from the theoretical levels. They also depend on the application. Values quoted in the literature [28.8] are reproduced in Table 28.7.

- *The rates of neutralisation* vary considerably. Limestones with greater than 10 % of $MgCO_3$ are not included in Table 28.7, because their reaction rates are

Table 28.7. Economically achievable pH levels

Alkali	Economically achievable pH level
High calcium limestone	
– without aeration	6.5
– with aeration	7[a]
Soda ash	7[b]
Dolomitic hydrated limes	9[c]
High calcium hydrate	12.4
Caustic soda	14

[a] Aeration assists the emission of carbon dioxide under acidic conditions.
[b] Above pH 7, the production of bicarbonate reduces the effective neutralising value.
[c] Above pH 9, $Mg(OH)_2$ precipitates, reducing the effective neutralising value.

too slow to be of commercial interest. In contrast, caustic soda solutions react as quickly as the acid and alkali can be mixed together. Although soda ash is also used as a solution, the achievable reaction rates are limited by the need to evolve gaseous carbon dioxide without excessive foaming.

Reaction rates of high calcium limes are relatively high at all pH levels up to 12.4 and increase with finer particle size distributions, subject to the above comments regarding blinding by insoluble reaction products. It is for this reason that, when preparing milks of lime from quicklime, it is usually beneficial to slake under conditions which give a finely divided milk of lime (see chapter 22). The solution rates of dolomitic hydrated limes up to pH 9 also depend on particle size, but are appreciably lower than those of high calcium limes.

- *The amounts of excess alkali required* reflect the extent to which the alkali is able to react within the residence time of the reaction vessel. In the case of lime, coarser particles may not react, either because of their relatively low solution rates, or because their dissolution is inhibited by insoluble reaction products.

- *The amount of sludge produced* depends on how much of the reaction products are insoluble and the ease with which they de-water. Thus the neutralisation of sulfuric acid with caustic soda, or soda ash produces virtually no sludge (all sodium salts are soluble), whereas its neutralisation with high-calcium lime produces significant quantities of calcium sulfate sludge, which may contain unreacted calcium hydroxide. Where dolomitic hydrated lime is used, the amount of sludge depends on the final pH — above pH 9, the magnesium is re-precipitated, thereby increasing the amount of sludge and reducing the effective neutralising value.

Some typical neutralisation reactions, producing low, intermediate and high levels of sludge are given in Table 28.8.

Where metals are precipitated, the production of relatively large quantities of sludge can be an advantage, as it can improve both the *settling/filtration characteristics* and *the acceptability of the sludge for disposal* (see section 28.3.6).

Table 28.8. Some neutralisation reactions

Reactions producing low levels of sludge:

$$2NaOH + H_2SO_4 \rightarrow Na_2SO_4 + 2H_2O$$
$$Na_2CO_3 + H_2SO_4 \rightarrow Na_2SO_4 + H_2O + CO_2\uparrow$$
$$Ca(OH)_2 + 2HCl \rightarrow CaCl_2 + 2H_2O$$
$$Ca(OH)_2 \cdot MgO + 4HCl \rightarrow CaCl_2 + MgCl_2 + 3H_2O \qquad \text{(pH below 9)}$$

Reactions producing intermediate levels of sludge:

$$Ca(OH)_2 \cdot MgO + 2H_2SO_4 \rightarrow CaSO_4\downarrow + MgSO_4 + 3H_2O$$
$$Ca(OH)_2 \cdot MgO + 2HCl \rightarrow CaCl_2 + Mg(OH)_2\downarrow + H_2O \qquad \text{(pH above 10)}$$

Reactions producing high levels of sludge:

$$Ca(OH)_2 + H_2SO_4 \rightarrow CaSO_4\downarrow + 2H_2O$$
$$Ca(OH)_2 \cdot MgO + H_2SO_4 \rightarrow CaSO_4\downarrow + Mg(OH)_2\downarrow + H_2O \qquad \text{(pH above 10)}$$

• *The capital costs* reflect the size of reaction vessels and of settling/filtration equipment required and need to be considered in conjunction with the *operating costs*. Thus a low cost neutralising agent (e.g., limestone), which has a low reaction rate and may have to be added in excess, may require a large plant with a high capital cost, which more than offsets any saving in operating costs.

28.3.3 Precipitation of Dissolved Metals

Waste waters frequently contain dissolved metals, which may have to be removed or reduced to acceptable levels before discharge to water courses or sewage plants.

Many metals are precipitated at the appropriate pH. Indeed, for most, there is an optimum pH at which the residual concentration in solution is at a minimum (Table 28.9) [28.2, 28.8, 28.9] (N.B., as mentioned on page 317, interference from other ions can cause the residual metal concentrations to be significantly higher than quoted).

Table 28.9. Precipitation pH ranges for metals

Metal ion	Precipitation pH range	Optimum pH	Approx. conc. at optimum pH (mg/l)[a]
Ferric iron	> 4.0	7–8.5	< 0.3
Aluminium	4.5–10	5.5–8	< 0.03
Chromium (III)	> 5.3	9.5	0.3
Copper	> 6	7.5	0.07
Ferrous iron[b]	> 9	> 10	high
Lead	> 6.0	10.0–10.3	0.03
Nickel	> 6.7	9.8–10.2	0.1
Cadmium	> 6.7	10.5	0.1
Zinc	8.0–11	10.5–11.0	0.01
Manganese[c]	> 6.5	9.5	< 0.01

[a] See note in para 2 of section 28.3.3.
[b] Fe (II) is generally oxidised to Fe (III).
[c] Under oxidising conditions.

To achieve the required reductions in metal ion concentrations, it may be necessary to carry out a multi-stage treatment. In other cases, the treatment may require the oxidation state of the metal to be changed (e.g., oxidation of ferrous iron to ferric, which is more readily removed at low pH levels, and reduction of soluble Cr(VI) to Cr(III), which can readily be precipitated).

The effective removal of relatively small quantities of precipitated metal hydroxides/basic carbonates generally requires the use of coagulating agents (e.g., ferric, aluminium, or magnesium hydroxides).

28.3.4   Precipitation of Sulfate and Fluoride

Waste waters often contain undesirable levels of sulfate and fluoride.

High levels of sulfate (in excess of about 500 mg/l) cause the water to be aggressive towards concrete. Lime can be used to precipitate calcium sulfate, but the residual sulfate concentration in solution may still be too high. Treatment with sodium aluminate and lime at pH 9.5 to 10.0 precipitates a $Ca:Al:SO_4:OH$ complex and reduces the sulfate level to well below 500 mg/l [28.2].

Fluoride can be precipitated by lime as the calcium salt. At pH 10, the residual level is in the range 10 to 20 mg/l (as F) [28.2]. The precipitate should be removed by adsorption on magnesium hydroxide, calcium phosphate, or activated alumina, but not on aluminium hydroxide, as alum and aluminate form a soluble complex with fluoride.

28.3.5   Reduction of Nutrients

Phosphate is generally considered to be the main cause of excessive algal growth, which can lead to eutrophication of lakes and streams. Where necessary, it is usually removed, at neutral pH levels, to below 1 mg/l (as P) by treatment with ferric salts, or alum/sodium aluminate, which respectively precipitate ferric and aluminium phosphates.

Phosphate can also be reduced to 2 to 3 mg/l (as P) by treatment with lime at ambient temperatures and at pH levels of above 10, in which case the precipitate is the hydroxyapatite. Levels below 1 mg/l can be obtained with lime treatment at higher temperatures [28.2]. Such high pH levels, however, produce an unacceptable taste and the pH of the treated water requires to be reduced to below 9 by the addition of carbon dioxide.

Recent research into the removal of phosphate from sewage [28.10] indicates that phosphorous concentrations can be reduced to below 2 mg P/l at pH levels in the range of 8.0 to 9.2, and to below 1 mg P/l at pH levels above 10.5.

Dissolved nitrogen, in the form of ammonium and nitrate salts, can also cause excessive algal growth. Where ammonia has to be reduced, it is generally oxidised by bacterial action, initially to nitrite and then to nitrate. Lime can be used to remove ammonia by raising the pH to 10 followed by degasification. Alternative techniques include ion exchange and adsorption on clays (e.g., clinoptilolite) [28.2].

Nitrate can be converted to nitrogen by nitrifying bacteria. Alternatively, it can be removed by ion exchange.

28.3.6   Modification of Sludge Characteristics

Some sludges are slow settling, or may not filter readily. In many cases, the use of lime results in the co-precipitation of a calcium compound (e.g., $CaCO_3$, or $CaSO_4$), which improves their settling and filtration characteristics.

Other sludges may be unsuitable for disposal to landfill, or for spreading on agricultural land, either because of their mechanical properties or because of the risks of leaching of heavy metals. In such circumstances, the use of lime products frequently improves the handling characteristics of the sludge, and, by raising the pH in the sludge, reduces leaching. Indeed, quicklime can also improve handling characteristics by reacting with some of the moisture and thereby reducing the moisture content .

28.3.7 Specifications

Commercial quick- and hydrated limes are suitable for the treatment of waste water. In some situations, due consideration will need to be given to the permissible levels of trace elements in the lime, in the context of consent levels for discharges to water courses.

28.4 Sewage Treatment

28.4.1 Introduction

Sewage consists of domestic and trade effluent. It contains suspended solids, dissolved colloidal organic matter and nutrients (phosphate and ammonia). It may also contain heavy metals, particularly from trade effluents.

Lime products have been used extensively in the treatment of sewage for over one hundred years. Its roles have included:

a) the adjustment of the pH of the incoming sewage,
b) coagulation and flocculation of the solids,
c) removal of metals and nutrients,
d) disinfection, and
e) conditioning the sludge for agricultural use or as a landfill.

Sewage treatment uses mechanical, biological and chemical processes to produce, at the lowest possible cost, a discharge of the required standard and a sludge that can be disposed of in a safe and acceptable manner. In most countries, the requirements for the discharged water and the disposal of sludge are becoming more stringent. This is causing radical changes in the industry which may well present opportunities for increased lime sales.

28.4.2 Treatment Processes

Although sewage treatment processes vary widely in terms of the equipment, treatments and sequence of unit operations, most fit into one of six categories [28.2]. Only one of these uses substantial quantities of lime [28.11].

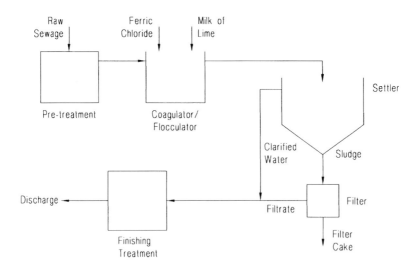

Figure 28.3. Diagram of a lime-based sewage treatment process

In this process (Fig. 28.3), milk of lime is used in conjunction with a coagulating agent (e.g. ferric chloride) to coagulate and flocculate the suspended and colloidal solids. The suspension is then fed to a clarifier.

The concentrated sludge from the clarifier is de-watered in a filter to produce a filter cake which, if a filter press is used, may contain 35 to 50 % solids. The filtrate is added to the clarified water.

The clarified water then passes into the finishing treatment plant, which includes filtration to remove residual solids, passage through a carbon bed, which absorbs dissolved organic matter, and disinfection with chlorine. The water is then discharged.

The lime-based process produces a readily filterable sludge and removes much of the colloidal organic matter in the clarifier, thereby reducing the load on the carbon bed. It produces an increased mass of dry solids, but, because of its improved de-watering characteristics, the volume of the filter cake may be no greater than with the alternative processes.

28.4.3   Removal of Metals and Nutrients

An advantage of the lime treatment process is that it can precipitate dissolved metals (see section 28.3.3) and remove them at the clarification stage. Increasingly, however, industry is required to treat its waste-water before discharge to sewer and this includes the removal of heavy metals.

Sewage typically contains 10 to 30 mg/l of phosphate (as P) and a similar amount of ammonia (as N) [28.2]. The EU Directive of 21 May 1991 requires that sewage plants serving more than 100,000 people reduce the level of phosphate to less than 1 mg/l (as P) by 2005.

Lime can be used to reduce the concentration of phosphate to 2 to 3 mg/l (as P) at a pH of 8.8 to 9.2 and to below 1 mg/l by the use of higher pH levels which cause the co-precipitation of carbonate hardness [28.10]. Phosphate can also be removed by treatment with aluminium and ferric salts [28.2].

Ammonia removal is generally achieved by biological action. However, in the lime-based process, ammonia can be removed by air-stripping at pH levels above 10 [28.2].

28.4.4   Disinfection

In most sewage plants, the final operation is disinfection using chlorine, although the use of ozone is becoming more feasible.

However, there are circumstances in which raw sewage or sewage sludge is disinfected, or "stabilised", by raising the pH using lime [28.1]. The purification process depends on:

a) the nature and concentration of the pathogens,
b) pH,
c) the reaction time, and
d) the temperature.

Some types of pathogen (which include bacteria, viruses and parasite eggs) can be reduced to an acceptable level at a pH as low as 9.5 in hours, whereas others may require a pH of over 12.0 and exposure times of weeks. For a given pathogen, the required reaction time decreases with increasing pH and increasing temperature. Many pathogens are considerably more resistant at temperatures below 15 °C. Conversely, their resistance is greatly reduced at temperatures above 50 °C.

The role of lime in the process is two-fold. Firstly, it creates the required high pH which attacks the pathogens and, secondly, it removes pathogens by physical absorption on the flocs produced by the lime as a result of removal of temporary hardness and precipitation of dissolved metals.

Lime is used to disinfect/stabilise the waste from septic tanks [28.12, 28.13] and to render sludge safe to spread on the land [28.14].

28.4.5   Disposal

Increasingly, attention is being given to how best to dispose of sewage sludge in a safe and acceptable way.

In some countries, this involves stabilisation and partial disinfection. In others, spreading disinfected sludge on agricultural land is the favoured option. There is concern that, where heavy metals are present in the sludge, this may result in the metals leaching into the ground water, or being taken up by crops. In Germany, for example, it is proposed to stop the agricultural use of sewage sludge and to incinerate most of it by 2005 [28.14].

Notwithstanding the above concerns, it is believed that lime will increasingly be used to improve the characteristics of the sludge. As described in section

28.4.2, the lime-based process, coupled with the use of filter presses, can produce a filter cake with 35 to 50 % dry solids (m/m). Cakes from the alternative processes typically contain 15 to 35 % dry solids. Such filter cakes do not have the necessary mechanical properties for landfill or for disposal to tips. However, the addition of ground quicklime:

a) removes water by hydration,
b) raises the pH to 12.4 and can raise the temperature to above 50 °C (resulting in disinfection within 2 hours), and
c) can give the necessary strength required for disposal to landfill or to tips [28.10].

Where sewage sludge is incinerated, the resulting ash also requires disposal. It has been suggested [28.10] that, as the ash contains SiO_2, Al_2O_3 and Fe_2O_3, it could be used as a raw material for the production of Portland cement, particularly if the lime-based process were used.

28.4.6   Specifications

Commercial quick- and hydrated limes are suitable for the treatment of sewage.

28.5   References

[28.1]   N.G. Pizzi, "Hoover's Water Supply and Treatment", 12th ed., Bulletin 211, National Lime Association, Kendall/Hunt Publishing Co., 1995, ISBN 0-8403-9625-2.
[28.2]   F.N. Hemmer, "The NALCO Water Handbook", McGraw-Hill, 1988, ISBN 0-07-045872-3.
[28.3]   prEN 12518: "Chemicals used for treatment of water intended for human consumption — High-calcium lime" (in preparation).
[28.4]   prEN 1017: "Chemicals used for treatment of water intended for human consumption — Half-burnt dolomite" (in preparation).
[28.5]   prEN 12485: "Chemicals used for treatment of water intended for human consumption — Calcium carbonate, high-calcium lime and half-burnt dolomite — Test methods" (in preparation).
[28.6]   AWWA Standard B202-93.
[28.7]   "CRC Handbook of Chemistry and Physics", 77th ed., 1996–97.
[28.8]   C.J. Lewis, R.S.Boynton, "Acid Neutralisation with Lime", National Lime Association, Bulletin No. 216, 1976.
[28.9]   L. Hartinger, "Taschenbuch der Abwasserbehandlung", Vol. 1, Carl Hanser Verlag, Munich, 1976.
[28.10] N. Peschen, "Phosphate precipitation and use of sewage sludge", Proc. of the 8th International Lime Congress, 1994.
[28.11] C.J. Lewis, K.A. Gutschick, "Lime in Municipal Sludge Processing", National Lime Association, Bulletin No. 217, 1980.
[28.12] A. Akyarli, N. Ozture, "Lime Usage in Treatment of Wastes from Septic Tanks", presented to the European Lime Association's 2nd Technical Conference, Oct. 1996.
[28.13] M. Montruccoli, "Hygenisation of sewage sludge by lime amendment", presented to the European Lime Association's 2nd Technical Conference, Oct. 1996.
[28.14] H.-P. Thomas, "Lime Use in Waste Water Engineering – Trends in Germany", presented to a Meeting of the European Lime Association, Sept. 1995.

29 Gaseous Effluents

29.1 Introduction

In the late 1970s, environmental control began to include the removal of oxides of sulfur from power station flue gases [29.15, 29.16]. It has since been extended to the removal of other acid gases, notably hydrogen chloride, hydrogen fluoride and the oxides of nitrogen. More recently, concern has been focused on the emission of dioxins, furans and volatile heavy metals. While some countries, notably the USA, Japan and Germany, have made considerable progress, on a global basis, this market segment is still in the initial stage of its growth curve.

Apart from the removal of oxides of nitrogen, lime has a part to play, either directly or indirectly in reducing the emission of the above pollutants. Lime, as the cheapest alkali, is widely used in removing acidic gases. A variety of abatement techniques have been designed to suit particular applications and new developments are continually being reported.

Lime-based techniques for the abatement of acid gases can be divided into five groups (four of which are illustrated in Fig. 29.1):

a) wet scrubbing, in which the gases are treated with milk of lime, principally to remove SO_2, with the neutralisation products being removed as a suspension,
b) semi-dry scrubbing, using either spray drier, or circulating fluidised bed technology, in which milk of lime is sprayed into the gases (principally to remove SO_2) with the reaction products being removed in a dust collector,
c) high temperature (over 850 °C) dry injection of hydrated lime, in which the hydrated lime calcines and the resulting calcium oxide reacts with the acid gases (principally SO_2), with the reaction products being removed as in (b),
d) low temperature (below 300 °C) dry injection of hydrated lime to remove HCl, HF and, to a lesser extent, SO_2, in which the reaction products are removed as in (b),
e) low temperature (below 300 °C) absorption by hydrated lime in a fixed bed, principally to remove HF from kilns calcining ceramic products.

Some of the above processes efficiently remove heavy metals, while others minimise the re-formation of dioxins and furans. In addition, hydrated lime may be used in conjunction with activated carbon or lignite coke for removal of dioxins, furans and volatile heavy metals.

The processes are described in some detail below, with an indication of the applications in which they are being used. The reader should, however, remember that this market segment is developing rapidly.

Abatement techniques using lime compete with processes using other absorbents such as limestone, sodium carbonate/bicarbonate, magnesium hydroxide,

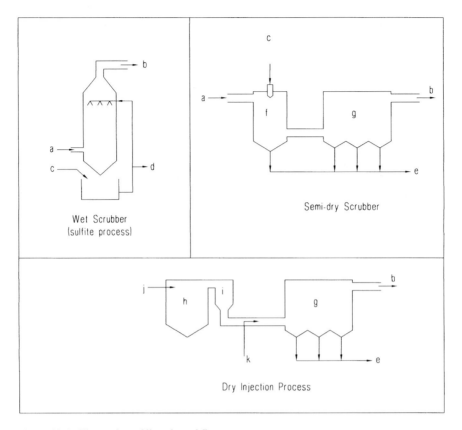

Figure 29.1. Illustration of lime-based flue gas treatment processes.
(a) untreated flue gas; (b) treated flue gas; (c) milk of lime; (d) disposal (wet); (e) disposal (dry);
(f) spray drier; (g) bag filter or ESP; (h) furnace; (i) economizer; (j) hydrated lime/limestone;
(k) hydrated lime

caustic soda and sea water. The use of limestone for desulfurisation and removal of HF is described in section 12.5.4.

Increasingly, the cost and acceptability of disposal of the reaction products is a major factor affecting the choice of abatement process. With lime and limestone, most of the reaction products are insoluble in water. Where landfill is the favoured disposal method, that is a significant advantage.

29.2   Wet Scrubbing

29.2.1   General

Of the five techniques described in section 29.1, wet scrubbing is the most efficient. It can reliably remove over 95 % of the SO_2, and can utilise over 95 % of the absorbent.

After being de-dusted, and before entering the scrubber, the flue gases are often cooled in a heat exchanger, or quenched with a water spray. Various techniques are used to minimise the capital and operating costs of the scrubber. One design uses recycled slurry from the sump to perform the initial scrubbing of the gases and to maximise the conversion of absorbent, while fresh milk of lime is used for the final stage of scrubbing the gases. Another uses co-current initial scrubbing and counter-current final scrubbing to minimise pumping costs.

As the processes operate at about 60 °C, volatile heavy metals such as mercury, cadmium and thallium are removed efficiently. Because of the low operating temperature, a heat exchanger is often used to transfer heat from the incoming gases to the scrubbed gases (see Fig. 12.1). Where, however, the required level of SO_2 removal is relatively low, part of the gases may be by-passed around the scrubber and used to pre-heat the scrubbed gases before discharge to atmosphere.

29.2.2   The Sulfite Process

This process is similar to the wet scrubbing process using limestone (section 12.5.2), but there are several important differences.

The flue gases are scrubbed with milk of lime in the absorber. The reactions between slaked lime and the oxides of sulfur are:

$$Ca(OH)_2 + SO_2 \rightarrow CaSO_3 \cdot 0.5H_2O\downarrow + 0.5H_2O \qquad (29.1)$$

$$Ca(OH)_2 + SO_3 + H_2O \rightarrow CaSO_4 \cdot 2H_2O\downarrow \qquad (29.2)$$

The rates of absorption of sulfur oxides in milk of lime and of solution of the slaked lime are considerably faster than the analogous reactions with limestone. For a given removal efficiency, therefore, the absorber is considerably smaller than that for limestone. This helps to reduce capital costs and the power consumed in recirculating the slurry. However, because of the rapid precipitation of calcium sulfite hemi-hydrate with some gypsum, the particles tend to be finely divided and form a thixotropic sludge which is difficult to de-water for disposal [12.1].

The milk of lime-sulfite process was widely used for early power station flue gas desulfurisation projects. It had a lower capital cost than the limestone-sulfite process and gave high absorption and reagent efficiencies. Subsequently, the problem of disposal of the calcium sulfite sludge led to three variants based on lime to be adopted — the gypsum process, the dual alkali process and the maglime process.

29.2.3   The Gypsum Process

This is analogous to the limestone–gypsum process (section 12.5.2). Oxides of sulfur are absorbed by the milk of lime, forming crystallites of calcium sulfite and sulfate (equations 29.1 and 29.2). Compressed air is blown into the scrubber sump to oxidise the calcium sulfite to sulfate, and results in the growth of gypsum crystals (29.3).

$$CaSO_3 \cdot 0.5H_2O + 0.5O_2 + 1.5H_2O \rightarrow CaSO_4 \cdot 2H_2O \downarrow \qquad (29.3)$$

The suspension of gypsum is removed from the absorber and classified in hydroclones, which remove crystals of the required size. Finer particles are re-cycled to the scrubber, where they act as seed crystals. Any unreacted slaked lime is recycled with the fines to the scrubber.

The gypsum produced from hydrated lime tends to be whiter than that from limestone [29.2] and can command premium prices.

29.2.4 The Dual Alkali Process

In this process, oxides of sulfur are absorbed in caustic soda. The reactions in the scrubber (29.4–29.7) produce soluble reaction products. The solution from the sump of the absorber is treated in a separate reaction vessel with lime (Fig. 29.2).

$$2NaOH + SO_2 \rightarrow Na_2SO_3 + H_2O \qquad (29.4)$$

$$Na_2SO_3 + SO_2 + H_2O \rightarrow 2NaHSO_3 \qquad (29.5)$$

$$2NaOH + SO_3 \rightarrow Na_2SO_4 + H_2O \qquad (29.6)$$

$$Na_2SO_4 + SO_3 + H_2O \rightarrow 2NaHSO_4 \qquad (29.7)$$

The lime precipitates calcium sulfite hemi-hydrate and gypsum and regenerates caustic soda (29.8, 29.9).

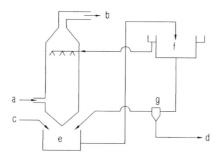

Figure 29.2. Illustration of the dual-alkali FGD process
(a) untreated flue gas; (b) treated flue gas; (c) milk of lime; (d) disposal (wet);
(e) causticiser and crystalliser; (f) clarifier; (g) hydrocyclone

$$2NaHSO_3 + 2Ca(OH)_2 \rightarrow 2NaOH + 2CaSO_3 \cdot 0.5H_2O\downarrow + H_2O \quad (29.8)$$

$$2NaHSO_4 + 2Ca(OH)_2 + 2H_2O \rightarrow 2NaOH + 2CaSO_4 \cdot 2H_2O\downarrow \quad (29.9)$$

By controlling the reaction, relatively coarse crystals can be grown, or, alternatively, the calcium sulfite can be oxidised to gypsum. In either case, the crystals can readily be de-watered for disposal or sale.

29.2.5   The Maglime Process

This is an elegant variant of the dual alkali process, which exploits the fact that many magnesium salts are soluble.

Quicklime, with about 7 % MgO, is slaked to produce a milk of lime. This is fed to the scrubber and removes oxides of sulfur (29.10, 29.11). The liquor that collects in the sump of the scrubber is a suspension of calcium sulfite and sulfate and a solution of magnesium hydrogen sulfite/sulfate.

$$MgO + 2SO_2 + H_2O \rightarrow Mg(HSO_3)_2 \quad (29.10)$$

$$MgO + 2SO_3 + H_2O \rightarrow Mg(HSO_4)_2 \quad (29.11)$$

The suspension is then treated in an external reactor with fresh milk of lime containing further magnesium oxide. The magnesium hydrogen sulfite and sulfate react with calcium hydroxide, forming magnesium sulfite and sulfate, which are recycled, and precipitate calcium sulfite hemi-hydrate and gypsum (29.12, 29.13).

$$Mg(HSO_3)_2 + Ca(OH)_2 \rightarrow CaSO_3 \cdot 0.5 H_2O\downarrow + MgSO_3 + 1.5H_2O \quad (29.12)$$

$$Mg(HSO_4)_2 + Ca(OH)_2 \rightarrow CaSO_4 \cdot 2H_2O\downarrow + MgSO_4 \quad (29.13)$$

As in the gypsum process, the calcium sulfite may be oxidised by injecting compressed air into the reaction vessel. Also, by classifying the precipitate in hydroclones, fines can be recycled to the reactor and crystals can be grown to the required size for de-watering for disposal or sale.

The concentration of magnesium in the recycle to the absorber is controlled by a purge through the solids de-watering system.

29.3   Semi-dry Scrubbing

Semi-dry scrubbing uses the principle of the spray drier. Milk of lime is atomised at the top of the spray drier chamber into relatively hot flue gases (e.g. 220 °C). The gases carry the spray towards the base of the chamber. The water in the milk of lime evaporates, cooling the gases and, at the same time, acid gases (SO_2, SO_3, together with any HCl/HF present) dissolve and react with the lime.

When all of the water in the spray has evaporated, the dry solids consist of the reaction products, together with unreacted calcium hydroxide. They are carried by the gases from the scrubber into the dust collector (preferably a bag filter) at a temperature of about 120 °C.

Some secondary absorption of acid gases (particularly SO_2) by the calcium hydroxide occurs en route to the bag filter, but the tertiary absorption on the filter medium is generally more significant.

The process is capable of removing up to 90 % of SO_2 and over 99 % of HCl and HF, using a Ca÷S stoichiometric ratio of about 2 [29.3]. By recirculating part of the collected dust, the Ca÷S ratio can be reduced to around 1.25.

The advantages of these techniques relative to wet scrubbing are lower capital costs, lower operating costs and relative ease of disposal of the products. The major disadvantages are the higher cost of absorbent, the lower SO_2 removal efficiency, the lack of applications for the product and the disposal costs. In addition, the absorbent utilisation decreases as the concentration of SO_2 increases, limiting its applicability to fuels with up to 2.5 % S [29.4], or to low SO_2 removal efficiencies.

29.4   High Temperature Dry Injection

In this technique, hydrated lime is fluidised in air and injected into the boiler or kiln at temperatures in excess of 850 °C. A system designed for injecting hydrated lime is described in [29.1]. The hydrated lime decomposes within 30 milli seconds [29.5] to a porous and extremely reactive form of quicklime.

$$Ca(OH)_2 \rightarrow CaO + H_2O \tag{29.14}$$

In the presence of excess oxygen (and in the absence of carbon monoxide), the quicklime reacts with oxides of sulfur at temperatures below 1200 °C to form calcium sulfate. The quicklime also reacts with any HCl or HF present.

$$CaO + SO_2 + 0.5O_2 \rightarrow CaSO_4 \tag{29.15}$$

$$CaO + SO_3 \rightarrow CaSO_4 \tag{29.16}$$

Early work, in which hydrated lime was injected into the firing end of the boiler, gave low SO_2 removal and poor utilisation of absorbent. It now appears that this was due to "over-burning" of the quicklime. More recent research shows that the temperature at the point of injection should be below 1200 °C and above 850 °C, with an optimum range of 1000 to 1100 °C [29.5].

The reaction products, together with ash from the combustion process, are removed in a dust collector and disposed to landfill sites. As with the dust from the low temperature injection process, the lime content can be used to solidify other wastes (see also section 29.8).

With commercial hydrated limes (BET surface area about $15 \, m^2/g$), and under optimum conditions, the high temperature injection process can remove 50 to 65 % of the sulfur dioxide, using a Ca to S stoichiometric ratio of 2.0, and given a residence time of about 500 milli seconds at above 850 °C [29.5].

A pilot plant investigation [29.5] showed that the efficiency of desulfurisation depended on the fuel, burner operation and injection mode, but not on the surface area of the hydrated lime. However, a strong correlation was found between the development of internal surface area during calcination and sulfation, and the capture of sulfur.

It should be noted that quicklime is not favoured for this desulfurisation technique, presumably because of:

a) the difficulty in grinding it sufficiently finely and
b) the absence of in-situ calcination, which restricts the surface area on which sulfation can occur.

Improved absorbents have been developed — see section 20.6, which describes the production of hydrated limes with high specific surface areas. [29.6] mentions a calcium lignosulfonate-modified hydrated lime for the high temperature injection process (presumably a high surface area hydrate).

The principle advantages of this technique are that it requires relatively little capital expenditure and can readily be retro-fitted. However, it has relatively high absorbent costs and is only suitable where partial desulfurisation is required.

29.5   Low Temperature Dry Injection

Acidic gases (HCl, HF, and SO_2) can be removed from flue gases by injecting hydrated lime into the exhaust ducting. Acceptable removal efficiencies can be obtained by providing the necessary residence time after the injection point and by using a bag filter to remove the neutralisation products and excess hydrate.

Applications of the technique include municipal and hazardous waste incinerators, small to medium sized industrial burners and brick kilns. The injection system used is similar to that for the high temperature injection process [29.1].

HCl and HF react more readily with hydrated lime than does SO_2. The efficiency of removal of a given gas depends on its initial concentration, the temperature, the stoichiometric ratio of Ca to the acid and the effective surface area of the hydrated lime. The required removal efficiency depends on the initial concentration and the emission limit.

The following paragraphs are based on typical information for incinerator flue gases [29.7] and on the use of a standard commercial hydrated lime, with more than 94 % $Ca(OH)_2$, a BET surface area of at least $12 \, m^2/g$ and a median particle size of less than $8 \, \mu m$. Table 29.1 gives typical concentrations of HCl, HF and SO_2 in the untreated gases from municipal incinerators and an indicative range of emission limits (which vary from one country to another and with the type of installation).

Table 29.1. Typical unabated acid gas concentrations from municipal incinerators

Pollutant	Typical concentrations mg/Nm^3	Range of emission limits mg/Nm^3
HCl	400–1,200	10–50
HF	2–20	1–10
SO$_2$	200–600	50–750

It will be noted that, while it is generally necessary to remove most of the HCl, the required percentage reductions in HF and SO$_2$ are often much lower. The neutralisation reactions are (29.17–29.19):

$$Ca(OH)_2 + 2HCl \rightarrow CaCl_2 + 2H_2O \qquad (29.17)$$

$$Ca(OH)_2 + 2HF \rightarrow CaF_2 + 2H_2O \qquad (29.18)$$

$$Ca(OH)_2 + SO_2 \rightarrow CaSO_3 + H_2O \qquad (29.19)$$

Some of the calcium sulfite is oxidised by the excess air to calcium sulfate: both compounds may form hydrates. In addition, some of the excess hydrated lime reacts with carbon dioxide to form calcium carbonate.

The effect of temperature and stoichiometric ratio of Ca to acid gas is indicated in Table 29.2 [29.7]. Generally, over 99 % of the HCl, over 95 % of the HF and over 90 % of the SO$_2$ can be removed, if required over the quoted ranges. Figure 29.3 illustrates how the removal efficiency varies with stoichiometric ratio in the temperature range 120 to 160 °C for HCl and SO$_2$ [29.7].

Table 29.2. Effect of temperature on required stoichiometric ratio

Temperature (°C)	Stoichiometric ratio	
	HCl (99 % removal)	SO$_2$ (90 % removal)
120–160	1.5	3.0
160–220	1.8	3.5
220–280	2.0	5.0

As removal efficiencies increase at lower temperatures [29.17], it is common practice to cool the gases either by using a gas to air heat exchanger, or by spraying controlled amounts of water into the gases before injecting the hydrated lime. The penalty of such "conditioning" is that it increases the density of the steam plume when the gases are discharged to atmosphere and may necessitate re-heating the gases before discharge.

Because of its simplicity and wide applicability, considerable effort has been put into ways of improving and optimising the technique. This has focused on improved absorbents and operation closer to the dew point.

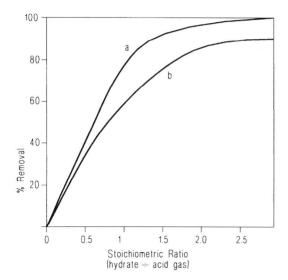

Figure 29.3. Effects of the amount of hydrate addition on acid gas removal
(a) hydrogen chloride; (b) sulfur dioxide

Section 20.6 describes two processes for the production of high surface area hydrated lime, with BET areas of 30 to 50 m^2/g, and median particle sizes of about 2 µm. One uses a methanol–water mixture for the hydration process, while the other uses water containing an additive. The higher surface areas enable the required acid gas emission concentrations to be achieved at lower levels of excess hydrate and with reduced amounts of waste material.

It is understood that other approaches are being investigated to increase the effective surface area, based in part on reducing the particle sizes. The development of a calcium silicate (advanced silicate or ADVACATE) absorbent has also been mentioned in the literature [29.6].

It is generally recommended that, for optimum removal of acid gases, the exhaust gases should be cooled to below 120 °C [29.8], or to within 20 to 30 °C of the dew point. Operation at lower temperatures can result in uncontrolled condensation leading to corrosion, high pressure drops across bag filters, caking of dust on filter bags and blinding of the filter medium.

Where a higher level of SO_2 removal is required than can be obtained with hydrated lime, sodium bicarbonate may be used. However, a recent proposal, based on laboratory work with hydrate involves operation so close to the dew point that the collected dust contains controlled amounts of moisture [29.9]. This results in improved capture of SO_2, particularly if the gases contain HCl which forms deliquescent calcium chloride.

Apart from calcium chloride, most of the components in the collected dust are insoluble in water and are, therefore, suitable for landfill. Moreover, as they contain some unreacted calcium hydroxide, they can be used to solidify other wastes before tipping (see also section 29.8).

29.6   Reduction of Dioxins and Furans

Although dioxins and furans are destroyed at above 850 °C, they re-form in the temperature range 200 to 450 °C [29.3]. It is understood that, without abatement techniques, municipal incinerators may emit at levels in the range 1 to 10 ng/Nm^3. Emission limits range from 0.1 to 1 ng/Nm^3.

Emission levels can be reduced by two techniques — rapid cooling between 450 and 200 °C, to minimise their re-formation, and adsorption on activated carbon.

The semi-dry scrubbing technique can be used to achieve the required rapid cooling from 450 °C, but at a cost, as this reduces the amount of recoverable heat (boiler flue gases are normally cooled to about 200 °C in a waste heat boiler).

Alternatively, activated carbon (or a less expensive equivalent such as lignite coke) can be injected into the exhaust gases and collected in a bag filter. An addition of 1 kg of activated carbon per tonne of organic matter is generally sufficient to reduce the dioxin and furan level to below 0.1 ng/Nm^3 [29.10]. A patent application describes the use of a blend of hydrated lime and activated carbon/lignite coke [29.11] uses the hydrated lime to effectively disperse the carbon and to eliminate the fire/explosion hazard associated with finely divided activated carbon.

29.7   Reduction of Heavy Metals

Industrial and municipal waste incinerators often emit heavy metals (e.g., Cd, Hg, Tl, Sb, As, Pb, Cr, Co, Cu, Mn, Ni, V and Sn) arising from items such as batteries, inks, pigments, plated metal and circuit boards. Many of the metals vaporise at the temperatures of 930 to 1200 °C required to destroy toxic organic compounds, but those with higher boiling points generally form sufficiently large particles when cooled to be removed in dust collectors to less than 1 mg/Nm^3 [29.12].

Mercury and cadmium are more volatile (boiling points 356 and 765 °C respectively) and are consequently more difficult to remove. They are reduced to about 0.2 mg/Nm^3 in total in the semi-dry scrubbing and pre-cooled dry injection processes and to about 0.1 mg/Nm^3 in the wet scrubbing processes [29.10]. Where lower emission levels are required, the injection of activated carbon can reduce the level to 0.05 mg/Nm^3 [29.10].

29.8   Disposal of Solid Residues

Disposal of the solid residues from the treatment of waste gases is an issue that needs to be addressed, particularly when they contain heavy metals and toxic organic compounds. One possibility is to convert the residues into an alinite cement, composed mainly of calcium silicates and calcium chloride [29.14]. When set, the concrete effectively locks up the toxic components (see also section 28.4.6).

29.9   References

[29.1]   K. Schneider, "Mopping-up Pollution", Gypsum, Lime and Building Products, April 1996, 23–26.

[29.2]   B Oppermann, M. Mehlmann, N. Peschen, "Products from the Lime Industry for Environmental Protection", Zement Kalk Gips **6**, 1991, 265–270 (English: **8**, 1991, 159–162).

[29.3]   J.T. Foster, P. Bisbjerg, D.W. James, "Best Available Technology for Air Pollution Control", Wheelabrator Report S1B/4.

[29.4]   D.H. Stowe, "Flue Gas Desulfurisation", Proc. 6th International Lime Association Congress, London, June 1986.

[29.5]   P. Flament, M. Morgan, "Fundamental and Technical Aspects of SO_2 Capture by Ca-based Sorbents in Pulverised Coal Combustion", International Flame Research Foundation, Doc. No. F 138/a/8, 1987.

[29.7]   "Lime for a Clean Environment", European Lime Association Brochure, 1995.

[29.6]   W. Jozewicz, B.K. Gullett, "The Effect of Storage Conditions on Handling and SO_2 reactivity of $Ca(OH)_2$ – based sorbents", Zement Kalk Gips **5**, 1991.

[29.8]   "The Dry Alternative", Process Engineering, June 1989, p. 57.

[29.9]   B. Naffin, U. Werner, "Removal of Harmful Acid Gases by Passage Through Moistened Lime Filter Layers", Part 3, Zement Kalk Gips **9**, 1996, 494–508.

[29.10]   Brochure by GEC Alsthom, NEU Process International.

[29.11]   "Verfahren zur Reinigung von Gasen und Abgasen", FTU GmbH, European Patent No. 0496 432 (Application No. 9210 3175.3).

[29.12]   A. A. Teeuwen, F. B. Laarkamp, "Dry Flue Gas Scrubbing for Waste Incinerators", report of DCE Benelux B. V.

[29.14]   R. Oberste-Padtberg et al., "Alinite cement, a Hydraulic Binder Made from Refuse Incineration Residues", Zement Kalk Gips **9**, 1992, 451–455 (English: **11**, 1992, E278–E290).

[29.15]   J. R. Cooper, I. A. Johnston, "The Engineering Requirements of the CEGB's Programme for Flue Gas Desulfurisation", IMechE Seminar on "Fossil Fired Emissions", 6 Dec. 1988.

[29.16]   J.R. Cooper et al., "Sulfur Removal from Flue Gases in the Utility Sector; Practicalities and Economics", Proc. Inst. Mech. Engrs. Vol. 211, Part A, 1997.

[29.17]   Th. Hunlich, R. Jeschar, R. Scholz, "Sorption Kinetics of SO_2 from Combustion Waste Gases at Low Temperatures", Zement Kalk Gips **5**, 1991 (English: **7**, 1991).

30 Agriculture, Food and Food By-products

30.1 Introduction

This chapter and chapters 31 and 32 describe a large number of applications of lime products, most of which represent a small proportion of the total production. Because of the difficulty of categorising many of them, they have been divided between the three chapters in what may appear to be a somewhat arbitrary way. The contents of each chapter are therefore summarised at the start of each chapter (see Table 30.1).

Table 30.1. Uses of lime products in agriculture, food and food by-products

Section	Application	Section	Application
30.2	Arable land and pasture	30.6	Leather processing
30.3	Miscellaneous agricultural uses	30.7	Glue and gelatin
	chicken litter	30.8	Dairy products
	animal carcases		butter
	compound fertilisers		casein
	acidic ponds and lakes		calcium lactate
	contaminated land	30.9	Fruit and vegetables
30.4	Sugar		storage
30.5	Pesticides		processing
	Bordeaux mixtures		
	lime-sulfur spray		
	calcium arsenate		
	dispersing agent		
	starfish control		

30.2 Arable Land and Pasture

30.2.1 General

As mentioned in section 10.2.1, the importance of soil pH and of calcium and magnesium for soil fertility was not fully appreciated until the second half of the 19th century. Thereafter, for about 100 years, most of the liming requirements were met with quick- and slaked limes. In the USA, for example, the use of lime in agriculture peaked in 1914 at almost 700,000 t/a [30.1], representing some 18 % of the total production.

During the past four decades, however, the increasing availability of ground agricultural limestone at relatively low prices has resulted in the growth of its sales at the expense of lime products. In many countries, the sales of lime products for use in agriculture amounts to 1 to 2 % of the total produced.

The lime requirements of various crops and the loss of calcium and magnesium from the soil are described in sections 10.2.2 and 10.2.3.

30.2.2   Effectiveness

When quicklime products are applied to the land, the first reaction is slaking. This produces finely divided calcium hydroxide and enables relatively coarse "kibbled" lime (a screened grade) to be used effectively. Some of the hydrated lime dissolves and produces a rapid increase in soil pH, within a week or two. The remainder of the lime carbonates, reacts relatively slowly, and keeps the pH in the required range for a prolonged period. The duration of that period depends on the rate of application, the nature of the soil and the rate of lime loss.

30.2.3   Application

Most lime products currently used for liming have a top size of 6 mm or less. This suits modern mechanical spreading equipment and helps to ensure an even distribution. Because the neutralising values of lime products are often almost double those of agricultural limestone, the application rates are correspondingly lower.

30.2.4   Lime Specification

When selecting a product for agricultural liming, a critical factor is the delivered cost per unit of neutralising value [30.2]. In consequence, when lime is used, inexpensive grades of lime are favoured. These include the fines screened from run-of-kiln lime, which contain more impurities (such as fuel ash, sulfur and clay) than the coarser fractions (see section 17.1.3). Hydrator rejects (section 20.4.3) and carbide lime (section 22.10) are also used.

Some of the lime products complying with the UK's Fertiliser Regulations [30.3] are listed in Table 30.2. While they are characterised in terms of grading and MgO content, the only declaration required is the neutralising value.

Table 30.2. Requirements for some agricultural lime products

Material	Form	% MgO	Grading requirements
Quicklime	Burned lime	< 27	None
(calcium)	Ground burned lime	< 27	100% passing 6.3 mm
	Kibbled burned lime	< 27	100% passing 45 mm
Quicklime	Burned lime	> 27	None
(magnesian)	Ground burned lime	> 27	100% passing 6.3 mm
	Kibbled burned lime	> 27	100% passing 45 mm
Slaked lime	Hydrated lime	[a]	95% passing 150 μm

[a] No limitation — either calcium or magnesian lime can be used.

30.3   Miscellaneous Agricultural Uses

- *Compost* — the process of composting organic matter can be accelerated by alternately adding about 5 % of hydrated lime and 5 % of a nitrogenous fertiliser to layers of compost. The hydrated lime provides an alkaline environment in which the fertiliser catalyses the conversion of the compost into humus.

- *Poultry* — hydrated lime is added to chicken litter, used in intensive poultry farming, to extend the life of the litter and to provide a degree of protection against parasites and disease.

- *Animal carcasses* can be decomposed and disinfected using quicklime. They are placed in a pit, quicklime is spread over them and water is added incrementally. The combination of high pH and elevated temperature causes the organic matter to decompose. It also disinfects the remains.

- *Compound fertilisers* (including calcium nitrate and calcium ammonium nitrate) sometimes contain 1 to 2 % of dolomitic hydrated lime. The lime confers similar benefits to pulverised dolomitic limestone (see section 10.3).

- The *pH of acidic ponds and lakes* has been raised using hydrated lime. The effect is similar to that of adding limestone (see section 10.6.3). However, as hydrated lime can produce a much higher pH than limestone, care must be taken to avoid over-liming, as that can cause as much damage as the initial acidity.

- *Restoration of contaminated* land has involved the use of lime products to immobilise cadmium and other heavy metals, thereby enabling soils to be used for growing forage crops and grasses [30.4]. See also section 32.18, which describes the use of lime products to treat land contaminated with organic matter.

30.4 Sugar

30.4.1 Sugarcane

Slaked lime is used in both the production and refining of sugar. Sugarcane is cut, shredded and treated with water to produce raw juice with a pH of 4 to 5. The juice is treated with lime to raise the pH and heated to approaching 75 °C to destroy invertase and other enzymes [30.5]. Typically some 2 to 5 kg of hydrated lime is used per tonne of sugar produced.

The lime also reacts with dissolved inorganic and organic compounds, forming insoluble calcium salts. A flocculating polymer is added to help the lime to coagulate suspended and colloidal matter (bagasse, soil, calcium salts, waxes and gums) in a process called defacation. Excess lime is precipitated either by adding phosphoric acid or by injecting carbon dioxide. The settled solids are pumped to filters from which residual sugar solution is recovered.

30.4.2 Sugarbeet

The production of sugar from sugarbeet requires approximately 200 kg of quick-lime per tonne of sugar. The raw beet is washed and chopped and treated with hot water to extract the sugar. The sugar solution also contains dissolved, suspended and colloidal matter.

The extract is first treated with excess lime, which raises the pH and precipitates insoluble calcium salts of both organic and inorganic acids. Carbon dioxide is then passed through the suspension to precipitate excess lime as calcium carbonate. The precipitation process removes suspended and colloidal matter as a "carbonation sludge", which is removed by filtration. The filtrate is a solution of sugar and is recycled. The composition of carbonation sludge is given in Table 30.3 [30.5]. It is frequently sold as a "lime" fertiliser (see also page 89).

Many sugarbeet processing plants operate lime kilns on-site to produce both the quicklime and the carbon dioxide required for the purification process. Shaft kilns are generally used, burning lump limestone. A few plants calcine the dried carbonation sludge, using rotary and circular multiple hearth kilns [30.6].

Table 30.3. Composition of sugarbeet carbonation sludge

Component	% by weight (dry)
$CaCO_3$	52–59
CaO in other compounds	11–17
MgO	0.9–1.2
$K_2O + Na_2O$	0.3
$PO_4^{3-} + SO_4^{2-}$	1.7–2.1
SiO_2	2–5
Organic substances	10–15

30.4.3   Sugar from Other Plants

Sugar is also extracted commercially from certain species of palm, from the sugar maple and from sweet sorghum. Lime is used to purify the extracts before evaporation.

30.4.4   Lime Specification

High calcium lime is generally specified, although some users require dolomitic lime [30.1]. A low level of impurities is desirable to minimise the amount of sludge produced. Some impurities such as arsenic can pass into the final product and limits may be imposed (e.g., less than 2 mg/kg of arsenic).

30.5   Pesticides

Bordeaux mixture is produced by mixing hydrated lime (9 to 12 g/l), copper sulfate (5 to 7 g/l) and cold water to produce a colloidal suspension of tetracupric sulfate. When sprayed on foliage, the mixture covers the leaves with a tenacious film which slowly releases soluble copper [30.7, 30.8]. It is effective against many fungi and insects.

Lime-sulfur sprays contain one part of hydrated lime to two parts of sulfur. They may be applied as a powder or as an aqueous spray [30.1].

The insecticide calcium arsenate consists of the powdered salt (produced by neutralising arsenic acid with milk of lime) and hydrated lime. The mixture is dispersed in water before applying as a spray [30.1].

Hydrated lime is used as a carrier for many insecticides. As it carbonates, it produces an adhesive film on the foliage which helps to retain the insecticide on the leaves.

Hard-burned quicklime with a preferred particle size of 2 mm, is used in the United States to control infestations of starfish on oyster beds. When a particle contacts a starfish, it causes a lesion which kills it. The quicklime is applied at approximately 150 kg per hectare [30.1].

30.6   Leather

"Liming" of hides to remove hair normally involves soaking the hides for 18 hours in a 2 % solution of sodium sulfide in water containing 3 % of hydrated lime. Wool is removed by applying a de-wooling paint, containing hydrated lime, sodium chloride, alkaline proteases and water to the flesh side of the skin [30.9].

30.7   Glue and Gelatin

Gelatin and animal glues (glutines) are extracted from waste bones and hides [30.10, 30.11]. Type A gelatin is produced by an acid extraction, while Type B uses an alkali — usually slaked lime. Demineralised bone (ossein) and cattle hide pieces are soaked in lime water at ambient temperature for up to 20 weeks. Additional milk of lime is added to the liquor to maintain the alkalinity.

The liming process hydrolyses collagen protein into single chains to produce a mixture of protein fragments of varying molecular weight. This degradation product, glutine, disperses in water to form a colloidal solution.

30.8   Dairy Products

In the production of butter, slaked lime may be used to neutralise the acidity of cream prior to the pasteurisation process [30.1].

Casein is the principle protein fraction of milk. "Acid casein" is precipitated at pH 4.6 by the addition of acid. Calcium caseinate is produced by treating acid casein with slaked lime [30.12], and is used in coffee whiteners, pasta, bread and cakes.

Calcium lactate may be produced by adding slaked lime to fermented skimmed milk [30.1] (see also section 31.14).

30.9   Fruit Industry

As apples and other fruit ripen, they emit carbon dioxide. In storage, the carbon dioxide lowers the level of oxygen in the atmosphere and accelerates the rate of deterioration of the fruit. By circulating air around the fruit and over hydrated lime, the level of carbon dioxide is reduced and the fruit remains marketable for longer [30.1].

Residues from processing citrus fruits are mixed with lime, dried, and sold as cattle feed. Lime is also used to neutralise waste citric acid and to raise the pH of fruit juices to stabilise the flavour and colour [30.1].

30.10   References

[30.1]   R.S. Boynton, "Chemistry and Technology of Lime and Limestone", John Wiley & Sons, 1980.
[30.2]   "Agricultural Lime and the Environment", The Agricultural Lime Producers' Council, published by the British Aggregate and Concrete Materials Industries, 1994.
[30.3]   The U.K. Fertiliser Regulations (S.I. No. 2197), 1991.

[30.4]   "Montana Hazardous Waste Seminar", Lime-Lites, Vol. LX, National Lime Associa-
         tion, July–Dec. 1993.
[30.5]   H. Schiweck, M. Clarke, "Sugar", Ullmann's*.
[30.6]   G.F. Kroneberger, Sugar Technology Rev. *4*, 1976/77, 3–47.
[30.7]   R.L. Metcalf, "Insect Control", Ullmann's*.
[30.8]   H.W. Richardson, "Copper Compounds", Ullmann's*.
[30.9]   E. Heidemann, "Leather", Ullmann's*.
[30.10] W. Haller et al., "Adhesives", Ullmann's*.
[30.11] J. Alleavitch et al., "Gelatin", Ullmann's*.
[30.12] J.L. Stein, K. Imhof, "Milk and Dairy Products", Ullmann's*.

---

* Ullmann's Encyclopedia of Industrial Chemistry, 5th ed. on CD-ROM, WILEY-VCH, 1997.

31 Use of Quick- and Slaked Lime in the Production of Chemicals

31.1 Introduction

Lime products are used in the manufacture of a wide range of chemicals. Some of those chemicals relate to agriculture, food and food by-products (chapter 30), while some will be found in chapter 32. The more important of the remainder are described in this chapter and are listed in Table 31.1.

Table 31.1. Use of lime products in the manufacture of chemicals

Section	Chemical	Section	Chemical
	Inorganic calcium compounds	*Inorganic chemicals*	
31.2	Precipitated calcium carbonate (PCC)	31.16	Aluminium oxide
31.3	Calcium hypochlorite bleaches	31.17	Potassium carbonate
31.4	Calcium carbide	31.18	Sodium chloride
31.5	Calcium phosphates	31.19	Sodium carbonates
31.6	Calcium chloride	31.20	Sodium hydroxide
31.7	Calcium bromide	31.27	Strontium carbonate
31.8	Calcium hexacyanoferrate		
31.9	Calcium silicon	*Organic chemicals*	
31.10	Calcium dichromate	31.21	Alkene oxides
31.11	Calcium tungstate	31.22	Diacetone alcohol
		31.23	Polyhydricalcohol esters
	Organic calcium compounds	31.24	Pentaerythritol
31.12	Calcium citrate	31.25	Anthraquinone intermediates
31.13	Calcium soaps	31.26	Trichloroethylene
31.14	Calcium lactate		
31.15	Calcium tartrate		

The roles of quick- and slaked lime in the production of industrial chemicals are many and varied. They include [31.1]:

a) source of calcium
b) alkali
c) desiccant
d) causticising agent
e) saponifying agent

f) bonding agent
g) flocculant and precipitant
h) fluxing agent
i) glass-forming product
j) degrader of organic matter
k) lubricant
l) filler and
m) hydrolysing agent

No doubt several others could be added, particularly in the field of organic chemistry.

31.2   Precipitated Calcium Carbonate (PCC)

31.2.1   General

Virtually all commercial PCC is produced by reacting a milk of lime with carbon dioxide. Some producers calcine limestone on site, slake the quicklime and then react the milk of lime with the kiln gases. Others buy-in lime (generally quicklime but sometimes hydrated lime) to produce a milk of lime, which they react with carbon dioxide.

Precipitated calcium carbonate is made as a by-product of causticisation (see section 31.20). While it has been shown that commercial grades of PCC can be made by this route, it has not yet been exploited on the full scale [31.2].

The uses of PCC are described in section 12.9 and listed in Table 12.4.

31.2.2   The Process

Figure 31.1 gives a simplified line diagram of the CO_2-based process.

Milk of lime is produced either by slaking quicklime, or by dispersing hydrated lime in water. The particle size of the slaked lime is an important parameter in the PCC process. Where quicklime is slaked, its reactivity and the quality of the water should be controlled. Where hydrated lime is used, its particle size distribution, and the quality of water used to disperse it, should be controlled. Milk of lime should be screened to remove coarse particles before use.

The process is complex and involves simultaneous dissolution of calcium hydroxide and carbon dioxide, and crystallisation of calcium carbonate. Carbonation is generally carried out in a series of reactors under closely controlled pH, temperature and degree of supersaturation, to produce the required PCC morphology and particle size distribution (see section 31.2.3). Crystallisation can occur on the surface of the calcium hydroxide particles (producing scalenohedral crystals), in the aqueous phase (producing rhombohedral crystals) and at the gas-liquid interface.

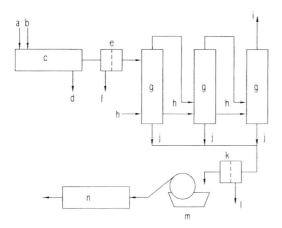

Figure 31.1. The CO_2-based PCC process
(a) quicklime; (b) water; (c) hydrator; (d) coarse grit; (e) filter; (f) fine grit; (g) reactor;
(h) carbon dioxide; (i) waste gases; (j) PCC suspension; (k) filter; (l) grit; (m) rotary filter;
(n) drying system

The resulting suspension of PCC is screened at about 50 µm and is either used as an aqueous suspension containing 20 to 50 % of solids, or de-watered using rotary vacuum filters, pressure filters or centrifuges. The resulting moist cake is dried in rotary, spray or flash driers.

Some grades of PCC are surface coated to improve handling characteristics and dispersability in, for example, plastics. Additives include fatty acids, resins and wetting agents. They help to reduce the surface energy of the calcium carbonate and improve dispersion in organic materials.

31.2.3   Morphology and Particle Size

Precipitated calcium carbonate can occur in five forms [31.2].

a) The scalenohedral (or needle-shaped) form of calcite is the one required by most customers. For a given median particle size, it has a high surface area (e.g. 10 to 15 m^2/g for a median particle size of 1 to 3 µm).
b) The rhombohedral form of calcite is primarily used as a coating material, but is also used as a paper filler.
c) The "amorphous" form of calcite does not have any applications but can be produced when control over the process is lost.
d) The aragonite form (which consists of fine needles) is used in specialised applications requiring a very high surface area.
e) The vaterite form is unstable and does not have any applications.

31.2.4   Lime Requirements

Reactivity of quicklime affects the particle size of the slaked lime, which in turn influences the PCC process. However, as the process parameters can be varied considerably to control the precipitation conditions, and as various particle sizes and forms of PCC are required, consistency of reactivity is as important as the absolute level of reactivity. Some producers prefer quicklimes with "moderate" reactivities, while others prefer "high" reactivities (see section 13.2 for an explanation of those terms).

A high available lime content is desirable, not only for economic reasons, but because it reflects low levels of impurity. The % $CaCO_3$ should be as low as practicable, while meeting the reactivity requirements. Much of the $CaCO_3$ can be removed by screening the milk of lime, albeit at the cost of losing some slaked lime.

The levels of MgO, SiO_2 and Al_2O_3 in the lime should be low, because most of the particles are less than 50 µm and therefore become incorporated as oversize in the PCC.

The level of Fe_2O_3 should be as low as possible, because 80 to 95 % of it passes into the PCC and adversely affects the brightness of the product. An increase from 0.1 to 0.5 % of Fe_2O_3 can reduce the brightness (R457) from 97 to 92 units [31.2].

31.3   Calcium Hypochlorite Bleaches

31.3.1   Bleaching Powder

Bleaching powder, also known as chloride of lime, is produced by reacting hydrated lime with chlorine. It consists of a mixture of calcium hypochlorite ($Ca(OCl)_2$ — the active ingredient), calcium chloride, calcium hydroxide and water. Some processes convert 40 % of the calcium hydroxide, giving about 25 % of "available chlorine". Others continue chlorination to completion, which corresponds to 60 % conversion and about 36 % of "available chlorine" [31.3]. The reaction (31.1) is surprisingly complex. It is exothermic and drives off excess water. The product is dried under vacuum at temperatures below 85 °C.

$$10Ca(OH)_2 + 6Cl_2 \rightarrow Ca(OCl)_2 \cdot 2Ca(OH)_2 +$$
$$2Ca(OCl)_2 \cdot 0.5Ca(OH)_2 + CaCl_2 \cdot Ca(OH)_2 \cdot H_2O +$$
$$2CaCl_2 \cdot H_2O + 3H_2O \qquad (31.1)$$

Tropical bleach is produced by blending bleaching powder with finely divided quicklime which converts any free water into calcium hydroxide. The resulting product is stable at temperatures up to 100 °C.

31.3.2   High-percentage Hypochlorite

"High-percentage" hypochlorite is produced by reacting chlorine with either milk of lime, or an aqueous suspension of bleaching powder. The reaction produces a mixture of calcium hypochlorite and calcium chloride (31.2), which is dried. "Available chlorine" values of greater than 70 % are obtained [31.3].

$$2Ca(OH)_2 + 2Cl_2 \rightarrow Ca(OCl)_2 + CaCl_2 + 2H_2O \qquad (31.2)$$

31.3.3   Lime Specifications

ASTM C911 (formerly ASTM C433) specifies the required quality of quick- and hydrated lime for chemical uses including the production of hypochlorite bleach. High purity products are required with high available lime contents, low levels of sludge-forming impurities and a low iron oxide content (iron oxide is understood to catalyse the decomposition of calcium hypochlorite) [31.1].

31.4   Calcium Carbide

31.4.1   Production

Technical grade calcium carbide is produced by reacting quicklime with carbonaceous matter at 1800 to 2100 °C.

$$CaO + 3C \rightarrow CaC_2 + CO \qquad (31.3)$$

Typically 950 kg of lime and 550 kg of coke or anthracite are used to produce a tonne of "300 litre" calcium carbide, consisting of about 80 % of CaC_2 and 13 % of unreacted CaO.

The most widely used process employs a short electrothermal furnace. Electrical energy is supplied via electrodes, which causes the lime and carbon to react, forming molten calcium carbide, which then acts as a reaction medium. An alternative process uses a shaft furnace, which is charged with a mixture of lime and coke/anthracite [31.4]. The temperatures required for the reaction to proceed are generated by the partial combustion of the carbonaceous matter in oxygen, or oxygen-enriched air. The residual carbonaceous matter then reacts with the lime to produce calcium carbide.

Most of the quicklime used is in the size range 6 to 50 mm to provide the necessary voidage for the gases produced in the reaction to escape. However, modern electrothermal furnaces have hollow electrodes through which fine material (less than 6 mm) can be blown. This enables less expensive fine fractions of coke and quicklime (including any recycled lime) to be introduced. Up to 25 % of the charge to the furnace may be added in this way.

31.4.2   Carbide Lime

Acetylene is produced from calcium carbide by reacting it with water in either "wet" or "dry" generators.

$$CaC_2 + 2H_2O \rightarrow Ca(OH)_2 + C_2H_2\uparrow \qquad (31.4)$$

In wet generators, the carbide is added to an excess of water and an impure milk of lime is produced. The milk may be used on site (e.g. for effluent treatment), or it may be fed to settling pits. The resulting lime putty, or "carbide dough" is then sold as an inferior grade of slaked lime (see also section 22.9).

In dry generators, the carbide is reacted with a controlled excess of water to produce acetylene, steam and hydrated lime with 1 to 2 % of excess water. Part of the hydrated lime may be calcined to quicklime and recycled to the carbide furnace. The remainder, which acts as a purge for impurities, can be used in many of the applications described for hydrated lime (see also section 20.10). Few details of kilns used for the dehydration of carbide lime appear to have been published. Kampmann [31.4] mentions granulation and briquetting of the hydrate, presumably for feeding into a rotary kiln. Other techniques, such as fluidised bed calcining would also appear to be appropriate (see section 16.4.11).

31.4.3   Lime Specification

ASTM C911 [31.5] (formerly ASTM C 258) specifies the quality of quicklime required by the process. The particle size requirements are described in section 31.4.1. From the chemical viewpoint, a high calcium lime is required with a MgO content of less than 2 % as it volatilises and can produce troublesome deposits. The level of minor impurities should be limited ($SiO_2 < 2$ %, $Fe_2O_3 + Al_2O_3 < 1$ %) and phosphorous should be less than 0.02 %. The loss on ignition (combined CO_2 and H_2O) should not exceed 4 %.

31.5   Calcium Phosphates

31.5.1   Monocalcium Phosphate

Commercial monocalcium phosphate ($Ca(H_2PO_4)_2 \cdot H_2O$, or hydrated calcium dihydrogen phosphate) is crystallised from aqueous solution by the partial neutralisation of phosphoric acid with a milk of lime (31.5). A controlled excess of lime is usually used, so that the product contains up to 20 % of dicalcium phosphate ($CaHPO_4$) [31.6].

$$2H_3PO_4 + Ca(OH)_2 \rightarrow Ca(H_2PO_4)_2\downarrow + 2H_2O \qquad (31.5)$$

Monocalcium phosphate is used in at least of one formulation of self-raising flour, in mineral enrichment of foods, as a stabiliser for milk products and as a feedstuff additive. The production process, therefore, requires a particularly pure slaked lime (Table 31.2).

31.5.2 Dicalcium Phosphate

Dicalcium phosphate dihydrate ($CaHPO_4 \cdot 2H_2O$, or hydrated calcium hydrogen-phosphate) is precipitated by reacting phosphoric acid with slaked lime. The reaction mixture is cooled to avoid the formation of the anhydrous phosphate [31.6].

$$H_3PO_4 + Ca(OH)_2 \rightarrow CaHPO_4 \cdot 2H_2O\downarrow \qquad (31.6)$$

The dihydrate is used as a component of toothpastes for its polishing properties and low abrasiveness. It is also used in the food industry for mineral enrichment and in pharmaceuticals as a pelletising aid and thickening agent. The anhydrous phosphate is produced in a similar way but precipitation is carried out at above 80 °C. It is used in the formulation of fertilisers for tropical soils, and in toothpaste formulations.

Because of the food and pharmaceutical applications, the slaked lime should be low in impurities (Table 31.2).

31.5.3 Tricalcium Phosphate

Commercial tricalcium phosphate consists mainly of hydroxyapatite ($Ca_5(PO_4)_3OH$) and is formed by the complete neutralisation of phosphoric acid with slaked lime (31.7). The precipitate is isolated by filtration, and is dried and ground to a fine powder. The molar ratio of Ca ÷ P depends on the operating conditions [31.6].

$$3\,H_3PO_4 + 5Ca(OH)_2 \rightarrow Ca_5(PO_4)_3OH\downarrow + 9H_2O \qquad (31.7)$$

Hydroxyapatite is added in amounts of 1 to 2 % as an anti-caking agent in table salt, sugar and fertilisers. It is also used in toothpaste in conjunction with dicalcium phosphate to modify the polishing properties. The slaked lime should, therefore, meet strict quality requirements (Table 31.2).

Tricalcium phosphate with a Ca ÷ P ratio of 1.5 is produced by heating hydroxyapatite to above 900 °C [31.6].

31.5.4 Lime Specifications

High-calcium quick and hydrated lime is used with low levels of impurities, particularly of lead, arsenic, heavy metals and fluoride (Table 31.2). The slaking characteristics of quicklime should be suitable for the design of the installed slaker. Hydrated lime should be free of grit which might contaminate the phosphates.

Table 31.2. Typical trace element limits in slaked lime for calcium phosphate production

Element	$Ca(OH)_2$ limit (mg/kg)
Arsenic (as As)	7
Lead (as Pb)	13
Zinc (as Zn)	75
Copper plus zinc (as Cu + Zn)	130
Heavy metals (as Pb)	70
Fluoride (as F)	75

31.6   Calcium Chloride

Calcium chloride ($CaCl_2$) is produced as a by-product of the ammonia-soda process (see section 31.19). Milk of lime is used to distil ammonia from the ammonium chloride stream and in so doing, produces a solution of calcium chloride. The solution is then concentrated by evaporation to 30 to 45 % $CaCl_2$, part of which is sold as a liquor. The remainder is concentrated to 75 % $CaCl_2$ (corresponding to $CaCl_2 \cdot 2H_2O$), flaked and dried. The dihydrate is dehydrated by heating to produce various anhydrous products containing 94 to 97 % $CaCl_2$ [31.7]. Calcium chloride is used for de-icing roads, and suppressing dust, as an additive for concrete, and a heat-transfer medium in food processing. The specification for the limestone used in the ammonia soda process is given in section 31.19.

$$2NH_4Cl + Ca(OH)_2 \rightarrow CaCl_2 + 2NH_3\uparrow + 2H_2O \qquad (31.8)$$

31.7   Calcium Bromide

Calcium bromide ($CaBr_2$) is produced by reacting slaked lime, quicklime or calcium carbonate with either hydrobromic acid or bromine and a reducing agent (e.g. formic acid, formaldehyde).

$$Ca(OH)_2 + 2HBr \rightarrow CaBr_2 + 2H_2O \qquad (31.9)$$

The resulting solution is then evaporated under reduced pressure to crystallise the anhydrous salt which is filtered and dried. Alternatively, the concentrated solution is spray-dried to produce a powder [31.8]. Calcium bromide is highly soluble in water (53 % m/m) and produces a dense solution (1.7 kg/l), which is used in oil well "packs" and "completion fluids".

31.8   Calcium Hexacyanoferrate

Calcium hexacyanoferrate (II) is produced by mixing liquid hydrogen cyanide with an aqueous solution of ferrous chloride and milk of lime. The resulting solution is filtered and evaporated to crystallise the hydrated calcium salt [31.9].

$$3Ca(OH)_2 + FeCl_2 + 6HCN \rightarrow Ca_2[Fe(CN)_6] + CaCl_2 + 6H_2O \quad (31.10)$$

Calcium hexacyanoferrate is usually converted into either the sodium and potassium salts, or into the hexacyanoferrates (III) (otherwise known as ferricyanides, $[Fe(CN)_6]^{3-}$). It is used in the production of blue pigments and salt (as an anti-caking agent).

31.9   Calcium Silicon

Calcium silicon ($CaSi_2$) is produced by reacting lime, quartz and coal at above $1200\,°C$ in an electric arc furnace. The raw materials are heated to the required temperature and the carbon reduces the lime and silica (31.11) [31.10].

$$CaO + 2SiO_2 + 5C \rightarrow CaSi_2 + 5CO \quad (31.11)$$

Calcium silicon is used in the iron and steel industry as a de-oxidizer and desulfurizer and for the modification of non-metallic inclusions. A high calcium quicklime is required (e.g. 94 to 97 % CaO) with controlled levels of combined CO_2 and water.

31.10   Calcium Dichromate

Calcium dichromate ($CaCr_2O_7$) is made industrially by oxidative roasting of chromium-containing ores with quicklime (or limestone). The dichromate is extracted by leaching with sulfuric acid [31.11].

31.11   Calcium Tungstate

The mineral scheelite (calcium tungstate) is beneficiated by fusion with soda ash and separation of the soluble sodium tungstate from insoluble impurities (31.12). Purified calcium tungstate is precipitated by the addition of slaked lime (31.13). It is used in the production of ferrotungsten and phosphors for items such as lasers, fluorescent lamps and oscilloscopes [31.12].

$$CaWO_4 + Na_2CO_3 \rightarrow Na_2WO_4 + CaO + CO_2\uparrow \qquad (31.12)$$

$$Na_2WO_4 + Ca(OH)_2 \rightarrow CaWO_4\downarrow + 2NaOH \qquad (31.13)$$

31.12 Calcium Citrate

Calcium citrate is produced as an intermediate in the purification of citric acid ($C_6H_8O_7$, or 2-hydroxypropane-1,2,3-tricarboxylic acid). A broth of citric acid is produced by the fermentation of sugar using a strain of Aspergillusniger mould. Suspended solids are removed by filtration. The solution is reacted with slaked lime to precipitate calcium citrate, which is separated by filtration.

The calcium citrate is stirred in dilute sulfuric acid to regenerate citric acid, the calcium being precipitated as the sulfate, which is removed by filtration. Citric acid is then crystallised from the solution [31.13].

The slaked lime should be of food-grade quality (e.g., [31.30] and Table 31.2).

31.13 Calcium Soaps

Calcium soaps [31.14] are generally the neutral soaps of:

a) saturated straight-chain aliphatic acids (C_8–C_{22}),
b) montan wax acids,
c) unsaturated straight-chain carboxylic acids, (e.g. oleic and linoleic acids),
d) branched-chain aliphatic carboxylic acids, (e.g. ethylhexanoic acids),
e) naphthenic acids, and
f) resin acids.

Some of the more common calcium soaps are the stearate, sulfonate, phenate, myristate, arachidate, 12-hydroxy-stearate, laurate, oleate, 2-ethyl-hexanoate, ricinoleate and naphthenate. They are produced either by double decomposition or direct reaction with slaked lime or calcium chloride.

Calcium soaps are also produced by reacting molten natural resins at temperatures above 200 °C with quick- or slaked lime. Calcium rosinate is used in the formulation of printing inks, principally for gravure and screen printing [31.16].

Calcium soaps are used as:

a) lubricants and stabilisers in the production of plastics to improve flow properties and prevent caking,
b) lubricants and as an additive to oils and greases,
c) mould-release agents,
d) waterproofing agents used for building protection and the surface treatment of fillers (including PCC), and
e) in the paper industry for coating slips.

The calcium hydroxide should have a high available lime content, a low level of residual $CaCO_3$ (typically less than 2 %) and be free of grit.

31.14   Calcium Lactate

Calcium lactate (CH_3–$CH(OH)$–$COO)_2Ca \cdot 5H_2O$) is produced as an intermediate in the production of lactic acid by the reaction of slaked lime with raw lactic
acid. It acts as a protein plasticiser (coagulant) and is used as an egg extender and
foaming agent. It is also used in pharmaceuticals as a source of calcium [31.17],
and as a buffer and source of calcium in bread. Its production requires food-grade
slaked lime (e.g., [31.30] and Table 31.2).

31.15   Calcium Tartrate

Calcium tartrate is produced by extracting wine lees (deposited on the bottom of
vats containing 19 to 38 % of potassium bitartrate) and grape marc (the residue
from pressing grapes), filtering the solution and then adding slaked lime and calcium chloride [31.18].

$$2KHC_4H_4O_6 + Ca(OH)_2 + CaCl_2 \rightarrow 2CaC_4H_4O_6\downarrow + 2KCl + 2H_2O \qquad (31.14)$$

The calcium tartrate is removed by filtration and treated with dilute sulfuric
acid to produce a solution of tartaric acid and a precipitate of calcium sulfate.
The tartaric acid is then concentrated by evaporation under reduced pressure and
separated by crystallisation. Tartaric acid is widely used in the food and drink
industries, in pharmaceutical preparations, as a retarding agent in plaster and gypsum formulations, and in the polishing and cleaning of metals. Food grade slaked
lime is generally specified (e.g., [31.30] and Table 31.2).

$$CaC_4H_4O_6 + H_2SO_4 \rightarrow H_2C_4H_4O_6 + CaSO_4\downarrow \qquad (31.15)$$

31.16   Aluminium Oxide

The Bayer process for the production of alumina uses caustic soda to extract
aluminium oxide from bauxite. The resulting solution of sodium aluminate also
contains dissolved sodium silicate. The silicate is normally removed by allowing
the aluminate and silicate to react to form precipitates of the zeolite structure
having a composition of approximately $Na_8Al_6Si_6O_{24}(OH)_2$. Slaked lime is
added, when very low levels of silicate are required, to precipitate the silicate as
cancrinite, which is removed, prior to the formation of the zeolite structures
[31.19].
Slaked lime is also added to the sodium aluminate liquor to control carbonate
levels by causticising sodium carbonate (31.16). It also helps to control the level
of phosphates by precipitating hydroxyapatite (31.17) — phosphates can interfere
with the subsequent precipitation of aluminium hydroxide.

$$Ca(OH)_2 + Na_2CO_3 \rightarrow CaCO_3\downarrow + 2NaOH \qquad (31.16)$$

$$5Ca(OH)_2 + 3Na_3PO_4 \rightarrow Ca_5(PO_4)_3OH\downarrow + 9NaOH \qquad (31.17)$$

The causticisation of soda ash by slaked lime is also used to generate the caustic soda required for the bauxite extraction when the price of lime relative to that of caustic soda justifies the additional handling and disposal costs.

31.17   Potassium Carbonate

Hydrated lime is used in the amine process for the production of potassium carbonate from potassium chloride [31.20].

Potassium chloride is reacted with carbon dioxide in precarbonated isopropylamine solution under pressure in an autoclave. Potassium hydrogencarbonate precipitates and the amine is converted into isopropylamine chlorohydrate. The potassium salt is isolated by filtration, washed free of amine and heated to convert it to the carbonate. Unreacted amine present in the filtrate is recovered by distillation. Hydrated lime is then added to convert the isopropylamine chlorohydrate back to the amine, which is also recovered by distillation. The main uses of potassium carbonate are the production of glass and sodium silicate.

31.18   Sodium Chloride

Rock salt is generally contaminated with calcium and magnesium sulfates and carbonates and with polyhalite ($K_2SO_4 \cdot 2CaSO_4 \cdot MgSO_4 \cdot 2H_2O$). Crude brine, therefore, contains calcium, magnesium and sulfate ions. Purification of the brine is necessary before it is used in many processes, for example, the ammonia soda process (section 31.18), the electrolytic production of chlorine/caustic soda, and the production of purified salt.

Magnesium ions are removed from the crude brine as a precipitate of magnesium hydroxide by adding milk of lime (or caustic soda). Calcium ions are removed by the addition of either sodium carbonate solution, or by injecting carbon dioxide in the form of combustion gases.

31.19   Sodium Carbonate

The ammonia-soda process [31.21, 31.22], is a long-established route for the production of soda ash (Na_2CO_3), washing soda ($Na_2CO_3 \cdot 10H_2O$) and sodium bicarbonate (Na_2HCO_3), as well as several co-products (including caustic soda, calcium chloride and ammonium chloride). Despite intense competition from naturally occurring sodium carbonate (e.g. trona — $Na_3H(CO_3)_2$), the process is still operated in many countries, and the operators are major (captive) producers of quicklime.

The primary raw materials are purified saturated brine (see section 31.18) and high-calcium limestone. The overall simplified equation is:

$$2\,NaCl + CaCO_3 \rightarrow Na_2CO_3 + CaCl_2 \qquad (31.18)$$

However, because of the low solubility of the calcium carbonate, the reaction does not proceed as shown. Ammonia is, therefore, used to "drive" the process, which consists of seven stages.

1. Limestone is calcined (usually in a mixed feed kiln — see section 16.4.3 — using coke as fuel) to produce quicklime (used in stage 6) and an exhaust gas rich in carbon dioxide (used in stage 3).

$$CaCO_3 \rightarrow CaO + CO_2\uparrow \qquad (31.19)$$

$$C + O_2 \rightarrow CO_2 \qquad (31.20)$$

2. The brine is saturated with ammonia (from stage 7).
3. Carbon dioxide, produced in stages 1 and 5, is injected into the solution from stage 2, precipitating sodium bicarbonate.

$$NaCl + H_2O + NH_3 + CO_2 \rightarrow NH_4Cl + NaHCO_3\downarrow \qquad (31.21)$$

4. The precipitated sodium bicarbonate is filtered and washed.
5. The bicarbonate is heated to produce sodium carbonate and carbon dioxide (which is used in stage 3).

$$2NaHCO_3 \rightarrow Na_2CO_3 + CO_2\uparrow + H_2O \qquad (31.22)$$

6. Quicklime from stage 1 is slaked to produce a milk of lime.

$$CaO + H_2O \rightarrow Ca(OH)_2 \qquad (31.23)$$

7. The ammonium chloride solution produced in stage 3 is heated with the milk of lime to liberate ammonia, which is used in stage 2.

$$2NH_4Cl + Ca(OH)_2 \rightarrow 2NH_3\uparrow + CaCl_2 + 2H_2O \qquad (31.24)$$

High purity limestone is required with a particle size to suit the lime kiln and with low levels of inerts (less than 3 % SiO_2 and less than 1.5 % $Fe_2O_3 + Al_2O_3$). It should also contain less than 2 % of $MgCO_3$ which produces unwanted by-products in the ammonia-soda process.

31.20   Sodium Hydroxide

The traditional route to sodium hydroxide involved the causticisation of soda ash (Na_2CO_3) with slaked lime [31.22]. This route has largely been superseded by the electrolysis of brine to produce chlorine and caustic soda. However, several companies operating the ammonia soda process or with access to natural sodium carbonate still use the causticisation route [31.23].

The process involves reacting a hot solution containing about 12 % of sodium carbonate with quicklime.

$$Na_2CO_3 + CaO + H_2O \rightarrow 2NaOH + CaCO_3\downarrow \qquad (31.25)$$

The precipitated calcium carbonate is removed by filtration and the resulting caustic soda solution is concentrated by evaporation. Impurities that precipitate during the concentration process, mainly sodium chloride and sulfate are removed. Most of the caustic soda is marketed as a 50 % solution.

In most processes, the precipitated calcium carbonate is a waste product. It does not meet the stringent requirements for PCC. At least one company calcines the precipitate and re-uses the resulting quicklime [31.23].

The use of slaked lime to regenerate caustic soda by causticising sodium carbonate is a feature of a number of other processes (e.g. the Bayer alumina process — section 31.15 and the processing of carbolic oil [31.24]).

31.21   Alkene Oxides (the Chlorohydrin Process)

The chlorohydrin process uses slaked lime (or caustic soda) to saponify, or dehydrochlorinate propylene and butene chlorohydrins to produce the corresponding oxides. The oxides may then be converted to the glycols by acidic hydrolysis [31.24].

$$2CH_3{-}CHOH{-}CH_2Cl + Ca(OH)_2 \rightarrow 2CH_3{-}CH{-}CH_2 + CaCl_2 + 2H_2O \quad (31.26)$$

Large tonnages of ethylene oxide and glycol were also produced by this route, for use in anti-freeze, until the direct oxidation of ethylene to ethylene oxide, using silver catalysts, was developed on a commercial basis.

The process requires a high-calcium quicklime with low levels of MgO and SiO_2 [31.25]. The relative merits of lime and caustic soda for the dehydrochlorination stage are discussed in [31.25].

31.22 Diacetone Alcohol

Hydrated lime is used as an alkaline catalyst to promote the self-condensation of acetone to form diacetone alcohol (4-hydroxy-4-methyl-2-pentanone) [31.26]. Diacetone alcohol is used as a solvent for natural and synthetic resins. It is also used as an intermediate in the production of mesityl oxide, methyl isobutyl ketone and hexylene glycol.

31.23 Hydroxypivalic Acid Neopentyl Glycol Ester (HPN)

HPN is produced by the disproportionation of two molecules of hydroxypivaldehyde in the presence of a basic catalyst such as calcium hydroxide [31.27]. HPN is used as a diol modification agent in polyesters, polyurethanes and plasticisers. It is also used as a component of polyester-varnish systems.

31.24 Pentaerythritol

Pentaerythritol is produced by reacting acetaldehyde with formaldehyde in the presence of calcium hydroxide (or caustic soda) [31.27]. Pentaerythritol is used in the production of alkyd resins.

31.25 Anthraquinone Dyes and Intermediates

Some 80 % of anthraquinone dyes are prepared via the anthraquinone sulfonic acids. In producing certain dyes, the sulfonic acid group is replaced by a hydroxyl group using high pressure fusion with lime [31.28].

31.26 Trichloroethylene

Trichloroethylene (widely used as a vapour degreasing solvent in the metal-working industries) may be produced from tetrachloroethane by cracking, using milk of lime (33.27).

$$2CHCl_2\text{--}CHCl_2 + Ca(OH)_2 \rightarrow 2CCl_2\text{=}CHCl + CaCl_2 + 2H_2O \qquad (31.27)$$

The heat of reaction may be used for the distillation of the trichloroethylene. Calcium chloride crystallises, settles to the base of the reactor and is withdrawn. It may be purified to produce a saleable product [31.15].

This route to trichloroethylene is not widely used, but provides an outlet for carbide lime in works which use the calcium carbide process to generate acetylene (see section 31.4.2).

31.27 Strontium Carbonate

Ground calestite ore ($SrSO_4$) can be extracted using ammonium carbonate solution to produce strontium carbonate and ammonium sulfate (31.28) [31.29]. The ammonium sulfate is treated with slaked lime to generate ammonia, which may be reacted with combustion gases (e.g., the exhaust gases from the lime kiln) to regenerate the ammonium carbonate (31.29). Strontium carbonate is the most important industrial compound of strontium, and is used in the manufacture of X-ray absorbing glass for cathode ray tubes.

$$SrSO_4 + (NH_4)_2CO_3 \rightarrow SrCO_3\downarrow + (NH_4)_2SO_4 \qquad (31.28)$$

$$(NH_4)_2SO_4 + Ca(OH)_2 \rightarrow 2NH_3\uparrow + CaSO_4.2H_2O \qquad (31.29)$$

31.28 References

[31.1]   H.N. Lee, "The Specification of Lime for Industrial Uses", The Chalk, Lime and Allied Industries' Research Association, Information Note, IP 58, 1966.
[31.2]   S. Hansen, "Limestone for Precipitated Calcium Carbonate (PCC)", Proc. International Lime Congress, Berlin, 1994.
[31.3]   P. Gallone, "Chlorine Oxides and Chlorine Oxygen Acids", Ullmann's*.
[31.4]   F.W. Kampmann et al., "Calcium Carbide", Ullmann's*.
[31.5]   ASTM C911–94: "Specification for Quicklime, Hydrated Lime and Limestone for Chemical Uses", 1994.
[31.6]   K. Schröder et al., "Phosphoric Acid and Phosphates", Ullmann's*.
[31.7]   R. Kemp, S.E. Keegan, "Calcium Chloride", Ullmann's*.
[31.8]   M.J. Dagani et al., "Bromine Compounds", Ullmann's*.
[31.9]   H. Klenk et al., "Cyano Compounds, Inorganic", Ullmann's*.
[31.10] W. Zulehner et al., "Silicon", Ullmann's*.
[31.11] G. Anger et al., "Chromium Compounds", Ullmann's*.
[31.12] R.S. Boynton , "Chemistry and Technology of Lime and Limestone", John Wiley & Sons, 1980, ISBN 0-471-02771-5.
[31.13] F.H. Verhoff, "Citric Acid", Ullmann's*.
[31.14] A. Szczepanek, G. Koenen, "Metallic Soaps", Ullmann's*.
[31.15] M. Rossberg, "Chlorinated Hydrocarbons", Ullmann's*.
[31.16] K. Fiebach, "Resins, Natural", Ullmann's*.
[31.17] S.P. Chahal, "Lactic Acid", Ullmann's*.
[31.18] J.-M. Kassaian, "Tartaric Acid", Ullmann's*.
[31.19] L.K. Hudson et al., "Aluminium Oxide", Ullmann's*.
[31.20] H. Schultz et al., "Potassium Compounds", Ullmann's*.

[31.21]   C. Thieme, "Sodium Carbonate", Ullmann's*.
[31.22]   T.-P. Hou, "Manufacture of Soda, with Special Reference to the Ammonia Process", Hafner Publishing Co., 1969.
[31.23]   F.-R. Minz, "Sodium Hydroxide", Ullmann's*.
[31.24]   G. Collin, H. Höke, "Tar and Pitch", Ullmann's*.
[31.25]   D. Kahlich et al., "Propylene Oxide", Ullmann's*.
[31.26]   S. Sifniades, "Acetone", Ullmann's*.
[31.27]   J. Schossig et al., "Alcohols, Polyhydric", Ullmann's*.
[31.28]   H.-S. Bien et al., "Anthraquinone Dyes and Intermediates", Ullmann's*.
[31.29]   J.P. MacMillan et al., "Strontium and Strontium Compounds", Ullmann's*.
[31.30]   British Pharmacopoeia, Pharmaceutical Press, London, 1993.

---

* Ullmann's Encyclopedia of Industrial Chemistry, 5th ed. on CD-ROM, WILEY-VCH, 1997.

32 Other Uses of Quick- and Slaked Lime

32.1 Introduction

This chapter describes a variety of applications of lime products (Table 32.1) not included in "Agriculture, food and food by-products" (see Table 30.1), or "Production of chemicals" (see Table 31.1).

Table 32.1. Other uses of quick- and slaked lime

Section	Application	Section	Application
32.2	Magnesium hydroxide	32.13	Drilling muds
32.3	Dead-burned dolomite	32.14	Oil-well cement
32.4	Silica/silicon carbide/	32.15	Oil additives & greases
	zirconia refractories	32.16	Paper & pulp
32.5	Lime refractory	32.17	Pigments & paints
32.6	Glass	32.18	Contaminated land
32.7	Whiteware pottery &		– heavy metals & mineral oils
	vitreous enamel	32.19	Destruction of organic wastes
32.8	Calcium aluminate cement		– oxidisable chemicals & PCBs
32.9	Flotation of metal ores	32.20	Briquetting of fuels
32.10	Refining non-ferrous metals	32.21	Soda lime
	Mg, Ca, Hg	32.22	Desiccant
32.11	Fluxing agent for SiO_2 & Al_2O_3	32.23	Non-explosive demolition
32.12	Lubricant for casting		agent
	& wire-drawing	32.24	Self-heating food cans

32.2 Magnesium Hydroxide

32.2.1 General

Magnesium hydroxide is precipitated from a solution of magnesium chloride solution, using slaked lime. It is used as an intermediate in the manufacture of:

a) caustic (or light-burned) magnesia,
b) sintered magnesia, and
c) metallic magnesium.

32.2.2   The Seawater Process [32.1]

Seawater contains dissolved magnesium chloride — about $400\,m^3$ are required to produce $1\,t$ of $Mg(OH)_2$. Dissolved carbon dioxide is first removed by acidifying the sea water to pH 4 to prevent the precipitation of calcium carbonate when milk of lime is added.

$$Ca(HCO_3)_2 + H_2SO_4 \rightarrow CaSO_4 + 2H_2O + 2CO_2\uparrow \qquad (32.1)$$

Reactive, low carbonate quicklime, or dolomitic lime, is then slaked with decarbonated fresh water to produce a milk of lime. Addition of the milk of lime precipitates the magnesium in the seawater as $Mg(OH)_2$. Where dolomitic lime is used, up to half of the $Mg(OH)_2$ arises from the lime.

$$Ca(OH)_2 + MgCl_2 \rightarrow Mg(OH)_2\downarrow + CaCl_2 \qquad (32.2)$$

$$Ca(OH)_2 \cdot Mg(OH)_2 + MgCl_2 \rightarrow 2Mg(OH)_2\downarrow + CaCl_2 \qquad (32.3)$$

The initial precipitation is done in the presence of excess calcium hydroxide at pH 12 to reduce the level of boron from about 0.2 % B_2O_3 to ca. 0.05 % and thereby limit contamination of the magnesium hydroxide with boron. The level of excess lime may subsequently be reduced by the addition of further quantities of sea water.

The precipitated magnesium hydroxide is concentrated in settlers and isolated by filtration. It is washed thoroughly before further processing. The clarified water is discharged back into the sea.

32.2.3   Production from Brines [32.1]

Magnesium brines are produced by solution mining of bischofite ($MgCl_2 \cdot 6H_2O$) and carnallite ($KCl \cdot MgCl_2 \cdot 6H_2O$). Magnesium chloride is also present in spent liquor from salt (NaCl) production. Any sulfate present is precipitated as calcium sulfate by addition of slaked lime. Magnesium hydroxide is then precipitated using slaked dolomitic lime (or if that is not readily available, slaked high-calcium lime), and removed by settling and filtration.

32.2.4   Quicklime Specifications

The lime should be light-burned, particularly in the case of dolomitic lime. It should have a particularly low residual CO_2 content, of less than 0.2 %, to minimise contamination of the magnesium hydroxide.

32.3   Dead-burned Dolomite [32.2]

Dead-burned dolomite is used in the production of refractory bricks, shaped refractories and for monolithic refractories. High purity, low iron dolomite for brickmaking is generally sintered at temperatures of 1800 °C or higher (see section 16.9). A lower purity product (which is often pre-blended with 5 to 10 % of iron oxide to assist sintering) is used for fettling purposes. It is sintered at 1400 to 1600 °C.

Even after sintering, dolomite is susceptible to hydration and should be protected from undue exposure to atmospheric moisture. Hydration resistance is increased by impregnating the bricks with pitch, resin or tar. 1 to 3 % of zirconia is sometimes added to increase the resistance to thermal shock.

Dolomite used in the production of refractories should be of high purity and, in particular, should contain less than 0.6 % of SiO_2 to limit the amount of dicalcium silicate formed in the sintering process, which can cause "dusting" owing to phase changes [32.2].

32.4   Silica, Silicon Carbide and Zirconia Refractories

Refractory silica bricks, containing 96 to 98 % of SiO_2, are bonded using 1 to 3 % of CaO added as a milk of lime or as hydrate, together with a small quantity of a finely divided sodium-iron-silicate flux [32.3].

Lime is also used as a bonding and stabilising agent in the production of silicon carbide and zirconia refractories [32.4].

32.5   Calcium Oxide Refractory

Calcium oxide has excellent refractory properties and many efforts have been made to commercialise its use (e.g., [32.48]). However, hydration by atmospheric water vapour has proved to be a major problem. Despite this, the nuclear industry is reported to use dead-burned calcium oxide crucibles, with an apparent density of 3.15 g/cm^3 [32.4, 32.49].

32.6   Glass

Relatively little lime is used in glass manufacture as limestone is generally more cost-effective (see section 12.2). However, dolomitic lime and occasionally high calcium lime are used in finely ground forms under specific circumstances.

Burned lime imparts greater brilliance and transparency to the glass than limestone on account of:

a) its lower content of organic matter, and
b) the iron oxide being present in the ferrous rather than the ferric form.

This reduces the requirement for costly decolouriser additives.

In glass processes using medium to fine-grained materials, the replacement of limestone by burned lime has been reported to increase solution rates and reduce heat requirements. This can increase the production capacity of a furnace [32.5]. Consistency of quality is particularly important with regard to grading, loss on ignition, calcium/magnesium oxide content and the level of sulfur. The iron content should be compatible with the quality of glass being produced (e.g. 0.1 to 0.2 % for optical glass and 0.5 % for standard glass).

For glass fibre production, the number and size of acid insoluble refractory particles (mainly grains of silica sand) are of particular importance as they can cause breakages of the fibres as they are drawn out to the required diameter.

32.7   Whiteware Pottery and Vitreous Enamel

Slaked lime is sometimes blended with the kaolin and ball clays used in the production of whiteware pottery. It helps to bind the materials and also increases the whiteness of the fired product [32.4].

The fluxing and glass-forming properties of lime are exploited in various formulations for vitreous enamel.

32.8   Calcium Aluminate Cement

While most calcium aluminate cements are produced from limestone and alumina (see section 9.5), high purity, refractory-grade material is made by at least one producer using ground quicklime as the source of calcium. The quicklime and alumina are blended and fed to a rotary kiln, where they sinter and melt at over 1500 °C. The molten calcium aluminate (about 70 % Al_2O_3 and 28 % CaO) is cooled in a rotating cylinder to produce a clinker, which is subsequently ground to substantially less than 90 μm [32.6].

Refractory grade calcium aluminate contains less than 0.8 % of SiO_2 and 0.4 % of Fe_2O_3 and, in consequence, is particularly white. This has led to its use in grouting and similar applications. The quicklime used should have low SiO_2 and Fe_2O_3 contents.

32.9   Flotation of Metal Ores

Lime is widely used in the benefication of non-ferrous metal ores to control the pH and acts as a depressant [32.7]. Its use, combined with other materials (frothers, collectors, activators, depressants and other reagents) enables, for example, copper pyrites (Cu_2S) to be separated from arsenopyrite (As_2S_5) [32.8, 32.9].

Lime is also used in producing concentrates by flotation of molybdenum [32.10], zinc, and lead ores [32.4]. It is used to maintain the slightly alkaline conditions (approx. pH 10) required for the dissolution of gold, silver [32.11, 32.12], and nickel [32.13] in cyanide extraction.

The quality of lime for such applications is specified in [32.14].

32.10   Refining of Non-ferrous Metals

32.10.1   Magnesium

The various processes used for the production of metallic magnesium are based on the reduction of magnesium oxide with ferrosilicon (FeSi). Calcined dolomite is used as the source of magnesium oxide and the reaction is carried out at 1200 to 1600 °C under reduced pressure (13 to 670 kPa). Gaseous magnesium distils from the other materials and is condensed at about 450 °C [32.15].

$$2CaO + 2MgO + Si \rightarrow 2Mg\uparrow + Ca_2SiO_4 \tag{32.4}$$

32.10.2   Calcium

Calcium is produced by the thermal reduction of lime with aluminium. High purity ground quicklime and aluminium powder are briquetted and heated in a retort to 1200 °C at a pressure of 0.1 Pa or less. Calcium vapour is formed which is condensed in a cooled section of the retort [32.16].

$$3CaO + 2Al \rightarrow Al_2O_3 + 3Ca\uparrow \tag{32.5}$$

32.10.3   Mercury

The most important minerals used for the production of mercury are cinnabar and cinnabarite (HgS). Quicklime can be used to flux the ore and combine with the sulfur. At above 300 °C and in the absence of oxygen, the reaction proceeds as in (32.6). In the presence of oxygen, reaction (32.7) occurs. The mercury distils from the retort and is condensed [32.17].

$$4HgS + 4CaO \rightarrow 3CaS + CaSO_4 + 4Hg\uparrow \tag{32.6}$$

$$HgS + CaO + 1.5O_2 \rightarrow CaSO_4 + Hg\uparrow \tag{32.7}$$

32.10.4   Quicklime Specification

A South African Standard specifies the qualities of lime for metallurgical purposes [32.14].

32.11   Lime as a Fluxing Agent

Quicklime is used as a fluxing agent in the production of:

a) beryllium compounds [32.18],
b) ferroboron from boric oxide and boric acid [32.19],
c) ferrochromium [32.20],
d) ferromanganese [32.21],
e) nickel [32.13],
f) tin [32.22], and
g) titanium [32.23].

The Högenäs process is widely used for the production of *reduced metal powders* [32.24]. In the production of iron powder, for example, pure magnetite ores are comminuted, screened and purified using magnetic separators. They are then mixed with a blend of coke and quicklime, and heated in a silicon carbide lined retort at 1050 to 1200 °C for 24 to 40 hours. The iron oxide is reduced by the coke to iron and the quicklime reacts with sulfur, silica and alumina in the ore and coke, forming a slag.

The reaction product is a strongly sintered sponge, which includes the slag. The sponge is ground into a powder, screened and non-metallic inclusions are removed using magnetic separation.

High-calcium quicklime is used with a similar chemical requirement to those for steelmaking (see section 27.4.2). As in steelmaking, the required particle size depends on the method of addition/injection of the quicklime.

32.12   Lubricant for Casting and Wire Drawing

This section refers to the use of quicklime as a component of continuous casting lubricants and of slaked lime as a lubricant carrier in wire drawing (the use of lime in greases and calcium soaps is described in section 31.13).

Continuous casting is used for materials based on iron, aluminium, copper, steel and noble metals. The molten strand passes through a cooled mould, which produces a shell of solidified metal. The skin of the metal is in contact with the mould. High temperature lubricants perform the essential role of preventing the solidified metal from sticking in the mould, which could then allow the molten core to escape.

Lubricants consist of intimate blends of lime, silicic acid, alumina and fluxing agents (such as fluorspar and alkali oxides) and carbon [32.25].

In wire-drawing, a drawing compound lowers friction and wear. Where the reduction in diameter is considerable, the compound tends to be stripped from the surface of the metal and lubricant carriers are required. Milk of lime is widely used as a lubricant carrier. The calcium hydroxide bonds to the surface of the wire, increases surface roughness and improves the adhesion of the drawing compound [32.25].

The milk of lime should be grit-free and should have as high a surface area as is practicable. Traditionally this has been produced by slaking a low carbonate quicklime under conditions favouring the production of finely divided calcium hydroxide (see chapter 22). More recently, ultra-fine milks of lime (see section 22.8) have proved to be particularly effective lubricant carriers.

32.13   Drilling Muds

When drilling through rock for oil or gas, "cuttings" (i.e. the fragments of rock produced by the drill bit) are carried to the surface by drilling mud. The mud is pumped to the bit through the hollow drill tube and returns to the surface on the outside of the drill tube. The mud helps to prevent blow-outs and lubricates the drill pipe. Formulations for drilling muds contain many materials including hydrated lime and clay. The hydrated lime maintains a high alkalinity, which keeps the clay in a non-plastic state [32.26].

The principal properties of hydrated lime for this purpose are its available lime content and its particle size distribution, which help to produce the required rheological characteristics.

32.14   Oil-well Cement

When it is required to seal the steel casing of gas and oil wells to the walls of the borehole and to seal porous formations, a grout mixture is used. Both Portland and pozzolanic cements are used as setting agents. The latter consists of slaked lime and a pozzolan. The setting time of the pozzolanic cement is shortened by the addition of an accelerator (e.g. soda ash which is immediately causticised by the lime) and by the relatively high temperatures at depth [32.27].

32.15 Oil Additives and Lubricating Greases

Certain oil additives are produced by reacting hydrated limes with alkyl phenates or organic sulfonates. The resulting calcium soaps act as wear inhibitors, help to prevent sludge build-up and neutralise acidity from products of combustion.

Lubricating greases frequently contain 10 to 25 % of calcium soaps dispersed in naphthenic or aromatic mineral oils. While greases containing "simple" soaps are only suitable for use at temperatures up to 60 °C, "complex" soaps can be used at up to 160 °C [32.50].

32.16 Paper and Pulp

The principle use of lime in the paper and pulp industry is in the *sulfate (or Kraft) process*. Kraft "black liquor" is dehydrated and burnt in a furnace to produce a smelt which consists primarily of sodium carbonate and sodium sulfide [32.28]. The smelt is causticised with slaked lime (see section 31.20). The products are insoluble calcium carbonate, which is removed by filtration, and sodium hydroxide liquor, which is recycled. Some 90 % of the calcium carbonate, so produced, is calcined in a suitable kiln (see section 16.4.11), slaked and re-used. About 250 kg of quicklime per tonne of pulp are required for causticisation.

The *sulfite pulp process* formerly used considerable amounts of lime. Other alkalis are now used e.g. ammonia, magnesia and soda ash, which do not create the waste disposal problems that arise with the use of lime or limestone.

Pulp mills use considerable quantities of *hypochlorite* for bleaching. Both sodium and calcium hypochlorites are used, produced by reacting chlorine with caustic soda and milk of lime respectively. It is reported, however, [32.29] that the trend is towards the use of sodium hypochlorite which produces soluble wastes, whereas calcium hypochlorite produces a calcium carbonate sludge.

Commercial *lignosulfonates* (also called lignin sulfonates and sulfite lignins) are by-products of the treatment of pulp. One of the methods of isolating and purifying them is to add excess slaked lime to spent sulfite liquor. This precipitates calcium lignosulfonates, which are removed by filtration [32.29].

32.17 Pigments and Paints

- *Precipitated calcium carbonate* is produced by the controlled carbonation of milk of lime (see section 31.2). The use of PCC as a pigment is mentioned in section 12.9.

- *Satin white*, a white pigment, is produced by reacting aluminium sulfate with milk of lime. The precipitate, which is reported to have the formula $3CaO \cdot Al_2O_3 \cdot 3CaSO_4 \cdot 3H_2O$, is passed through a fine mesh screen to remove grit and de-watered in a filter press. It is stored as a damp filter cake. Satin white is blended with PCC, china clay and a binder (e.g. casein or resin) to form a slurry, which is used to coat paper [32.30].

● *Limewash (or whitewash)* is a traditional wall decoration and, despite the availability of a wide range of alternative paints, there are still circumstances where it is both the best and cheapest option [32.31]. Five formulations are given in [32.32], which were evaluated in a 5-year trial and which proved to be generally satisfactory.

Limewash is particularly suitable for the internal decoration of buildings with solid walls and without damp-proof courses. The moisture content of such walls is frequently high and varies with the seasons. In such circumstances, any wall decoration must be porous. Limewash is also widely used in agricultural buildings, owing to its mild germicidal qualities coupled with ease of application and low cost. It has been recommended by the Building Research Establishment for use on bituminous surfaces, such as flat roofs to reduce radiant heat absorption from sunlight.

32.18   Lime Treatment of Contaminated Land

32.18.1   Heavy Metal Contaminants

Lime products have been used to raise the pH of soils contaminated with cadmium and other heavy metals. This treatment immobilises the metals and enables the land to be used for growing forage crops and grasses [32.33].

32.18.2   Contamination with Mineral Oils and Similar Substances

A patented process, known as the DCR process, [32.34] describes the use of coated ground quicklime to disperse mineral oils and similar substances. The coating delays the hydration of the quicklime, which subsequently carbonates. Because the organic matter is finely dispersed, it is claimed to be biodegradable. The process is reported to be suitable for the treatment of contaminated soils and sludges.

The process is understood to have been applied successfully to many sites in Europe. The treatment undoubtedly improves the characteristics of the soil and sludge and immobilises the organic matter to a degree. However, there is concern as to whether the treated material meets the requirements of the appropriate leaching tests [32.35, 32.36]. The latter simulates the effects of acidic rain, which is much more aggressive than normal groundwater.

32.19   Destruction of Organic Wastes

32.19.1   Oxidisable Chemicals

A patented process [32.51] describes the treatment of water, sludge and soil, contaminated with "small to moderate" concentrations of organic substances. According to [32.37], the organics are oxidised in 20 to 40 min. by oxygen dissolved in saturated limewater at 90 to 170 bars pressure and 220 to 300 °C. The role of the lime is presumably to maintain the alkaline conditions favouring the hydrolysis of the organic matter and to react with inorganic oxidation products such as sulfuric acid.

32.19.2   Polychlorinated Biphenyls (PCBs)

Considerable interest was generated in 1990/91 by reports arising from the US Environmental Protection Agency stating that the use of quicklime to stabilise PCB-contaminated soil had (apparently) led to the disappearance of the PCBs. However, it subsequently appeared likely that the PCBs had not been destroyed [32.38]. Indeed, the suggestion was made that, if quicklime were able to replace chlorine atoms in PCBs by hydroxyl groups, the reaction products might be more hazardous than the PCBs [32.39].

Subsequent trials were reported [32.40, 32.41], in which PCBs were mixed intimately with quicklime using a ball mill as a reactor, which also resulted in their apparent destruction. It is understood, however, that the expected reaction products were not detected in the predicted quantities, casting doubt on the success of the treatment.

As of August 1997, the National Lime Association of America was not aware of any current research into the subject [32.42].

32.20   Briquetting of Fuels

Slaked lime has frequently been proposed as a binding agent for the briquetting of fuels (e.g. [32.43]). Its potential for binding some of the sulfur in the fuel made it a particularly attractive binding agent [32.44, 32.45]. However, the general conclusion was that the strength of the resulting briquettes was not adequate.

Alternative approaches have been suggested for improving the strengths of briquettes [32.45], including the conversion of the slaked lime into:

a)   calcium carbonate by treatment with carbon dioxide, or

b)   calcium silicate by adding siliceous materials (such as pulverised fuel ash) and steam curing at 100 °C.

A more recent publication [32.46] described the production of dense refuse-derived fuel pellets. Calcium hydroxide was selected as the most effective binding agent from over 150 potential binders and binder combinations. The future use of refuse derived fuel pellets, however, is likely to depend on political factors, as much as technical feasibility.

32.21   Soda Lime

Soda lime consists of hydrated lime treated with caustic soda solution. It is used for absorbing carbon dioxide from air in medical, deep sea diving and military applications. It is manufactured by mixing hydrated lime with about 3 % of caustic soda and sufficient water to enable the paste to be extruded into pellets some 4 mm in diameter and 8 to 15 mm long. The pellets are then dried, screened and re-hydrated to a moisture content of about 20 %.

The caustic soda dissolves in the free water and is the active component for the absorption of carbon dioxide at the surface of the pellet (32.8). The sodium carbonate so formed then reacts with calcium hydroxide to form calcium carbonate and regenerate sodium hydroxide (32.9).

$$2NaOH + CO_2 \rightarrow Na_2CO_3 + H_2O \tag{32.8}$$

$$Na_2CO_3 + Ca(OH)_2 \rightarrow CaCO_3\downarrow + 2NaOH \tag{32.9}$$

The hydrated lime should have a high active lime content and a particle size distribution that gives the paste the required rheological properties. It should be sound and not cause expansion in the pellets which could lead to low strength or dusting.

32.22   Use as a Desiccant

Quicklime's high affinity for water has resulted in its use as a desiccant for both gases and organic liquids. In some applications, finely ground quicklime is used. In others, the quicklime is mixed into a paste with an inert organic liquid and possibly other components. The paste is then formed into blocks or granules and the solvent removed by evaporation.

A high-calcium quicklime is preferred. Reactivity to water does not appear to be particularly important, but, where blocks or granules are produced, the particle size distribution should be such that it gives the required rheological properties.

32.23   Use as a Non-explosive Demolition Agent

The hydration of quicklime is associated with an increase in volume of over 2.5 times. This property has been exploited to develop products, which, when moistened with water and confined in drill holes, can split rocks and concrete [32.37, 32.47].

The product was developed in Japan in the 1970s [32.37]. Initially, mixes containing hard-burned quicklime were used. However, the expansive stress, which can exceed 5000 t/m^2 after 48 hours was sometimes relieved by "blow-outs" (i.e. ejection of the lime from the hole), rather than causing splitting.

Currently, the product is made by calcining mixtures of limestone and a flux (e.g. a clay) at temperatures of 1400 to 1500 °C. The resulting clinker is ground and used either as a powder or as cylindrical capsules. The formulation, and/or the degree of sintering, is varied to produce a range of products for use at ambient temperatures ranging from –5 to +35 °C.

32.24 Self-heating Food Containers

The heat of hydration of quicklime has been used to produce a range of "self-heating" food containers [32.48]. The food can is packed in a larger can and medium to low reactivity quicklime, in the form of fine grains (e.g. 2 to 5 mm), is packed into the annular space around the inner can. Small, sealed bags of water are placed on top of the lime. The outer container is sealed to the inner using an annular lid.

When hot food is required, the water bags are punctured, releasing the water on to the quicklime. After 5 to 10 min. the food in the inner can is sufficiently hot to be eaten. The cans are used by emergency services and the military in situations when the convenience of the cans outweighs their extra mass, bulk and cost.

32.25 References

[32.1]   M. Seeger at al., "Magnesium Compounds", Ullmann's*.
[32.2]   G. Rotschka, K.-E. Granitzki, "Refractory Ceramics", Ullmann's*.
[32.3]   O.W. Flörke et al., "Silica", Ullmann's*.
[32.4]   R.S. Boynton, "Chemistry and Technology of Lime and Limestone", John Wiley & Sons, 1980, ISBN 0-471-02771-5.
[32.5]   M. O'Driscoll, "Burnt Lime/Dolime – Seeing Markets Green", Industrial Minerals, May 1988.
[32.6]   A. Beadle, Lafarge Aluminates, personal communication.
[32.7]   B. Yarar, "Flotation", Ullmann's*.
[32.8]   K. Hanusch et al., "Arsenic and Arsenic Compounds", Ullmann's*.
[32.9]   H. Fabian, "Copper", Ullmann's*.
[32.10] R.F. Sebenik, "Molybdenum and Molybdenum Compounds", Ullmann's*.
[32.11] H. Renner, M.W. Johne, "Gold, Gold Alloys and Gold Compounds", Ullmann's*.
[32.12] T. Voeste et al., "Liquid-Solid Extraction", Ullmann's*.
[32.13] D.G.E. Kerfoot, "Nickel", Ullmann's*.
[32.14] SABS 459-1955: "Specification for Lime for Chemical and Metallurgical Purposes", 1955.
[32.15] N. Høy-Petersen et al., "Magnesium", Ullmann's*.
[32.16] S.E. Hluchan, "Calcium and Calcium Alloys", Ullmann's*.

[32.17] M. Simon et al., "Mercury, Mercury Alloys and Mercury Compounds", Ullmann's*.
[32.18] G. Petzow, "Beryllium and Beryllium Compounds", Ullmann's*.
[32.19] R. Fichte, "Boron and Boron Compounds", Ullmann's*.
[32.20] J.H. Downing et al., "Chromium and Chromium Alloys", Ullmann's*.
[32.21] D.B. Wellbeloved et al., "Manganese and Manganese Alloys", Ullmann's*.
[32.22] G.G. Graf, "Tin, Tin Alloys and Tin Compounds", Ullmann's*.
[32.23] H. Sibum at al., "Titanium and Titanium Alloys", Ullmann's*.
[32.24] H. Cohrt, M. Enders, "Sintered Steel and Iron", Ullmann's*.
[32.25] D. Klamann, "Lubricants and Related Products", Ullmann's*.
[32.26] "Lime Products in the Oil Industry", ICI Technical Service Note, TS/E/22.
[32.27] F.W. Locher, J. Kropp, "Cement and Concrete", Ullmann's*
[32.28] R. Patt et al., "Paper and Pulp", Ullmann's*.
[32.29] H.U. Süss, "Bleaching", Ullmann's*.
[32.30] "Lime Products in the Paper Trade", ICI Technical Service Note No. TS/E/71.
[32.31] R. Bennett, "The Use of Limewash as a Decorative and Protective Coating", The Building Conservation Directory, 1997, 136–137.
[32.32] "Lime in Building", British Quarry and Slag Federation (now the Quarry Products Association), 1974.
[32.33] "Montana Hazardous Waste Seminar", Lime-Lites, Vol. LX, National Lime Association, July–Dec. 1993.
[32.34] "Process for the Safe Elimination of Mineral Oils and Substances similar to Mineral Oils", German Patent Application DE 3632337A1, 24 Sept. 1986.
[32.35] "Draft Guidelines for Performing the Eluate Tests", Federal Department for the Environment, Forestry and Land Use, Berne, Aug. 1988.
[32.36] "Overview of the Revised Toxicity Characteristic Leaching Procedure (TCLP)", US Federal Register, Vol. 55, No. 61, 29 March 1990.
[32.37] K. Soeda, "Non-explosive Demolition Agent", Proc. International Lime Congress, London, 1986.
[32.38] R.L. Einhaus et al., "The Fate of Polychlorinated Biphenyls (PCBs) in soil following stabilization with quicklime", EPA/600/2-91/052, Sept. 1991.
[32.39] W. Gruber, "Treating PCBs with Quicklime", EI Digest **6**, June 1991.
[32.40] S. Borman, "Novel Idea Developed to Destroy Toxic Chemicals", C & EN, Oct. 11, 1993.
[32.41] "Lime Used in Destroying PCBs", Lime-Lites, Vol. LX, Jan.–June 1994 (based on an article in Chemical Engineering, June 1994).
[32.42] J. Hammond of the National Lime Association, personal communication to M.B. Kenny, 8 Aug. 1997.
[32.43] P.L. Waters, "Binders for Fuel Briquettes — a Critical Survey", Commonwealth Scientific and Industrial Research Organisation, Australia, Technical Communication 51, May 1969.
[32.44] E.F. Maust, "Method for Enhancing the Utilisation of Powdered Coal", US Patent No. 4,230,460, Oct. 28, 1980.
[32.45] Davey McKee Corporation, "A Literature Review and Binder and Coal Selection for Research Studies on Coal Agglomeration", Febr. 26, 1982.
[32.46] O.O. Ohlsson, "Development of Lime Enhanced Refuse Derived Fuel (RDF) Pellets and its Potential Impact on the Lime Industry", Zement Kalk Gips **6**, 1995, 328–333.
[32.47] "Safe Exploding from Access Onoda Bristar", Construction News, 18 Dec. 1980.
[32.48] S. Aitoh, "Research and Development of a Self-heating Food Container Using Lime", Proc.International Lime Congress, Rome, 13–14 Sept. 1990.
[32.49] E. Marino, "The Use of Calcium Oxide as Refractory Material in Steelmaking Processes". In: "Refractories for the Steel Industry", Elsevier Applied Science, London, 1990, pp. 59–68.
[32.50] A. Szczepanek, G. Koenen, "Metallic Soaps", Ullmann's*.
[32.51] "Wet Oxidation for Pulp and Paper Industry Wastes", Environmental Science & Engineering, June 1996.

---

* Ullmann's Encyclopedia of Industrial Chemistry, 5th ed. on CD-ROM, WILEY-VCH, 1997.

Part 6   Safety, Health and Environment

33   Control of the Environmental Effects of Lime and Limestone Production

33.1   Introduction

The production of lime and limestone inevitably results in a number of environmental effects. A responsible producer will, therefore, have an environmental management system, which addresses the relevant areas listed in Table 33.1, to ensure that appropriate actions are taken to minimise the environmental impact of his operations [33.1, 33.2].

Table 33.1. Topics for an environmental management system

Air quality	– dust	Energy
	– SO_x	Transport
	– NO_x	Solid wastes
	– CO_2	Liquid discharges
	– CO	Public safety
	– other	Security
Blasting	– vibration	General noise
	– noise	Disturbance
Employee	– awareness	Lighting
	– training	Archaeology
Visual impact	– new developments	Ecology
	– existing operations	Restoration
Community relations		After-use

In the context of the production of lime and limestone the following effects merit detailed consideration:

a) dust,
b) sulfur dioxide,
c) oxides of nitrogen,
d) oxides of carbon,
e) general noise,
f) noise and vibration from blasting,
g) aqueous discharges, and
h) disposal of solid waste.

These effects are considered briefly in the following sections: for greater detail on individual effects, the reader is referred to the standard reference texts.

33.2 Standards

The control of environmental effects is an area in which acceptable standards and practices are changing rapidly as a result of public pressure, developing technology and improved management systems.

In the European Union, for example, the Integrated Pollution Prevention and Control (IPPC) Directive requires the operator to use the "Best Available Technique" (BAT) to protect the environment. This includes the concept that "the cost should not be excessive". The term technique covers both technology and operating/maintenance practices. Thus, a new plant is expected to comply with BAT, which will change as advances are made. An existing plant will be required to up-grade to the BAT (which may differ from that for a new plant) after an appropriate period.

IPPC recognises the need to "integrate" the approach to pollution prevention and control, ie. to consider the pollution potential with regard to air, land and water, and to select the solution which has the least net impact. The thorny problem of determining how to assess relative impacts of a process on more than one medium is still under consideration. A document for the guidance of national governments — the "BAT Reference Document" for cement and lime production — is in preparation [31.14].

33.3 Dust

33.3.1 Quarrying

In the quarrying process, drilling is the main potential source of dust. The air carrying pulverised rock up the hole may be de-dusted using multi-cyclones, but the use of bag filters is the preferred control method. The collected dust should be disposed to tip in sealed bags, or should be wetted to avoid its becoming air-borne when blasting.

Dust emission from blasting is controlled by sealing the top of the holes with "stemming" (section 4.3.4). Transport of limestone from the quarry face along the quarry roads, which are normally un-made, can raise a dust cloud in dry weather. Various ways of suppressing this dust have been investigated, but there does not appear to be an ideal solution. Waste oil and a bitumen–water emulsion are effective in binding the dust, but may lead to the contamination of ground water with organic matter. Calcium chloride which is deliquescent, is effective in binding the dust for prolonged periods in dry weather, but is expensive and appears to be washed away by rainfall. Spraying water is widely practised, but, in hot, dry weather, it evaporates quickly.

Un-published work by the author showed that the dust raised by vehicles travelling along unmade quarry roads is relatively coarse and that very little of it carries more than a few tens of metres from the road. Clearly any fine particles of (say) below 20 µm remain air-borne and drift away, leaving only coarse particles for the next vehicle to raise. Thus, the cost-effectiveness of dust suppression on quarry roads, in the context of reducing dust emission beyond the works boundary, merits further investigation.

33.3.2   Limestone Processing and Storage

Crushing, screening and conveying of limestone can give rise to dust. High speed impact rotary crushers, which displace large volumes of air, should be fitted with dust extraction, or, at the very least, be contained within a dust-tight building. Other crushers should be fitted with appropriate containment, dust suppression, or dust collection equipment.

In general, dust emission from screening dry, crushed limestone should be controlled by enclosure or extraction. An alternative approach is to suppress the dust using water, preferably containing a wetting agent. This, however, may introduce down-stream problems as a result of carry-forward of fines and contaminants.

Belt conveyors handling un-treated limestone should be partially, or totally enclosed to control wind-whip, unless the material has been pre-screened (e.g. at 3 mm) to remove fines. Transfer points between conveyors should be enclosed, and, where dry limestone is being handled, they should be fitted with dust extraction.

Where conveyors are used to feed to and from stockpiles the drop height of the stone should be minimised to limit the effects of wind whip. When water is used to condition stone, it should be added before the point of discharge of the conveyor. Use of a wetting agent increases the effectiveness of water addition and helps to reduce the amount of water required. All conveyors should be fitted with an effective means of cleaning to minimise dust dropping off the returning conveyor (for example belt scrapers should be fitted at all head drum returns and the scrapings should fall by chute so as to rejoin the main material run).

Plants producing dry ground limestone products require appropriate dust collection equipment — generally a bag filter.

All crushed limestone containing a substantial proportion of material below about 3 mm in size should, where practicable, be kept in covered storage (e.g., consisting of not less than three walls and a roof [33.3]) for routine use. Where this is not possible the material should be conditioned with water to control wind-whip.

Limestone dust should be stored in enclosed silos and hoppers which are vented through arrestment equipment.

With regard to the choice of dust collectors, bag filters are generally favoured where it is possible to use the collected dust. At transfer points the collected dust may be discharged on to the product for onward conveying. Where it is necessary to dispose of the collected dust, a wet scrubbing system is often preferred. The water from a wet scrubbing system may, after removal of solids, be recycled or discharged.

33.3.3   Production of Quicklime

Lime kilns produce relatively large volumes of exhaust gases (e.g., 3,500 to 4,000 Nm^3/t lime) at temperatures which may range from 100 °C to over 1000 °C. The dust concentration can be high, depending on the feedstone and the kiln design, and range from about 500 mg/Nm^3 to over 5,000 mg/Nm^3. The dust contains quicklime, limestone, and clay, as well as any ash from the fuel, and calcium sulfate produced by the reaction of any sulfur dioxide in the exhaust gases with quicklime. A number of alternative collection systems are used to cope with this range of conditions. They include cyclones, fabric filters, electrostatic precipitators, wet scrubbers and gravel bed filters.

The capital and operating cost of dust collectors, per tonne of lime produced, vary with the type of collector and the size of the installation. For kilns with capacities of over 250 t/d, the all-in cost (1997) for bag filters (charging capital at 10 % depreciation and 10 % interest charge) is probably in the range £1 to 3 per tonne. For similarly sized kilns fitted with electrostatic precipitator the cost is understood to be in the range £2 to 4 per tonne. The capital charges per tonne of lime tend to be significantly greater for smaller kilns.

- *Cyclones* are relatively inexpensive and have low operating and maintenance costs. Practical experience shows that, while they are capable of reducing dust concentrations in kiln exit gases to about 500 mg/Nm^3, they are generally not suitable for meeting emission limits of below 200 mg/Nm^3. Cyclones may, however, be used for pre-cleaning gases before treatment in other dust collectors.

- *Fabric filters* are widely used owing to their simplicity, reliability, efficiency and competitive capital cost. Their application is limited by both the humidity and the temperature of the gas streams. Condensation or collection of water droplets of the filter fabric will lead to blinding of the cloth. Normal filter cloths can operate up to 180 °C and fibreglass fabric can operate up to 250 °C. Where the limitation of humidity or temperature is likely to be a problem, precautions must be taken (such as the use of gas to air heat exchangers, the addition of cooling air/water, or the use of auxiliary heaters), or an alternative collection system must be sought.
  While fabric filters have relatively low capital coats, their operating costs are high. Well maintained fabric filters can reliably reduce dust concentrations to below 50 mg/Nm^3. Where incompletely combusted fuels are likely to be present the bags should be earthed, to prevent build-up of static electricity, and the filter housing should incorporate a suitably designed explosion vent.

- *Electrostatic precipitators* can be used over a wider range of temperature (up to 400 °C) and at higher humidities than bag filters. They have lower operating costs, owing to lower pressure drops, but their capital costs are considerably higher. When the gas temperature exceeds 400 °C, the gas must be cooled (e.g., by the injection of ambient air, of water, or by the use of gas-to-air heat exchangers). Electrostatic precipitators can reliably achieve dust loadings below 50 mg/Nm^3.

  Sparking in the precipitator could cause explosions of partially burned fuel. Where there is a risk of significant levels of combustibles, a combustibles meter should be fitted and used to de-activate the precipitator when the level of combustibles exceeds a pre-set value. As an additional precaution, the precipitator housing should be fitted with an explosion-relief vent.

- *Wet scrubbers* are particularly suitable for treating exhaust gases with high humidities and low temperatures. These conditions are obtained with highly efficient kilns, burning either small sizes of limestone, or stone, with a significant water content. Their low capital costs are offset by high operating costs. Wet scrubbers can also reduce the emitted dust level to below 50 mg/Nm^3.

  The disadvantages of wet scrubbers are that they produce a visible steam plume, and liquid and solid effluents, which can cause disposal problems. Where low sulfur fuels are used, the effluent may contain calcium hydroxide which in one installation has been found to give pH values as high as 11.3: the sludge is also alkaline and may present disposal problems. Where high sulfur fuels are used, the effluent contains dissolved salts, mainly calcium sulfate and possibly calcium bisulfite. The latter could lead to a biological oxygen demand.

- *Gravel bed filters* can operate at high temperatures in excess of 400 °C. It is understood that they cannot always meet an emission limit of 50 mg/Nm^3. Moreover, capital and operating costs are believed to be generally higher than those of the techniques described above. Condensation in the bed can give rise to problems and should be avoided.

33.3.4 Processing and Storage of Quicklime

Because quicklime is relatively sensitive to abrasion and impact, it is an inherently dusty product. *Run of kiln lime* typically contains 5 to 10 % of material finer than 6 mm, a significant proportion of which is sufficiently fine (less than 75 µm) to become airborne at transfer points. Such material is generally of marketable quality and, after collection, can usually be blended back into the product.

   Belt *conveyors* are widely used to transport run-of kiln quicklime to the processing plant. The conveyors should be enclosed to prevent wind-whip (as well as to keep off rain). Transfer points should be fitted with hoods and dust extraction. Bag filters are generally preferred, with the collected dust being discharged on to the down-stream conveyor. Conveyor discharge points should be arranged to minimise free-fall and should either be enclosed, or fitted with dust extraction.

Other conveying systems, such as elevators, drag chains, vibrating tubes, screw feeders, pneumatic conveyors, and air slides are generally enclosed. Any air displaced from them should be vented to a suitable dust collector.

Crushing and screening quicklime generate airborne dust. The units and their discharge chutes should be enclosed and vented to a dust collector. Quicklime *grinding plants* generally include bag filters to collect the product. The filters may be preceded by cyclones to reduce the dust loading on the filters and to recycle part of the air stream. *Bagging* of quicklime should be done using a purpose-designed plant fitted with dust extraction equipment.

All quicklime *bunkers* should be enclosed. Where air is used in the conveying operation, they should be fitted with dust collection equipment to filter the air discharged from the bunker.

Spillages should be cleaned up to prevent entrainment of dust. Use of vacuum equipment connected to a filter is generally preferable to shovelling or sweeping.

33.3.5   Production and Storage of Hydrated Lime

The exhaust gas from hydration of quicklime (ca. 800 m^3/t of hydrate) is generally the main source of dust emission from the process. Depending on the detailed design of the plant, it may contain dust concentrations of up to 2000 mg/Nm^3. As the gas consists principally of steam, and as the process requires a controlled feed of water, many operators use wet scrubbers to remove the dust. Fabric filters are also used. By integrating them into the production unit, problems caused by condensation can largely be avoided.

The processing stage of a hydration plant includes conveyors, elevators, air classifiers, chutes and mills. It is all effectively enclosed. Dust extraction and filtration equipment is provided for all discharges to atmosphere and for pressure relief, at, for example, the top of elevators. The collected dust is returned to the process. Hydrated lime and any rejects from the process are stored in enclosed bunkers fitted with filters to de-dust displaced air.

A significant proportion of hydrated lime is bagged, using purpose-designed plant fitted with dust extraction equipment.

33.3.6   Production of Milk of Lime

Slaking quicklime is not an inherently dusty operation and the volume of gaseous effluent is small. It is good practice to keep slakers under slight suction to ensure that dust and steam do not escape into the plant, or into the lime feed system. One technique employs a water spray in a pipe, which acts as an ejector and keeps the hydrator under slight suction. The spray removes the dust, and the collected dilute milk of lime is used for slaking.

33.3.7   Vehicle Loading and Movement

Loading of limestone into road vehicles should be done in such a way as to minimise the amount of airborne dust generated. The vehicle should be sheeted
before leaving the site [33.3].

Screened quicklime inevitably contains some dust produced by impact and
abrasion during conveying and storage. Loading-out is, therefore, potentially a
dusty operation and measures should be taken to minimise airborne dust emissions [33.4]. When loading screened, granular quicklime into open tipper lorries,
it is good practice to have a loading chute which can be extended to minimise the
free-fall distance. Dust extraction in such a situation is often of limited benefit
and partial enclosure of the tipping point to reduce wind-whip, is generally regarded as acceptable practice. All open tipper vehicles should be covered with a
waterproof sheet before leaving the works.

Finely divided products, such as ground quicklime and hydrated lime, are loaded into air pressure discharge tankers for transport. A flexible chute is used,
which fits the loading hatch of the vehicle, and incorporates an extraction hood
connected to a vacuum filter system. The hood removes both displaced and conveying air emerging from the hatch.

Dust emission caused by vehicle movement should be controlled by good
housekeeping practices. For example, spillages on the top of tankers should be
removed at the loading point, preferably using vacuum cleaning, and roadways
should be hard-surfaced and kept clean.

33.4   Sulfur Dioxide

The emission of sulfur dioxide in the exhaust gases from lime kilns depends on a
number of factors, the more important of which are:

a) the design of the kiln,
b) the sulfur content of the fuel,
c) the particle size of the limestone feed, and
d) the sulfur content of the limestone.

Generally factors (a) and (b) are dominant.

The emission of sulfur dioxide is only likely to be an issue with rotary kilns.
Straight rotary kilns readily emit most of the sulfur in the fuel as sulfur dioxide in
the exhaust gases. Thus, with a fuel containing 3 % of sulfur, emission levels can
be in excess of $750\,mg/Nm^3$ (expressed in terms of the specified reference conditions, e.g., $0\,°C$, 101.3 kPa, dry and at 11 % oxygen). The level of SO_2 emission
from such kilns is affected by:

a) the level of excess air and combustibles in the kiln gases,
b) the flame shape,
c) the temperatures to which the lime is subjected and its reactivity.

Emissions from rotary kilns with preheaters vary considerably and may correspond to between 40 and 80 % of the sulfur in the fuel.

Sulfur dioxide emission is generally not a problem with shaft lime kilns, which, because quicklime is an efficient absorber of the gas at temperatures between 900 and 1100 °C, typically retain from 70 % to over 90 % of the sulfur in the fuel. Typically, SO_2 emissions are less than 300 mg/Nm³, when fired with high-sulfur coal, and even lower with low-sulfur fuels.

If it is necessary to reduce sulfur dioxide emission, the lime producer has several options:

a) use of a fuel containing less sulfur,
b) adjusting kiln conditions and raising product reactivity to retain more of the sulfur in the product,
c) use of standard flue gas desulfurisation techniques (see chapter 29), and
d) use of "non-standard" desulfurisation techniques.

One non-standard desulfurisation technique, used on a rotary kiln by a producer of calcined dolomite, is to include finely divided dolomite in the feedstone. The fines calcine, become airborne as a result of the action of the kiln internal fittings (i.e., trefoils and lifters), and remove a significant amount of the sulfur dioxide. It is not known whether this technique would be as successful with high calcium limestone, which calcines at higher temperatures than dolomite. Other techniques will no doubt be evaluated, such as the injection of hydrated lime into the back-end of the kiln. The cost-effectiveness of such techniques in relation to alternatives, and their effects on kiln operation, would need to be assessed.

33.5   Oxides of Nitrogen (NOx)

For most lime kilns, the emission of oxides of nitrogen is not an issue.

The concentration of NOx in the exhaust gases reflects the temperatures in the flame and/or in the kiln. In kilns such as rotaries, the flame influences the NOx level, and a low NOx burner may be used to reduce the NOx level [33.5, 33.6]. Further experience is probably required to asses the effects of such burners on other aspects of kiln operation. Production of dead-burned dolomite in rotary kilns requires temperatures of up to 1900 °C. This results in very high NOx levels (Table 33.2).

In many shaft kilns, the fuel is burned as an ill-defined flame and, in consequence, the gas temperatures within the kiln are generally low. As a result, the NOx levels are well within required emission limits (Table 33.2). Low NOx burner technology may, however, be of some benefit in those shaft kilns which have combustion chambers, and in which there is a well-defined flame.

If it were necessary to reduce NOx levels by means other than low NOx burners, and lowering kiln temperatures/excess oxygen levels, standard NOx destruction techniques would have to be considered.

Table 33.2. NOx emission levels from various types of lime kiln

Kiln	Product[a]	NOx range mg/Nm^3
Rotary	soft-burned CL	100–700
	hard-burned CL	400–1700
	dead-burned DL	2000–5000
PFR[b]	soft-burned CL	< 300
Mixed-feed	medium-burned CL	< 300
	hard-burned DL	< 300
Other Shaft	medium-soft burned CL	< 500

[a] CL = calcium quicklime
   DL = calcined dolomite.
[b] PFR: parallel flow regenerative.

33.6 Oxides of Carbon

33.6.1 Carbon Monoxide

The emission of carbon monoxide is generally indicative of inefficient combustion, which the lime producer would normally seek to avoid.

Some designs of lime kiln, however, operate best at carbon monoxide concentrations of over 1 % (Table 33.3). At this level the carbon monoxide does not present an environmental hazard, although any leakage of the exhaust gases into plant building could present a health hazard.

When operating rotary kilns to produce a low sulfur, high reactivity, quicklime, it is necessary to maintain a controlled low level of carbon monoxide in the exhaust gases of up to 1 %, to assist with the reduction and decomposition of calcium sulfate in the calcining zone (see section 13.2).

33.6.2 Carbon Dioxide

In the context of global warming, the emission of large amounts of carbon dioxide from quicklime production merits careful consideration. The thermal decomposition of $CaCO_3$ and $MgCO_3$ inevitably produces up to 0.75 t of CO_2/t of lime. The quantity of CO_2 produced from combustion, is 0.2 to 0.45 t/t of lime, depending on the composition and calorific value of the fuel, and the specific heat requirement of the kiln.

For example, a natural gas-fired parallel-flow regenerative kiln produces about 0.95 t of CO_2/t of quicklime, of which 0.73 t arises from the dissociation of the quicklime and 0.22 t from the combustion of the gas. The use of less efficient kilns and fuels containing higher proportions of carbon would result in corresponding increases in the amount of CO_2 emitted per tonne of lime. Carbon dioxide as-

Table 33.3. Carbon monoxide emission levels from various types of lime kiln

Kiln	Conditions[a]	CO level %
Rotary	typical CL	0.1
	for elimination of S	< 1.0
PFR[b]	all conditions	< 0.1
Mixed-feed	typical CL	1–3
	hard-burned DL	3–5
Traditional shaft	typical CL	< 1.0
Modern shaft	typical CL	< 0.1

[a] CL = calcium quicklime.
    DL = calcined dolomite.
[b] PFR: parallel flow regenerative.

sociated with the consumption of electricity could result in an additional emission of 0.02 t of CO_2/t of lime.

The above information should be considered in the context of the facts that:

a) many lime kilns are among the most thermally efficient furnaces used by industry, (> 80 % relative to the heat of dissociation of 760 kcal/kg), following many decades of sustained improvement, and
b) a significant proportion (perhaps over 20 %) of the lime eventually absorbs atmospheric carbon dioxide and reverts to calcium carbonate.

33.7 Other Emissions to Atmosphere

To date, there has been little published information on the emissions of substances such as volatile organic compounds (VOC), dioxins (polychlorinated dibenzo-dioxins, or PCDD), furans (polychlorinated dibenzofurans, or PCDF) and heavy metals from lime kilns.

Emissions of VOCs are believed to be generally very low, with the rare exceptions of those kilns which emit dark smoke, or which are fed with limestone containing exceptionally high levels of organic matter. Fuels containing volatile chlorides (e. g., secondary liquid fuels containing chlorinated solvents) have the *potential* to form dioxins and furans. However, the presence of lime in the preheating and calcining zones appears to inhibit their formation, presumably because of the absoption of the gaseous chlorides by the lime. Emissions of heavy metals are understood to be insignificant, because of the high purity of most limestones used for the production of calcium and dolomitic limes.

33.8 General Noise

Noise emission from open-cast quarries and from industrial operations, such as lime kilns, is a complex issue. Various approaches have been adopted by different countries. In the UK, a number of guidance notes have been issued, which relate to noise from quarrying operations (e.g. [33.7–33.9]) Guidance in connection with the production and processing of lime is less readily available. This section indicates some of the main points to consider.

Handling of limestone at transfer points, and charging it into lime kilns, can generate high impact sound levels. Such operations are often carried out at elevated levels and can cause disturbance, particularly at night. Installing noise insulation to existing plant can often present difficulties, as, for example, the addition of rubber linings to the internal faces of chutes can lead to blockages. Such insulation is often best done at the design stage, or by re-design of the plant item. Exhaust fans and Rootes-type blowers can generate pure tones, which can be particularly irritating. They can generally be reduced by fitting outlet silencers.

Many lime kiln installations incorporate a screen to remove fines produced by breakage before charging the limestone into the kiln. When the particle size of the stone is large, a metal screen deck can produce high sound levels. Use of rubber decking reduces the problem.

Guidance for rating industrial noise and assessing the likelihood of complaint is given in [33.10]. Pure tones cease to be noticeable when they are 10 dB or more below the general sound level.

33.9 Blasting Noise and Vibration

33.9.1 Blasting Noise

The emission of noise from primary blasting is best limited by good blasting practices (see section 4.3.4). The use of down-the-hole initiation coupled with adequate stemming, helps to contain the gases produced by the explosion and to reduce sound levels. The use of unconfined explosive, or explosive cord, should be avoided wherever possible.

The sound level experienced at a distance from an explosion can be heavily depended on transient atmospheric conditions (e.g. temperature inversions, wind shear, and even gusts of wind). The low frequency air pressure from a blast can cause vibration in buildings, in addition to its startle effect. Such vibration can give rise to concern about possible building damage at relatively low over-pressure levels. The subject has been investigated in depth by the United States Bureau of Mines, whose conclusions are summarised in [33.11].

Guidance has also been published relating to human response to vibration in buildings [33.12].

33.9.2 Blasting Vibration

Primary blasts inevitably produce ground vibration. The level of vibration can be controlled by introducing delays (e.g. of 25 milli-seconds) between initiating the explosive in each hole, and by limiting the amount of explosive fired at any instant. In contrast to the propagation of sound, the propagation of ground vibration is generally quite predictable, once the characteristics of the site have been established. Blasting vibration can give rise to adverse reactions as a result of the subjective response [33.12]. At higher levels, there is a risk of the vibration causing damage to buildings. Guidance on this aspect is given in [33.13].

33.10   Discharges to Water

Aqueous discharges from limestone quarrying and processing do not, in general, present a significant environmental hazard. The most common contaminant is suspended matter, which can be removed by settling in sumps or slurry ponds.

Discharges from lime production and processing include drainage and effluent from wet scrubbers. These may be alkaline, as a result of dissolved calcium hydroxide, or may contain dissolved salts, such as calcium sulfate and bisulfite. Such effluents are generally of low volume and are best kept separate from those arising from quarrying. They should be processed to meet the discharge requirements for pH and biological oxygen demand.

33.11   Solid Wastes

The wastes from quarrying are generally of clay or limestone and can, therefore, be disposed to land with few constraints.

A considerable amount of effort has been, and still is being directed to minimising wastes from lime production. They may be blended into commercial products (e.g., lime for building, soil stabilisation, hydration and pelletisation). Such blending should be carefully controlled and the resulting products should be segregated from premium products.

Wastes from lime production and processing contain alkaline components such as calcium oxide and hydroxide. They may contain slightly soluble salts, such as calcium sulfate, as well as insoluble calcium carbonate, clay and fuel ash related products. Water draining through "lime tips" becomes alkaline and precipitates carbonate hardness in ground waters (33.1).

$$Ca(OH)_2 + Ca(HCO_3)_2 \rightarrow 2CaCO_3\downarrow + 2H_2O \qquad (33.1)$$

As a minium requirement, lime tips should, therefore, be built on sites which do not have any significant natural springs and the surface layer should be contoured so that it is self-draining.

33.12 References

[33.1]   "British Aggregate and Construction Materials Industries (BACMI) Environmental Code", March 1992.

[33.2]   ISO 14001: "Environmental Management Systems — Specification with Guidance for Use", 1996.

[33.3]   IPC Guidance Note PG3/8(96): "Secretary of State's Guidance — Quarry Processes", HMSO, London.

[33.4]   IPC Guidance Note S2 3.01: "Cement Manufacture, Lime Manufacture and Associated processes", The Environment Agency, Sept. 1996, HMSO, London.

[33.5]   C. Manias et al., "New Combustion Technology for Lime Production", World Cement, Dec. 1996.

[33.6]   P.J. Mullinger, B.G. Jenkins, "NOx Reduction Techniques for Rotary Kilns", Proc. AFRC/JFRC Pacific Rim Conference on Environmental Control of Combustion Processes, Haui, Hawaii, 16–20 Oct., 1994.

[33.7]   BS 5228, Part 1: "Noise Control on Construction and Open Sites", 1984.

[33.8]   MPG 11: "The Control of Noise at Surface Mineral Workings". Mineral Planning Guidance Note, HMSO, London, 1993, ISBN 0-11-752779-3.

[33.9]   W.S. Atkins Engineering Services, "The Control of Noise at Surface Mineral Workings", HMSO London, 1990, ISBN 0-11-752338-0.

[33.10] BS 4142: "Rating industrial noise affecting mixed residential and industrial areas", 1990.

[33.11] T. Wilton, "The Air Over-pressure Problem", Quarry Management, July 1991, 25–27.

[33.12] BS 6472: "Evaluation of Human Exposure to Vibrations in Buildings (1 Hz to 80 Hz)", 1992.

[33.13] BS 7385, Part 2: "Guide to Damage Levels from Groundborne Vibration", 1993.

[33.14] "Best Available Techniques Reference Document for the Cement and Lime Industries", in preparation (see Annex 2.1 for further details).

34   Toxicology and Occupational Health

34.1   Toxicology

34.1.1   Limestone

Limestone is a practically non-harmful material, providing its crystalline silica content is less than 1 % [34.1].

As an air-borne dust, its long term exposure limit (8-hour time-weighted average) is $10\,mg/m^3$ of total inhalable dust and $5\,mg/m^3$ of respirable dust (respirable being defined in [34.1]). More stringent standards apply if the crystalline silica content exceeds 1 %.

As with other "nuisance dusts", limestone may irritate the eyes and cause discomfort. It can also cause discomfort as a result of its drying effect on the mouth and upper respiratory tract. It is not-irritant to the skin.

Limestone is used in the treatment of water for human consumption. In such applications, limits are applied to the minor and trace element concentrations (section 12.10.1).

34.1.2   Quick- and Slaked Lime

Quicklime and slaked lime are caustic alkalis in the presence of water (pH 12.4). In addition, when quicklime comes into contact with water, it generates a considerable amount of heat.

As airborne dusts, the long term exposure limits (8-hour time-weighted average) for quicklime and hydrated lime are 2 and $5\,mg/m^3$ of dust respectively [34.1]. The dusts are irritating to the respiratory tract and may cause inflammation.

Contact of lime with the eyes can cause painful irritation and may result in serious damage unless immediate treatment is given.

Both quicklime and hydrated lime are classified as irritant and can cause "chemical burns" of the skin, when subject to abrasion in the presence of moisture, or perspiration. If proper care is neglected, prolonged and repeated contact may cause the skin to become dry and cracked, and may lead to dermatitis.

Ingestion can cause corrosion and damage to the gastrointestinal tract.

Both quick- and slaked lime are used in the treatment of drinking water. Limits for minor and trace elements for this application are given in section 28.1.8.

34.2 Precautionary Measures

The following information should be read in conjunction with specific guidance and safety data provided by the supplier.

34.2.1 Limestone

For limestone dust, the main precautions to be observed are the use of appropriate eye protection and respiratory protective equipment. Such equipment should be of an approved standard.

34.2.2 Quick- and Slaked Lime

The protective measures for quicklime, hydrated lime, lime putty and milk of lime are more comprehensive than for limestone [34.2]. The eyes are particularly vulnerable to damage and adequate eye protection should be worn at all times when handling lime products. Hot lime putties and milks of lime require special precautions as splashes can burn both physically and chemically.

Protection for the mouth and nose should be provided by the use of approved respiratory protective equipment. In some circumstances, the face and neck can be protected with barrier cream, but this may cause increased sweating, in which case a ventilated visor may be necessary. A cloth worn around the neck may reduce chaffing of the skin by clothing.

The hands should be protected by gloves with a tight-fitting wristband. Breathable canvas or leather is suitable for dry conditions, but, where lime putty or milk of lime is being handled, waterproof gloves should be used. Any exposed parts of the arms, wrists and hands should be protected with barrier cream.

Lime should be prevented from reaching the feet to avoid burns or irritation. In dry conditions, gaiters may be worn over the boot tops and trouser bottoms. In wet conditions, waterproof trousers should be worn over rubber boots.

Quicklime should not be allowed to come into contact with combustible matter, such as wood shavings, sawdust and straw, nor should it be stored on wooden floors as there is risk of fire when it comes into contact with water. Both quicklime and slaked lime are non-combustible and inhibit the spread of flame.

The following information is registered under the UK's Chemicals (Hazard Information and Packaging) (CHIP) Regulations 1993 for quick- and slaked limes [34.3]:

- Classification for conveyance: None
- Classification for supply: Irritant
- Risk phrases: Irritating to skin (R38)
  Risk of serious damage to eyes (R41)

- Safety phrases: Wear suitable gloves and eye/face protection.
                                          In case of contact with eyes, rinse immedia-
                                          tely with water and seek medical advice.
                                          Keep out of reach of children

The CHIP Regulations implement the 88/379/EEC Dangerous Preparations Directive (DPD) and subsequent amendments/adaptions plus the 91/155/EEC Safety Data Sheets Directive.

There is currently (January 1998) a debate regarding the classification of quick- and slaked lime. Most EU countries agree that the dry solids should be classified as irritant, although some use different risk phrases from the above. At least one non-EU country classifies quicklime as corrosive. Very recent evidence (as yet unpublished) is understood to support continuing to classify both quicklime and hydrated lime as irritant, with somewhat less arduous risk phrases than the above being appropriate for hydrated lime.

Milks of lime with solids contents above about 22 % are regarded by some as corrosive, based on the United States Soap and Detergents Association's test, which takes account of the pH and the "reserve alkali" content. Milks of lime with lower solids contents have a lower reserve alkali content and are classified as irritant.

34.3   First Aid Treatment [34.2]

34.3.1   Limestone

Limestone dust in the eye should be removed immediately by irrigation. Any particles remaining should be removed with extreme caution, using a cotton wool bud and irrigation with eyewash solution or gently flowing, clean, mains water.

In case of extreme inhalation of limestone dust, the nose and throat should be irrigated with water, taking care to avoid inhaling the water.

34.3.2   Quick- and Slaked Lime

Lime dust in the eye should be removed immediately. Speed is essential. Particles should be removed with extreme caution using a cotton wool bud. Irrigation with eyewash solution or gently flowing clean mains water should start at once and should continue until medical attention can be obtained.

If lime dust has been inhaled, the nose and throat should be thoroughly irrigated with water (for at least 20 min). It is important to avoid inhaling the water.

If lime has been swallowed, wash out the mouth with water and give copious quantities of water. Do not induce vomiting.

Milk of lime on the skin should be washed off without delay. Any lime entering footwear should be removed as quickly as possible. Where contact with lime removes the skin's natural oils, they should be replaced by the use of skin cream.

Whenever there is a risk of lime entering the eye, suitable eye irrigation bottles should be readily accessible. In all cases where the eye is affected, or in severe cases of skin contamination, the person should receive medical attention as soon as possible and, in all cases after first aid treatment, the person should consult a qualified medical practitioner.

34.4   Occupational Health

34.4.1   Limestone

The quarrying industry in general has a poor safety record relative to general manufacturing [34.4, 34.5]. In the UK, *mobile equipment* (shovels, dump trucks etc.) is the major cause of deaths. The second largest cause of deaths and over half of the major injuries is *stumbling, falling and slipping*. While the use of explosives is generally perceived to be a major hazard, it causes less than 5 % of the UK deaths and major injuries, presumably because relatively few, well-trained people handle explosives, and their use is well covered by legislation.

In limestone mines, *rock falls* are the major cause of deaths and serious injuries, with *mobile equipment* being the next most significant category. Special electrical precautions (e.g. high integrity earthing and the use of low voltage portable tools) are essential as water percolates into most limestone mines [34.6].

Stone processing involves a number of mechanical operations. The largest category of injuries caused by these operations generally arises from *inadequate guarding* of machinery and from failure to *electrically isolate* equipment when carrying out maintenance and clearance of blockages [34.7].

Many stone processing operations produce high sound levels. Operators' exposure to noise is controlled by a combination of reducing/containing it at source, excluding it from control rooms, and the use of remote cameras with monitors in control rooms. Effective personal hearing protection is still required when operators are required to enter noisy areas.

The control of workplace dust arising from quarrying and stone processing is considered in section 33.3.

34.4.2   Quick- and Slaked Lime

The lime industry generally has a lower accident rate than quarrying. Contributory factors are likely to be the reduced use of mobile equipment and the greater size of the works — there is a correlation between accident rate and the number of people employed on site [34.8]. Tripping and falling constitute the largest category of accidents.

Lime processing involves a number of mechanical operations. Adequate guarding of moving machinery and effective isolation procedures, when carrying out maintenance, are essential.

Both quicklime and hydrated lime are dusty, alkaline products. Not surprisingly, a common injury associated with them is grit or dust in the eye. The use of adequate eye protection in all lime production and handling areas is essential. Indeed, some firms have made the use of some form of eye protection compulsory throughout the lime area and have, as a consequence, dramatically reduced the number of eye injuries.

34.5 References

[34.1]   "Occupational Exposure Limits, 1996", EH40/96, Health and Safety Executive, HSE Books, London.
[34.2]   BS 6463: "Quicklime, hydrated lime and natural calcium carbonate", Part 101: "Methods for preparing samples for testing", 1996.
[34.3]   "Chemicals (Hazard Information and Packaging) Regulation", 1993.
[34.4]   Health and Safety Commission and Health and Safety Executive Reports 1986/87, HMSO.
[34.5]   Health and Safety Executive, "Quarries", Report 1985/86, HMSO.
[34.6]   Health and Safety Executive, "Mines", Report 1985/86, HMSO.
[34.7]   J.M. Hobbs, "Safety in Quarries (Part 1)", private publication, 1988.
[34.8]   H. Krummhaar, "The progress of Occupational Safety in the German Lime Industry", Zement Kalk Gips **1**, 1989, 37–40.

Annexes

Annex 1 Glossary of Terms*

A

AAV see aggregate abrasion value.

Absorption is the assimilation of molecules, or other particles, into the physical structure of a liquid or solid, without chemical reaction.

ACV see aggregate crushing value.

Adsorption is the physical adhesion of molecules or colloids to the surfaces of solids, without chemical reaction.

Agglomerate is to gather fine particles together to form a larger mass.

Aggregate consists of particles of rock, of a controlled particle size distribution, used in the construction of a building or civil engineering structure (see also coarse and fine aggregate).

Aggregate abrasion value (AAV) is the resistance of an aggregate to abrasion, as measured by the aggregate abrasion test. Low values indicate increased resistance to abrasion.

Aggregate crushing value (ACV) is the resistance of an aggregate to crushing, as measured by the aggregate crushing test. Low values indicate increased resistance to crushing.

Aggregate impact value (AIV) is the resistance of an aggregate to impact, as measured by the aggregate impact test. Low values indicate increased resistance to impact.

Agricultural hydrate is a relatively coarse, unrefined form of hydrated lime, used mainly for adjusting the acidity of soils.

Agricultural lime is a term which includes any limestone, quicklime, or hydrated lime product used to neutralize soil acidity.

Agricultural limestone is a ground limestone product used to neutralize soil acidity.

Air-entrainment occurs when a surfactant is added to a mortar to increase the level of entrained air, thereby improving its workability.

Air limes (translation of a term used in most European countries) consist mainly of calcium oxide or hydroxide, which, when incorporated into a mortar mix, slowly harden in air by reacting with atmospheric carbon dioxide. They do not harden under water as they do not have any hydraulic properties. They may be either quicklimes or hydrated limes.

---

* See also BS 6100, Part 6, Section 6.1, "Glossary of Building and Civil Engineering Terms — Binders", 1984 (1995).

Air-slaked lime is produced by excessive exposure of quicklime to the atmosphere. It contains varying proportions of the oxides, hydroxides, and carbonates of calcium and magnesium.

AIV see aggregate impact value.

Algal limestone is composed of biosparites, or biomicrites resulting from the activities of algae.

Alkali-carbonate reaction is a reaction which can occur under certain conditions between sodium and potassium hydroxides in cement and carbonate rocks (see section 8.3.6).

Alkali-silica reaction is an expansive reaction in concrete that can occur when a solution of sodium or potassium hydroxide reacts with a siliceous aggregate to form a gel of hydrated alkali silicate.

"All-in" aggregate has a defined top size, but includes fine fractions.

Aluminous cement consists predominantly of calcium aluminate.

Ammonium nitrate-fuel oil mixture see ANFO.

Amorphous limestone appears to the naked eye to be non-crystalline and to be without form or texture.

ANFO is a mixture of ammonium nitrate and fuel oil, used as a blasting agent.

Angle of repose is the angle (relative to the horizontal) above which a pile of material becomes unstable and begins to move under gravity.

Annular shaft kiln is a type of shaft kiln in which the burden occupies an annulus between a central column and the shell.

Anthracite is form of coal that has a volatiles content of between 3.5 and 10 %. It is suitable for use in mixed-feed kilns.

Aragonite is one of the less abundant crystalline (rhombic) forms of calcium carbonate. It slowly recrystallises to calcite (hexagonal structure) in the presence of water.

Arch is any bridging brick structure.

Arenaceous limestone contains significant amounts of particles of silica sand.

Argillaceous limestone contains significant amounts of clay or shale (i.e. of silica and alumina).

Artificial hydraulic limes consist mainly of calcium hydroxide, calcium silicates and calcium aluminates. They are produced by blending suitable powdered materials, such as natural hydraulic limes, fully hydrated air limes and dolomitic limes, pulverised fuel ash, volcanic ash, trass, ordinary Portland cement and blast furnace slag.

Asphalt is a mixture of bitumen and mineral matter, which may be natural or manufactured.

Asphaltic cement can be bitumen, or a mixture of asphalt with bitumen, or flux oils, or pitch, which has cementing qualities suitable for the manufacture of asphalt pavements.

Asphaltic limestone see carbonaceous limestone.

Autoclaved lime see Type S hydrated lime.

Availability is the percentage of time that an item of equipment is available for production.

Available lime is an analytical term for the calcium oxide content of quicklime or hydrated lime that is able to react with a sucrose solution in the available lime test (e.g., as specified in BS 6463 and ASTM C25).

B

Back end of a rotary kiln is the upper end into which the limestone is fed.

Ball-deck is a design of screen, which uses balls trapped between two decks to reduce pegging.

Ball mill is a grinding machine, consisting of a short horizontal rotating cylinder charged with steel balls. The material to be milled is fed into one end of the cylinder, broken between the tumbling balls, and emerges in a finely divided form from the other end of the mill.

Base is an alkaline substance.

Base course is the layer in a road below the wearing course and above the road base.

Bastard limestone see magnesian limestone.

Bedding is the way in which distinct layers, or beds, of sedimentary rock are laid on one another. The surface between successive beds is called a bedding plane.

Benches are the various quarry floor levels produced by benching.

Benching refers to the operation of a quarry as a number of stepped quarry faces.

Benefication is the processing of rocks and minerals to reduce impurities. In ore processing, the up-graded fraction is called a concentrate and the waste a gangue.

Bin is a bunker or silo used for storage.

Binder is any cementing agent (e.g., cement, lime, bitumen and asphalt).

Biomicrites are limestones consisting of skeletons or fragments of organic debris in a micrite matrix.

Biosparites are limestones consisting of calcareous fragments in a matrix of recrystallised calcite (sparite).

Bitumen is a viscous liquid, or a solid, which consists largely of hydrocarbons and their derivatives. It softens gradually on heating and is essentially non-volatile. It occurs naturally as bitumen and, in conjunction with mineral matter, as asphalt. It is also produced as a by-product of petroleum refining. It is used as a water-proofing and binding agent.

Bituminous macadam see macadam.

Bituminous limestone see carbonaceous limestone.

Blast furnace is the furnace in which iron ore is reduced by coke to form molten iron.

Bleeding is the release of water from a concrete or mortar after placing.

Blinding (re. construction) is a layer of mortar or concrete, used to cover a surface prior to construction.

Blinding (re. screening) is the obstruction of the apertures in a screen deck by an accumulation of fine material.

Blow-bar is the wear surface attached to the rotor of an impact crusher.

Bound aggregate is an aggregate which is coated and bound with a cementitious, or bituminous binder, as in bituminous macadam, or concrete.

Boundstone is a limestone in which the components were bound together during deposition, as in a coral reef.

Brachiopod is a fossil, which is frequently found in limestones, consisting of two bilaterally symmetrical shells.

Brecciated limestone was formed from coarse angular fragments.

Building limes are limes used in building and construction, the main constituents of which, on chemical analysis, are the oxides and hydroxides of calcium, with lesser amounts of magnesium (MgO, $Mg(OH)_2$), silicon (SiO_2), aluminium (Al_2O_3) and iron (Fe_2O_3). They include calcium limes, dolomitic limes and hydraulic limes (see EN 459).

Building sands are sands, which may be calcareous or siliceous, suitable for use in mortars.

Bulk density is the mass per unit volume of a solid, including the voids in a bulk sample of the material.

Bunker is a large storage vessel (sometimes called a bin — see also silo).

Burden, in quarrying, is the distance between blast holes and the quarry face. Burden also refers to the solid material in lime kilns (i.e., the lime and limestone).

Burning zone is that part of a kiln in which dissociation/reaction of the burden occurs.

Burnt limes are quicklimes mainly consisting of calcium oxide.

C

Calcareous ooze is a naturally occurring, unconsolidated mud, the main component of which is calcium carbonate.

Calcareous (or calc) tufa see tufa.

Calcia is the traditional name for calcium oxide.

Calcination is the heating of a substance so that a physical, or chemical change occurs. In the case of limestone, it refers to the dissociation of calcium and magnesium carbonates.

Calcining zone see burning zone.

Calcite is the most abundant crystalline form (hexagonal structure) of calcium carbonate.

Calcitic limestone refers to a high calcium limestone, with less than 5 % of magnesium carbonate.

Calcitic mudstone is cemented calcitic mud (micrite), with few coarse grains.

Calcium carbonate sludge is a precipitated by-product (e.g. from water softening, the sulfite paper-pulp process and from the purification of sugar).

Calcium limes are limes mainly consisting of calcium oxide or calcium hydroxide (see air limes).

California Bearing Ratio (CBR) is the measure of the load-bearing capacity of a soil, obtained using the California Bearing Ratio test.

Calorific value is the quantity of heat released when unit mass (or volume) of fuel is completely combusted.

Carbide lime is a by-product of acetylene production from calcium carbide. It is principally calcium hydroxide and occurs either as a wet sludge or a dry powder of varying degrees of purity and particle size.

Carbonaceous limestone contains organic matter as an impurity. It is often dark grey and has a musty odour.

Carbonate hardness is that hardness in a water caused by bicarbonates of calcium and magnesium.

Carbonate rocks are rocks formed predominantly from the carbonates of calcium, magnesium, iron, etc., occurring either singly or in combination, of which limestone is the most common.

Carbonatite is a limestone which is of magmatic (igneous) origin.

Carboniferous limestone was deposited in the Carboniferous period 286 to 360 million years ago.

Caustic soda is sodium hydroxide.

Causticisation refers to the reaction between alkali-metal carbonates and calcium hydroxide to produce the alkali-metal hydroxide and calcium carbonate.

CBR see California Bearing Ratio.

Cement is the natural or synthetic material which binds rock particles together. In sedimentary rocks, the cementing substances include silica, calcite, clay and iron oxide. See also ordinary Portland cement.

Cementation is the binding together of the components of a rock by interstitial crystals. Sparite is the most common cement in limestones.

Cementation Index (CI) is used to categorise hydraulic limes (see section 16.10.1).

Cementstone is limestone which contains sufficient amounts of silica, alumina and iron oxide for it to be converted into a cement, when fired at a sufficiently high temperature. (In practice, most deposits require additions to meet present-day specifications.)

CEN is the abbreviation for Comité Européen de Normalisation — the European standards organisation.

Chalk is a naturally occurring limestone, which has been only partially consolidated, and therefore, has a high porosity. It is generally relatively soft.

Charge is used to describe a batch of limestone fed into a lime kiln.

Chemical-grade (or -quality) limestone see high-calcium limestone.

Chert is a form of silica, which often concentrates in layers.

Cherty limestone is an impure form containing substantial levels of chert. It often has poor physical properties.

Classification is a method of sizing fine particles by exploiting differences in settling velocity in water or air.

Classifier is a particle-sizing device which uses the principle of classification.

Clay minerals are a complex group of finely crystalline to amorphous hydrated silicates, mainly alumino-silicates, which were largely formed by alteration, or weathering of silicate minerals.

Coagulation is the process whereby a suspension is de-stabilised. De-stabilisation involves neutralising the electrical charges on suspended particles by the oppositely charged particles of the coagulant. This enables the suspended particles to move close to one another and permits agglomeration to occur (see flocculation). Coagulants can be inorganic (e.g., $Al(OH)_3$ and $Fe(OH)_3$), or organic (e.g., polyamides and polyamines).

Coarse aggregate is aggregate which is mainly retained on a 5 mm BS 410 test sieve.

Coated macadam is a road material consisting of aggregate that has been coated with tar or bitumen.

Colloid is a dispersion in liquid of very fine particles, generally between 0.1 and 10 μm.

Colloid mill is a general term for mills which are able to reduce the particle size of a suspension into the colloidal size range.

Comminution is the reduction of particle size.

Compact limestone has a low porosity (typically less than 5 % by volume). It is generally fine-grained, homogeneous and hard.

Completely hydrated dolomitic limes see Type S hydrated lime.

Composite sample is a mixture of spot samples taken at different times, or places, from a larger mass of the same material. It is produced by thoroughly mixing the combined spot samples and, if necessary, sub-dividing the quantity of the resulting mixture, using approved techniques.

Concrete is a structural material made by mixing controlled amounts of sand, aggregate, binder and water. The binder is generally ordinary Portland cement, but may be a hydraulic lime, or hydrated lime plus a pozzolan.

Conductivity is the ability of a substance to conduct heat or electricity.

Conglomeratic limestone was formed by the cementation of relatively coarse fragments of calcium carbonate.

Cooler is a vessel in which hot lime discharged from the calcining zone of a rotary or other kiln is cooled by a stream of air, which is usually used as combustion air.

Cooling zone is that part of a kiln in which the lime emerging from the calcining zone is cooled before discharge.

Coquina is a fossiliferous shell deposit. It is generally soft.

Coral limestone is one in which the main fossil is coral.

Core describes the uncalcined limestone within an under-burned particle of lime.

Cretaceous limestone was deposited in the Cretaceous period 65 to 144 million years ago.

Crosses are cross-shaped metal inserts in the preheating zone of some rotary kilns which divide the kiln into four channels to improve heat transfer between the hot gases and the limestone.

Crotching refers to the sintering of particles of lime in a kiln to each other and/or to the lining of the kiln. The flux holding the particles together generally consists of calcium silicates, aluminates and ferrates. Several other terms are used for crotches, including "bears", "scaffolds" and "clinkering".

Crusher-run aggregate is the unscreened product from a single crushing operation.

Crystalline limestone is a highly crystalline form with a large average grain size (over 50 µm), e.g. marble.

Cyclone is a classifying device which creates a vortex to produce a size separation.

D

Dam is a band of thicker refractory in a rotary kiln, which holds back the burden and increases residence time.

Dead-burned dolomite is a highly sintered form of dolomitic quicklime which is used primarily as a basic refractory.

Dead-burned lime is sintered quicklime which does not slake readily under normal conditions.

Deck is a screening surface.

Decrepitation refers to the cracking, or breaking-up of lumps of limestone during heating. It is caused by differential thermal expansion.

Detonator is a device, containing a very small amount of sensitive explosive, which can be initiated remotely (by electric current, a fuse or detonating cord) and safely to detonate a larger mass of explosive (see primer).

Diagenesis is the process which converts sediments into rock.

Dimension stone is stone which has been cut into blocks of specific dimensions.

Disk attrition mills have rotor-stator configurations which produce narrow gaps in the order of 30 to 500 µm, through which the suspension to be milled is forced. The milled product may have a maximum particle size as low as 10 µm.

Disinfection is the use of heat or chemicals to kill pathogenic organisms.

Doloma and **dolime** refers to dolomitic lime.

Dolomite is strictly speaking the double carbonate containing 54 to 58 % of $CaCO_3$ and 40 to 44 % of $MgCO_3$. The mineral has a hexagonal-rhombohedral crystal structure. The term is, however, frequently used to describe dolomitic limestone.

Dolomitic limes are limes mainly consisting of calcium oxide and magnesium oxide (see calcined dolomite), calcium hydroxide and magnesium oxide (see semi-hydrated dolomitic lime), or calcium hydroxide and magnesium hydroxide (see Type S hydrated lime).

Dolomitic limestone is generally understood to contain 20 to 40 % of $MgCO_3$.

Dolomitisation is the process in which the passage of sea-water through calcitic limestone results in the partial replacement of calcium by magnesium ions.

Dolostone is rock consisting of the mineral dolomite.

Draw describes a batch of lime discharged from a lime kiln.

Drop see draw.

Drop-ball is a small crane designed to lift and then drop a heavy weight (the "ball"), on to over-sized rocks to break then into smaller pieces, which are suitable for feeding into the primary crusher.

Drowned lime is produced when lime is slaked in excessive amounts of water so that the temperature required for effective slaking is not produced. A skin of lime putty seals the surface of the particles and prevents further slaking.

E

Earthy limestone see marl.

Eminently hydraulic limes (also called Roman limes and hydraulic limes) are natural hydraulic limes, which have pronounced hydraulic properties.

Emissivity of a fuel is a measure of its ability to radiate energy, particularly in the infra-red wavelengths.

EN, or Europäische Norm, is the European Standard.

ENV, or Europäische Vornorm, is the European Prestandard, equivalent to the Draft for Development in British Standards.

Equivalent weight is the weight of a substance which combines with, or displaces, unit weight of hydrogen.

Eutrophication is the choking and de-oxygenation of water resulting from enrichment of water with nutrients, which causes excessive growth of aquatic plants and algae.

F

Fallen lime see air-slaked lime.

Fat lime is used to describe a quick- or hydrated lime, having a high volume yield and producing a plastic putty. It is also used for relatively pure quick- or hydrated lime, as opposed to impure and hydraulic limes.

Fattening of lime putties is the slow absorption of water, which increases volume yield and plasticity.

Fault is a surface along which rock has been fractured and displaced.

Feebly-hydraulic limes (also called semi hydraulic) are limes which possess slight hydraulic properties.

Ferriferous means containing iron.

Ferruginous limestone contains sufficient iron oxide to colour the deposit yellow or red.

Filler is inert material with a particle size less than 75 μm, e.g., limestone dust, which is used to fill voids and, in the case of asphalt, to modify its viscosity.

Fine aggregate is aggregate which mainly passes a 5 mm BS 410 test sieve and is retained on a 75 μm sieve.

Fine lime generally refers to screened products with a top size below 0.6 cm.

Finishing lime is a type of refined hydrated lime that is suitable for the finishing coat in plastering.

Flint is a variety of chert (silica), found in chalk.

Flocculation is the process in which de-stabilised suspended particles are induced to come together and form larger (and possibly denser) agglomerates and "flocs".

Flux refers to the reaction of solid substances to form a melt. Calcium oxide is frequently used to react with silica, alumina and iron oxides to form a molten slag.

Fluxstone see metallurgical-grade limestone.

Fossiliferous limestone is one which contains fossils which are readily visible to the naked eye.

Free lime is an analytical term for the calcium oxide component of quicklime or hydrated lime. It excludes calcium oxide in $CaCO_3$, $Ca(OH)_2$, and calcium silicates.

Front end of a rotary kiln is the firing end from which the lime is discharged.

G

Glass-stone is a high carbonate limestone, which is low in iron oxide and is suitable for the manufacture of glass.

Glauconitic limestone contains glauconite — a hydrated silicate of iron and potassium.

Grading strictly refers to the particle size distribution, which is normally determined using square mesh sieves, unless otherwise stated. When used to describe products, it refers to the nominal size (e.g., 10 to 20 mm, or "–20", +10 mm which indicates that, say, 90 % by weight of the particles are within that range).

Grainstone is a rock which consists mainly of grains in close contact, without a mud cement.

Granular quicklime usually refers to screened products with a top size of above 0.6 cm and below 2.5 cm.

Grey lime is used to describe some semi-hydraulic limes, which often have a warm grey colour.

Ground quicklime refers to powdered products produced by milling.

H

Half-burned dolomite ($CaCO_3 \cdot MgO$) is produced by the controlled calcination of dolomite at low temperatures to dissociate most of the $MgCO_3$ component and little of the $CaCO_3$ component.

Hard-burned quicklime has been sintered as a result of over-burning at high temperatures. It slakes very slowly.

Hardness approximates to the concentration of calcium and magnesium salts in water. Total hardness is the sum of carbonate and non-carbonate hardness. It may be expressed as degrees of hardness, millimoles per litre (expressed as calcium equivalent), or as parts per million of $CaCO_3$ equivalent.

High alumina cement see aluminous cement.

High-calcium lime is quick- or slaked lime produced from high-calcium limestone.

High-calcium limestone is a general term for limestone consisting of mainly $CaCO_3$ (at least 95 % m/m) and having less than 5 % (m/m) $MgCO_3$.

High-calcium quicklime contains mainly CaO and not more than 5 % (m/m) MgO.

High carbonate rock is a pure form of limestone, which contains more than 95 % (m/m) of calcium and magnesium carbonates. It may be may be high calcium, magnesian, or dolomitic.

High purity carbonate rock is limestone containing more than 95 % of ($CaCO_3 + MgCO_3$).

High reactivity see reactivity.

Hopper is an open-topped receptacle into which road or rail vehicles discharge their load.

Hydrated calcium limes are powdered slaked limes consisting mainly of calcium hydroxide.

Hydrated dolomitic limes are powdered slaked limes consisting of calcium hydroxide, magnesium hydroxide and magnesium oxide (see semi-hydrated lime and Type S hydrated lime).

Hydrated hydraulic lime is a hydraulic lime which has been treated with sufficient water to convert the free CaO into $Ca(OH)_2$, without hydrating the cementitous components.

Hydrated limes are hydrated air limes, calcium limes, or dolomitic limes, resulting from the controlled slaking of quicklimes to produce a dry powder.

Hydration describes the reaction of quicklime with a controlled excess of water to produce hydrated lime as a dry powder.

Hydraulic, as applied to cements and limes, refers to their ability to set under water.

Hydraulic limes have the property of setting and hardening under water — see natural hydraulic limes, special natural hydraulic limes and artificial hydraulic limes. The term is also used to describe "eminently hydraulic" and "Roman" limes.

Hydraulic limestone is an impure carbonate containing considerable amounts of silica and alumina. Calcination of hydraulic limestone at temperatures below 1250 °C produces natural hydraulic lime.

Hydrocyclone a cyclone which creates a vortex in water to produce a size separation.

Hydrothermal refers to processes which occur through the action of hot (generally superheated) water.

I

Iceland Spar is a rare and very pure form of calcite. It is transparent and has been used in optical instruments.

Increment is a quantity of material taken in a single operation of sampling equipment.

Initiation (in blasting) refers to the method of firing the explosive, and to the instant at which the explosion starts.

J

Joint is a discontinuity, fracture or parting in a rock, involving no displacement.

Jurassic limestone was deposited in the Jurassic period 135 to 180 million years ago.

K

Karst is a term describing the features produced in limestone and at limestone surfaces by weathering.

Kibbled is a traditional term used to describe a crushed and/or screened grade of agricultural lime, which has a defined top size, but may include fine fractions.

Kiln is a vessel or chamber in which solid materials are heated at intermediate temperatures (e.g., from 500 to 1800 °C — the term oven is used for lower temperatures and furnace for higher ones).

L

Laboratory sample is a sample prepared by thoroughly mixing, and, if necessary, sub-dividing the quantity of a spot or composite sample. The laboratory sample may be reduced in particle size if appropriate.

Lagoon (in waste disposal) see slurry pond.

Leaching is the process by which soluble matter is removed from soil or rock by the action of percolating water.

Lias limestone see hydraulic limestone.

Lifter is a device, made of steel or refractory, attached to the inside of a rotary kiln or cooler which lifts the burden and causes it to cascade through the hot kiln gases to increase heat transfer.

Light-burned quicklime is quicklime that is lightly sintered and has a high reactivity to water.

LIM see lime improved mixture.

Lime is a general term for the various forms of calcium oxide and/or hydroxide with lesser amounts of magnesium oxide and/or hydroxide. It is sometimes used *incorrectly* to refer to limestone.

Lime improved mixture (LIM) is a mixture, generally incorporating a soil, the handling characteristics of which improve immediately following the addition of lime. There does not need to be an enhancement of the characteristics in the medium or long-term.

Lime mortar is sand bound in a matrix of lime, or lime plus pozzolan (but not cement), used for laying bricks, blocks or stones in building.

Lime plaster is a form of lime mortar, which is applied to ceilings and internal walls.

Lime putties are slaked limes mixed with water to a desired consistence, mainly consisting of calcium hydroxide with or without magnesium hydroxide.

Lime render (or rendering) is a form of lime mortar, which is applied to external walls.

Lime slurry see milk of lime.

Lime stabilised mixture (LSM) is a mixture, generally incorporating a soil, for which the addition of lime significantly enhances, in the medium to long-term, the characteristics of the material and renders it permanently stable.

Lime treated mixture (LTM) is a mixture, generally incorporating a soil, that, following the addition of lime, has the required handling and strength characteristics for the intended purpose. LTMs demonstrate improvement and optionally stabilisation.

Limestone is a naturally occurring rock consisting principally of the carbonates of calcium and magnesium, in which the ratio by weight of $CaCO_3$ to $MgCO_3$ is not less than 1.2 to 1.0.

Limewash (or whitewash) is a type of paint based on lime.

Lime water is a solution of calcium hydroxide in water.

Lining is the refractory layer on the inner face of the shell of a kiln.

Los Angeles abrasion value is a measure of the resistance of a rock to a combination of impact and abrasion, as measured by the Los Angeles abrasion test. Lower values indicate higher resistance.

Loss on ignition is an analytical term for the loss on weight produced by heating a sample to a specified temperature. In the case of quicklime and limestone products, it generally refers to heating to $1000\pm50\,°C$ in air to decompose all carbonates and hydroxides to the oxides and to burn-off organic matter. In the case of hydrated lime, it may refer to $575\pm25\,°C$, at which temperature calcium hydroxide decomposes and any organic matter is burnt-off.

Low reactivity see reactivity.

LSM see lime stabilised mixture.

LTM see lime treated mixture.

Lump quicklime usually refers to products with a top size above 2.5 cm.

M

Macadam consists of aggregate, which has been mechanically locked together by rolling and cemented together by application of stone screenings and water. Bituminous macadam is crushed aggregate in which the fragments are cemented together by bitumen or asphalt.

Magnesia is the traditional name for magnesium oxide.

Magnesian limestone is generally understood to be mainly $CaCO_3$ with 5 to 20 % of $MgCO_3$.

Magnesium sulfate soundness value (MSSV) is the soundness of an aggregate as measured by the magnesium sulfate soundness test. In the BS 812 test (% retained), higher values indicate increased soundness. In the ASTM C88 test (% loss), smaller values indicate increased soundness.

Marble is a highly crystalline metamorphic limestone which may be calcitic or dolomitic limestone. It occurs in many colours with veined and mottled effects.

Marl is an impure, soft, earthy rock which contains 5 to 65 % of clay and sand. If it contains more that 50 % of calcium plus magnesium carbonates, it is classified as a limestone.

Masonry refers to building elements consisting of bricks, blocks, or stones. In some contexts, it excludes brickwork.

Masons' lime is a hydrated lime used in mortar.

Mat is a screening surface, which is often replaceable.

Maturing of lime putties refers to the process of storing them under water to fatten them and render them sound.

Metallurgical-grade limestone is a high purity carbonate rock used in metallurgical process to flux solid impurities.

Metamorphic limestone is one in which the carbonate has completely re-crystallised.

Metamorphism refers to the mineralogical and structural changes of rocks caused by the physical and chemical conditions.

Metasomatic limestone is one where the original character of the deposit has been modified by the intrusion of secondary impurities.

Micrite is a very fine-grained carbonate material, or a lime mud.

Micritisation is the partial or total alteration of grains to a homogeneous microcrystalline (micritic) fabric, generally following repeated boring of algae or fungi into the perimeter of a grain.

Milk of lime is a fluid suspension of slaked lime in water.

Mixed-feed describes the process in which the limestone and fuel are both charged into the top of a shaft kiln.

Moderate reactivity see reactivity.

Moderately hydraulic limes are hydraulic limes, which develop moderate strengths when combined into a mortar.

Mohs' scale of hardness is a scale of scratch resistance, graduated in terms of the relative hardness of 10 common minerals, which include talc (1), gypsum (2), calcite (3), quartz (7) and diamond (10).

Mole is a weight of a chemical corresponding to its molecular weight. A gram mole of carbon dioxide weighs 44 grams and occupies 22.4 litres at a temperature of $0\,^\circ C$ and a pressure of 1 atmosphere (760 mm of mercury).

Mortar a mixture consisting of a binder (cement and/or hydrated lime), sand and water and the hardened product of the mixture.

MSSV see magnesium sulfate soundness value.

Mudstone see calcitic mudstone.

N

Natural hydraulic limes are limes produced by burning, at below $1250\,^\circ C$, of more or less argillaceous or siliceous limestones, with reduction to powder by slaking with or without grinding. They consist of calcium silicates, calcium aluminates and calcium hydroxide.

Neomorphism is the process of re-crystallisation, e.g of aragonite into calcite and of calcite mud into larger crystallites.

Neutralisation generally refers to the reaction of an acid with an alkali to produce a solution, or a salt that is neither acid nor alkaline.

Neutralising value is an analytical term for that proportion of limestone, quicklime, or hydrated lime (expressed as CaO) that is capable of reacting with hydrochloric acid under specified conditions. It includes the contribution of $CaCO_3$, CaO, $Ca(OH)_2$, and the acid-soluble fraction of the calcium silicates, aluminates and ferrate.

Nodulise is to form nodules, or granules from powders, generally using the growth granulation process.

Non-carbonate hardness is hardness in water caused mainly by calcium and magnesium compounds other than the bicarbonates.

Nose-ring dam is a deep dam at the discharge end of a rotary kiln, used to hold back the lime in the kiln for an additional "soaking" period, before discharge into the cooler.

Nuisance is an undesirable, but non-hazardous environmental effect of an activity.

O

Ooids are spherical, accretionary calcareous grains of up to 3 mm in diameter, precipitated by algal action in turbulent waters.

Ooliths are limestones composed of ooids cemented in calcite.

Oolitic limestone see ooliths.

OPC see ordinary Portland cement.

Ordinary Portland cement is a controlled blend of calcium silicates, aluminates and ferrate, which is ground to a fine powder with gypsum and other materials (see chapter 9).

Overburden is the material that lies over the mineral deposit. It consists of top soil, sub-soil and any over-lying rocks.

Over-burned lime is quicklime that has a low reactivity to water, as a result of being calcined at high temperatures.

P

Packstone is limestone consisting mainly of coarse grains in close contact, with some calcitic mud cement.

Partial pressure is the pressure exerted by a particular gas in a mixture of gases. The particular gas may be produced by the thermal dissociation of a solid such as calcium hydroxide or calcium carbonate.

Pathogens are disease-producing organisms.

Pavement is the whole constructed thickness of a road or similar slab.

PCC see precipitated calcium carbonate.

Pebble quicklime usually refers to screened products with a top size in the range 1.5 to 6 cm.

Pegging is the blocking of the apertures in a screen deck by single particles which lodge in individual apertures.

Petrography is the description and classification of rocks, based mainly on the microscopic study of thin sections.

pfa is pulverised fuel ash, a pozzolan produced by burning pulverised fuel in power station boilers.

pH is a way of expressing the acidity, or alkalinity of an aqueous solution, based on the concentration of hydrogen ions. pH 0 corresponds to a 1 Normal acid solution (1 gram of hydrogen ions per litre), pH 7 corresponds to a neutral solution, with equal concentrations of hydrogen and hydroxyl ions, and pH 14 corresponds to a 1 Normal alkali solution (17 grams of hydroxyl ions per litre). As the scale is logarithmic, an increase of one unit corresponds to a ten-fold reduction of the hydrogen ion and a ten-fold increase in the hydroxyl ion.

Phosphatic limestone is derived from invertebrate marine creatures. It contains significant levels of phosphorous, typically up to 5 %, (largely derived from fish bones).

Photic zone is the upper layer of water through which sufficient light penetrates to support the growth of photosynthetic organisms.

Pisolitic limestone is similar to oolitic limestone, but with a larger grain size (about 5 mm).

Pitch is a residue from the distillation of tars, which is liquid when hot and almost solid when cold.

Pitting is caused by expansion of a particle in plaster, which causes the material between the particle and the surface to crumble, typically causing a conical pit.

Plasticiser is an agent, which may be added to mortar or concrete to improve its workability by increasing air entrainment.

Poke holes are small apertures through the kiln shell and lining, which enable the burden to be rodded to break up crotched material.

Polished stone value (PSV) is the resistance of an aggregate to polishing, using the accelerated polishing test in BS 812. High values indicate increased resistance to polishing.

Popping is caused by the expansion of a particle (e.g., of chalk in concrete as a result of freeze-thaw cycles, and of unsound slaked lime in lime mortar) and results in lifting of the surface.

Porosity is the volume of voids in a material expressed as a percentage of the total volume of the material.

Portland cement see ordinary Portland cement.

Pozzolana is a volcanic ash, which contains reactive silica, and which, when mixed with quicklime and water, sets to a hard mass.

Pozzolanic is used to describe materials, which, like pozzolana, contain reactive silica, and which, when mixed with quicklime and water, set to a hard mass. They include pulverised fuel ash, trass and burnt shale.

Precipitate is an insoluble reaction product, which is usually crystalline and which usually grows in size to become settleable.

Precipitated calcium carbonate (PCC) is produced by blowing carbon dioxide into milk of lime, thereby precipitating finely divided calcium carbonate, generally with a mean particle size of less than $5\,\mu m$.

Preheater is a vessel in which limestone is heated by the exhaust gases from a rotary or other kiln, prior to being fed into the kiln.

Preheating zone is that part of a kiln where the limestone is heated to just below its dissociation temperature.

prEN is a draft (or provisional) European Standard. At the appropriate stage, prENs are given a number, which transfers to the EN, or ENV document.

Primary blasting is used to fragment naturally occurring rock (compare with secondary blasting).

Primer is a relatively small amount of high explosive, into which a detonator is inserted, which is used to initiate blasting agents such as ANFO.

Producer gas is made by the partial combustion/gasification of coal in a gas producer. It has a lower calorific value than natural gas or liquid petroleum gas.

PSV see polished stone value.

Pulverised fuel ash see pfa.

Q

Quartering is a process for reducing the quantity of a sample, by dividing a cone into quarters and retaining opposite quarters. The process is repeated until the required reduction has been achieved.

Quicklimes are air-limes consisting mainly of calcium oxide and magnesium oxide produced by calcination of limestone and/or dolomitic rock. They include calcium limes and dolomitic limes.

R

Rank is used to describe the quality and degree of metamorphosis of a coal.

Reactivity of quicklime is a measure of the rate at which it reacts with water. There are many reactivity tests (see Fig. 13.1). The terms "very high", "high", "moderate", "medium", and "low" reactivity are used as broad classifications for quicklimes. Their reactivity ranges, as used in this publication, are given in section 13.2.

Recuperator is a heat exchanger used to recover surplus heat (generally in exhaust gases) and to transfer it to the combustion air.

Reef limestones are mounds and units of organically produced debris, cemented in situ and often consisting of complete fossils.

Refractory describes the ability to withstand high temperatures without damage. It is also used to describe bricks with refractory properties.

Regenerative kiln is a type of shaft kiln in which the limestone is used as a regenerative heat exchange medium to transfer heat from exhaust gases to the combustion air.

Regional subsidence refers to the lowering of an area of land, relative to neighbouring areas, as a result of tectonic activity.

Regional up-lift is the reverse of regional subsidence.

Respirable dust is dust which, when inhaled, is likely to pass into the lungs (see section 34.1.1).

Riffling is sample reduction using a riffle box, which is a device designed to divide a sample into two equivalent fractions.

Ring is used to describe a build up, generally of ash or calcium sulfate, around the inside of a rotary kiln.

Ringlemann test is a method of comparing the opacity of a stack plume with a scale on a chart to obtain a visual measure of the optical density of the plume. A scale value of 5 corresponds to 100 %, with 0 corresponding to 0 % obscuration.

Riprap refers to lumps of stone placed on the surface of slopes to protect against erosion by water.

Road base is the main structural element in a road pavement. It spreads concentrated loads from traffic over such an area that the subgrade is able to withstand them.

ROC see run-of-crusher.

ROK see run-of-kiln.

Roman limes see eminently hydraulic limes.

ROQ see run-of-quarry.

Rotary kiln is a long, slightly inclined rotating tube. Limestone is fed into the upper end and fuel is fired into the lower end, from which lime is discharged. Rotary kilns often have limestone preheaters and/or lime coolers.

Run-of-crusher (ROC) describes the material produced by a crusher and includes the fines and impurities.

Run-of-kiln (ROK) describes the material discharged from a kiln and includes the fines and impurities, such as fuel ash, flint, clay and fragments of refractory.

Run-of-quarry (ROQ) describes the material produced by quarrying activities for subsequent processing. It includes fines, clay and other impurities, but not material sent directly to tip.

S

Saccharoidal limestone is a coarsely crystalline grainstone.

Sample is a quantity of material taken at random, or in accordance with a sampling plan, from a larger quantity. It may consist of one or more increments.

Sand — Building sand is a granular material, generally in the size range 0.06 to 2 mm (e.g., BS 1377, 1975 & BS 5930, 1981). Concreting sand generally has a nominal upper size of 5 mm, with constraints on the particle size distribution (e.g., BS 882, 1983).

Sandlime (or calcium silicate) bricks are produced by autoclaving a mixture of sand and lime.

Scaffolding refers to the formation of columns of fused material within a shaft kiln. It can develop into arches and prevent the burden from moving uniformly down the kiln (see crotching).

Scale is a precipitate that forms on surfaces as a result of a physical or chemical change.

Scalping is the removal of the finer fraction of a feed to a stone processing plant to reject unwanted material.

Scrubber (in gas cleaning) is a device in which dust or gaseous impurities are removed using droplets of water.

Scrubber (in mineral processing) is a device in which rock is washed to remove fine particles of rock or clay from the surface of larger particles.

Sedimentary rocks have been formed from materials deposited in water. They generally show distinct layers or beds.

Semi-hydrated dolomitic limes are hydrated dolomitic limes mainly consisting of calcium hydroxide and magnesium oxide (see Type N).

Semi-hydraulic limes (also called feebly-hydraulic) are limes which possess slight hydraulic properties.

Settling pond see lagoon.

Sewage is the waste fluid in a sewer.

Shell limes are a form of calcium lime, produced by the calcination of shells.

Shell limestone is a consolidated limestone consisting mainly of shells and/or fragments of shells.

Shelly limestone see shell limestone.

Siliceous limestone see cherty limestone.

Silo is a large storage vessel which is generally filled pneumatically (see bunker).

Single-sized generally refers to aggregate, which passed through a screen deck with an aperture size of 1.4 units and is retained on one with an aperture size of 1.0 units (e.g., passing 14 mm and retained on 10 mm).

Size ratio is widely used to describe the size range of limestones. Thus a size ratio of 3:1 would describe a product that has passed through a screen mesh with an square aperture of 30×30 mm and has been retained on a screen with an aperture of 10×10 mm.

Slag (in metallurgical processing) refers to the solid or liquid impurities removed from molten metal.

Slaked limes are air limes mainly consisting of calcium and possibly magnesium hydroxides, resulting from the slaking of quicklime. They generally refer to a dispersion of calcium hydroxide in water, but may also include powdered hydrated lime. In this book, the term is used to include *both* aqueous dispersions and dry hydrated lime.

Slurry describes a high concentration of suspended solids in water, typically over 5 g/l.

Slurry pond (in waste disposal) is a contained area of water into which slurried solids are discharged to permit the solids to settle and the clarified water to be discharged or recycled.

Soda ash is sodium carbonate.

Soft-burned quicklime see light-burned quicklime.

Softening is the removal of hardness (calcium and magnesium) from water.

Solid-burned quicklime see hard-burned quicklime.

Soundness of aggregate is its resistance to chemical attack, or to repeated physical changes, such as freeze-thaw cycles.

Soundness of slaked lime refers to the absence of expansion potential (i.e., the lime is fully slaked and will not cause popping and pitting, when made into a mortar).

Sparite is clearly crystalline, interstitial calcite in carbonate rocks, which acts as a cement binding larger particles.

Special hydrated lime see Type S hydrated lime.

Special natural hydraulic limes are produced by blending natural hydraulic limes with up to 20 % of suitable pozzolanic products (e.g., pulverised fuel ash, volcanic ash and trass), or hydraulic materials (e.g., ordinary Portland cement and blast furnace slag).

Spot sample is a sample taken on the same occasion, and from the same place. It can consist of one increment, or may be obtained by combining two or more consecutive increments.

Stalactites are icicle-like deposits of calcium carbonate, which hang from the roofs of limestone caverns. They are formed by crystallisation from ground-water.

Stalagmites are conical deposits of calcium carbonate on the floors of limestone caverns, formed by crystallisation from groundwater.

Stoichiometric refers to the quantities of chemicals that will react together according to the theoretical chemical reaction.

Stucco is a form of plaster for coating walls.

Subgrade is the rock or soil immediately below a road pavement.

Subbase is a layer, generally of granular material, laid on the sub-grade and on which the road base is laid.

T

Table is a cone-shaped insert, usually of metal, placed just above the discharge of a shaft kiln to give more uniform movement of the burden down the shaft.

Tailings are the residue left after removing the useful components from an ore.

Tar is a viscous black liquid produced by the destructive distillation of coal, or other organic materials. Its adhesive properties are used to bind aggregates in the construction of roads etc.

Tarmacadam is a road-building material, consisting of aggregate bound with tar or a tar-bitumen mixture. It contains little fine material and consequently gives a relatively open surface.

Ten per cent fines value (TFV) is a measure of the resistance of an aggregate's resistance to crushing in kN, as measured by the ten per cent fines test. Larger values indicate increased resistance to crushing.

Terrigenous matter is silt originating from the land.

TFV see ten per cent fines value.

Total lime is an analytical term for the total CaO plus MgO content of a limestone or lime, expressed in terms of CaO equivalent. It includes the carbonates, oxides, hydroxides, silicates, aluminates, and ferrate.

Trass, or tras is a natural pozzolan occurring in Germany.

Travertine is a crystalline form of calcium carbonate formed at hot-water springs by the evaporation of carbon dioxide from calcium bicarbonate solutions. It has a banded appearance.

Trefoil is a refractory structure in the preheating zone of some rotary kilns, dividing the kiln into three channels, to improve heat transfer between the hot gases and the limestone.

Tufa is produced under similar conditions to travertine, but is typically deposited over rocks. It is softer than travertine.

Tuyère refers to a refractory-lined pipe, which projects into a shaft lime kiln to increase the penetration of combustion gases, which are passed through the pipe.

Type N or normal hydrated lime is defined in ASTM specification C-207 (2). It is generally produced by hydrating high-calcium quicklime at ca 100 °C.

Type S hydrated lime (also called special hydrated lime) is defined in ASTM specification C-207. It is produced by heating lime (generally dolomitic) in an autoclave at ca. 180 °C. It may contain up to 8 % of unhydrated oxide. It is required to meet specified plasticity, water retention and particle size requirements.

U

Ultra-high calcium limestone contains more than 97 % $CaCO_3$.

Under-burned lime contains a substantial core of residual limestone in many of the particles.

Unslaked lime is any form of quicklime.

Unsoundness indicates that a material does not have adequate resistance to chemical attack, or to repeated physical changes, such as moisture, freeze-thaw, temperature etc.

V

Vaterite is a metastable form of calcium carbonate with a hexagonal crystal structure, which can crystallise from highly alkaline lake waters. It is of no commercial interest.

Vertical kiln is a term used to differentiate shaft kilns from rotary kilns.

Volume yield is the volume of a putty of standard consistency produced by slaking a specified weight of quicklime.

Vugs are small cavities caused by the dissolution of limestone.

W

Wakestone is a limestone containing at least 10 % of coarse grains, distributed in a mud matrix.

Washing see scrubbing.

Water-burned lime is partially slaked as a result of having been slaked incorrectly — see drowned lime.

Water retentivity is the ability of a mortar to retain water against the capillary suction of the masonry units.

Water soluble lime is an analytical term for the component in hydrated lime, expressed as $Ca(OH)_2$, that is able to react with hydrochloric acid at a pH of above 9.2, using the test specified in prEN 12485.

Wearing-course is the top layer of a road.

Whitewash see limewash.

Whiting is a finely powdered product produced by milling and classifying limestone (generally chalk). The nominal top size varies from 10 to 75 µm.

Workability is a measure of the ease with which a fresh mix of concrete or mortar can be handled and placed.

Y

Yield see volume yield.

Annex 2   General References

Annex 2.1   Books and Journals

Arenas, A., H. Carcamo, and G. Coloma, "La Cal en el Benefico de los Minerales" (in Spanish), Inacesa — Universidad Catolica del Norte, Antofagasta, Chile, 1993.

Boynton, R.S., "Chemistry and Technology of Lime and Limestone", John Wiley & Sons, 1980, ISBN 0-471-02771-5.

"Chemical Lime Facts", Bulletin 214, 5th ed., National Lime Association, 1988.

Hernandez, M., A. Arenas, H. Carcamo, V. Conejeros, and G. Coloma, "La Cal en la Metalurigica Extractiva" (in Spanish), Inacesa — Universidad Catolica del Norte, Antofagasta, Chile, 1995, ISBN 956-7012-39-3.

Holmes, S., and M. Wingate, "Building with Lime: A practical introduction", Intermediate Technology Publications, 1997, ISBN 1-853393-84-3.

"Kirk Othmer — Encyclopedia of Chemical Technology" , 4th ed., Wiley & Sons, 1995.

Knibbs, N.V.S., "Lime and Magnesia", Ernest Benn, 1924.

"Lime: handling, application and storage", Bulletin 213, 5th ed., National Lime Association, 1988.

Schiele, E., and L.W. Berens, "Kalk" (in German), Verlag Stahleisen, 1972.

Schwarzkopf, F., "Lime Burning Technology — a Manual for Lime Plant Operators", 3rd ed., Svedala Industries, Kennedy Van Saun, 1994.

Searle, A.B., "Limestone and its Products", Ernest Benn, 1935.

Smith, M.R., and L. Collis, "Aggregates", The Geological Society, Bath, 1993, ISBN 0-903317-89-3.

Stowell, F.P., "Limestone as a Raw Material in Industry", Oxford University Press, 1963.

"Ullmann's Encyclopedia of Industrial Chemistry", 5th ed., VCH, 1990.

Wingate, M., "Small-scale Lime-burning", Intermediate Technology Publications, 1985, ISBN 0-946688-01-X.

A reference document describing the "Best Available Techniques" for the cement and lime industries is in preparation by the EIPPC Bureau in connection with the European Union's Council Directive on Integrated Pollution Prevention and Control (96/61/EC, Article 16.2).

Journals covering aspects of the lime and limestone industries include:
Ciments, Chaux et Plâtres
GLP — Gypsum, Lime, and Building Products
Industrial Minerals
Pit and Quarry
Quarry Management
Rock Products
World Cement
Zement, Kalk, Gips (German and English)

Annex 2.2   European (CEN) Specifications

EN 196: "Methods of testing cement".
   Part 2: "Chemical analysis of cement", 1993.
   Part 7: "Methods of taking and preparing samples of cement", 1992.
ENV 197-1: "Cement — Composition, specifications and conformity criteria — Common cements", 1993.
ENV 413-1: Part 1: "Masonry cement — Specification", 1995.
EN 459: "Building lime"
   Part 1: "Definitions, specifications and conformity criteria", 1995 (being revised).
   Part 2: "Test methods", 1995 (being revised).
   Part 3: "Conformity evaluation" (in preparation).
EN 771-4: "Specification for masonry units — Autoclaved aerated concrete masonry units" (in preparation).
EN 932: "Tests for general properties of aggregates".
   Part 1: "Methods for sampling".
   Part 2: "Methods for reducing laboratory samples".
   Part 3: "Procedure and terminology for simplified petrographic description".
   Part 4: "Methods for description and petrography – Quantitative and qualitative procedures".
   Part 5: "Common equipment and calibration".
   Part 6: "Definitions of repeatability and reproducibility".
   Part 7: "Conformity criteria for test results".
EN 993: "Tests for geometric properties of aggregates".
   Part 1: "Determinâtion of particle size — Sieving method".
   Part 2: "Determination of particle size — Test sieves, nominal size of apertures".
   Part 3: "Determination of particle shape of aggregates – Flakiness index".
   Part 4: "Determination of particle shape of aggregates – Shape index".
   Part 5: "Assessment of surface characteristics – Percentage of crushed or broken surfaces in coarse aggregates".

Part 6: "Determination of shape/texture — Flow coefficient for coarse aggregates".
Part 7: "Determination of shell content — Percentage of shells for coarse aggregates".
Part 8: "Assessment of fines — Sand equivalent method".
Part 9: "Assessment of fines — Methylene blue test".
Part 10: "Determination of fines — Grading of fillers (air jet sieving)".

EN 998: "Specification for mortar for masonry", 1997.
Part 1: "Rendering and plastering mortar with inorganic binding agents" (in preparation).
Part 2: "Masonry mortar" (in preparation).

EN 1017: "Chemicals used for treatment of water intended for human consumption — Half-burnt dolomite" (in preparation).

EN 1018 "Chemicals used for treatment of water intended for human consumption — Calcium carbonate" (in preparation).

EN 1097: "Tests for mechanical and physical properties of aggregates".
Part 1: "Determination of the resistance to wear (micro-Deval)".
Part 2: "Methods for the determination of resistance to fragmentation".
Part 3: "Determination of loose bulk density and voids".
Part 4: "Determination of the voids of dry compacted filler".
Part 5: "Determination of water content by drying in a ventilated oven".
Part 6: "Determination of particle density and water absorption".
Part 7: "Determination of the particle density of filler — Pycnometer method".
Part 8: "Determination of the polished stone value".
Part 9: "Determination of the resistance to wear by abrasion from studded tyres: Nordic test".
Part 10: "Water suction height".

EN 1367: "Tests for thermal and weathering properties of aggregates".
Part 1: "Determination of resistance to freezing and thawing".
Part 2: "Magnesium sulfate test".
Part 3: "Determination of volume stability (Sonnenbrand)".
Part 4: "Determination of drying shrinkage".
Part 5: "Determination of resistance to heat".

EN 1482: "Sampling of solid fertilizers and liming materials", 1996.

EN 1744: "Tests for chemical properties of aggregates".
Part 1: "Chemical analysis: (a) chloride; (b) acid soluble sulfate; (c) impurities affecting setting and hardening of cement; (d) impurities that affect surface finish; (e) water solubility, (f) loss on ignition; (g) slag unsoundness; (h) free lime; (i) fulvo acid test; (j) sodium hydroxide test".
Part 2: "Determination of resistance to alkali reaction".
Part 3: "Water susceptibility of fillers".

EN 12485: "Chemicals used for treatment of water intended for human consumption — calcium carbonate, high-calcium lime and half-burnt dolomite — Test methods" (in preparation).

EN: 12518 "Chemicals used for treatment of water intended for human consumption — High-calcium lime" (in preparation).

EN XXX "Tests for (bituminous bound fillers) filler aggregate" (in preparation).
Part 1: "Delta ring and ball test".
Part 2: "Bitumen number".

EN XXX: "Unbound and hydraulically bound mixtures for roads — specification for lime-treated mixtures for road construction and civil engineering — definitions, composition and laboratory mixture requirements" (in preparation).

EN XXX: "Hydraulic road binders — Composition, specifications, and conformity criteria" (in preparation).

Annex 2.3   American (ASTM) Specifications

C 5-92: "Specification for quicklime for structural purposes".

C 25-95: "Test methods for chemical analysis of limestone, quicklime and hydrated lime".

C 33-93: "Specification for concrete aggregates".

C 40-92: "Test method for organic impurities in fine aggregates for concrete".

C 50-94: "Practice for sampling, inspection, packing and marking of lime and limestone products".

C 51-95: "Terminology relating to lime and limestone (as used by the Industry)".

C 110-95: "Test methods for physical testing of quicklime, hydrated lime and limestone".

C 131-89: "Test methods for resistance to abrasion of small size coarse aggregate by use of the Los Angeles machine".

C 141-85: "Specification for hydraulic hydrated lime for structural purposes".

C 206-84: "Specification for finishing hydrated lime".

C 207-91: "Specification for hydrated lime for masonry purposes".

C 270-97: "Specification for mortar for unit masonry".

C 400-93: "Test methods for quicklime and hydrated lime for neutralization of waste acid".

C 586-92: "Test method for potential alkali-reactivity of carbonate rocks for concrete aggregates (rock cylinder method).

C 593-95: "Specification for fly ash and other pozzolans for use with lime".

C 602-95: "Specification for agricultural liming materials".

C 706-92: "Specification for limestone for animal feed use".

C 737-92: "Specification for limestone for dusting of coal mines".

C 821-78: "Specification for lime for use with pozzolans".

C 911-94: "Specification for quicklime, hydrated lime and limestone for chemical uses".

C 977-95: "Specification for quicklime and hydrated lime for soil stabilization".

C 1097-95: "Specification for hydrated lime for use in asphaltic concrete mixtures".

C 1164-92: "Specification for evaluation of limestone or lime uniformity from a single source".

C 1271-94: "Test method for X-ray spectrometric analysis of lime and limestone".

C 1301-95: "Test methods for major and trace elements in limestone and lime by inductively coupled plasma — atomic emission spectroscopy (ICP) and atomic absorption (AA)".

C 1318-95: "Test method for determination of total neutralizing capability and dissolved calcium and magnesium oxide in lime for flue gas desulfurisation (FGD)".

D 1199-86: "Specification for calcium carbonate pigments".

D 1863-93: "Specification for mineral aggregates used on built-up roofs".

E 11-95: "Specification for wire cloth and sieves for testing purposes".

Annex 2.4   British (BS) Specifications

(N.B., EN standards are also adopted as BS standards.)

BS 63: "Road aggregates", 1997.

BS 187: "Specification for calcium silicate (sandlime and flintlime) bricks", 1978.

BS 594, Part 1: "Hot rolled asphalt for roads and other paved areas — Specification for constituent materials and asphalt mixtures", 1992.

BS 812: " Testing aggregates".

    Part 2: "Methods for determination of density", 1995.

    Part 100: "General requirements for apparatus and calibration", 1990.

    Part 101: "Guide to sampling and testing aggregates", 1984.

    Part 103: "Methods for determination of particle size distribution", 1985/89.

    Part 105: "Methods for determination of particle size distribution", 1989/90.

    Part 111: "Methods for determination of ten per cent fines value (TFV)", 1990.

    Part 112: "Method for determination of aggregate impact value (AIV)", 1990.

    Part 113: "Method for determination of aggregate abrasion value (AAV)", 1990.

    Part 124: "Method for determination of frost-heave", 1989.

BS 882: "Specification for aggregates from natural sources for concrete", 1992.

BS 890: "Specification for building limes", 1995.

BS 1047: "Specification for air-cooled blastfurnace slag aggregate for use in construction", 1983.

BS 1199: "Specifications for building sands from natural sources — sands for mortars for plastering and rendering", 1986.

BS 1200: "Specifications for building sands from natural sources — sands for mortars for bricklaying", 1984.

BS 1377: "Methods of test for soils for civil engineering purposes".
Part 2: "Classification tests", 1990.
Part 4: "Compaction-related tests", 1990.

BS 1924, Part 2: "Stabilised materials for civil engineering purposes — Methods of test for cement-stabilised and lime-stabilised materials", 1990.

BS 3108: "Specification for limestone for making colourless glasses", 1980.

BS 3406, Part 2: "Methods for determination of particle size distribution — Recommendations for gravitational liquid sedimentation methods for powders and suspensions", 1984.

BS 4142: "Rating industrial noise affecting mixed residential and industrial areas", 1990.

BS 4887, Part 1: "Mortar admixtures — Specification for air-entraining (plasticizing) admixtures", 1986.

BS 4987, Part 1: "Coated macadam for roads and other paved areas — Specification for constituent materials and for mixtures", 1993.

BS 5224: 1995, "Specification for masonry cement".

BS 5228, Part 1: "Noise control on construction and open sites", 1984.

BS 5262: "Code of practice for external renderings", 1991.

BS 5309: "Methods for sampling chemical products".
Part 1: "Introduction and general principles", 1976.
Part 4: "Sampling of solids", 1976.

BS 5328: "Concrete", 1990.

BS 5492: "Code of practice for internal plastering", 1990.

BS 5628: "Code of practice for use of masonry".
Part 1: "Structural use of unreinforced masonry", 1992.
Part 2: "Structural use of reinforced and prestressed masonry", 1995.
Part 3: "Materials and components, design and workmanship", 1985 (being revised).

BS 5835, Part 1: "Compactability test for graded aggregates", 1980.

BS 5930: "Code of practice for site investigations", 1981.

BS 6073: "Precast concrete masonry units".
Part 1: "Specification for precast concrete masonry units" (in preparation).
Part 2: "Methods for specifying precast concrete masonry units" (in preparation).

BS 6100: "Glossary of building and civil engineering terms".
Part 6, Section 6.1: "Binders", 1984 (1995).

BS 6463: "Quicklime, hydrated lime and natural calcium carbonate".
   Part 101: "Methods for preparing samples for testing", 1996.
   Part 102: "Methods for chemical analysis" (in preparation).
   Part 103: "Methods for physical testing" (in preparation).
   Part 2: "Methods for chemical analysis", 1984.
   Part 3: "Methods of test for physical properties of quicklime", 1987.
   Part 4: "Methods of test for physical properties of hydrated lime and lime putty", 1987.
BS 6472: "Evaluation of human exposure to vibrations in buildings (1 Hz to 80 Hz)", 1992.
BS 7385, Part 2: "Guide to damage levels from groundborne vibration", 1993.
BS 7583: "Specification for Portland limestone cement", 1996.
BS 8007: "Code of practice for design of concrete structures for retaining aqueous liquids", 1987.

Annex 2.5   Other Standard Specifications

DIN 53163: "Testing of pigments and extenders: Determination of lightness of extenders and white pigments in powder form", 1988.
DIN 5033, Part 9: "Colorimetry: reflectance standard for colorimetry and photometry", 1982.
SABS 459-1955: "Specification for lime for chemical and metallurgical purposes".
ISO 14001: "Environmental management systems — Specification with guidance for use", 1996.

Annex 3

Appendix A
Reactivity Test for Aircrete Production

1. General

A number of reactivity tests are used to characterise quicklime for use in the production of aircrete. Perhaps the most widely used is that specified in EN 459-2. That test, however, suffers from excessive thickening of the milk of lime with reactive limes, leading to poor repeatability between laboratories. The method used to obtain the curve in Fig. 26.7 has been found to give acceptable with high (but not very high) reactivity limes.

2. Apparatus

The apparatus consists of a stirred 1 l Dewar flask with a lid (see Fig. A3.1). The stirrer impeller consists of angled blades and is rotated at a speed which ensures that the surface of the contents is agitated throughout the test. The level of insulation of the apparatus is such that the rate of heat loss is within the limits described in section A.5 below. The temperature of the contents is measured using a thermometer or recording thermocouple, immersed 60 mm below the static meniscus.

3. Procedure

3.1. Add 750 ± 5 ml of distilled water at $25 \pm 0.3 \,°C$ to the Dewar flask, switch on the stirrer and allow the apparatus to reach a steady temperature. If necessary, raise the temperature to $25 \pm 0.3 \,°C$ by inserting a heated rod into the water.

3.2. Weigh 150.0 ± 0.1 g of the quicklime, which should be at a temperature of 15 to 30 °C. Switch off the stirrer and open the lid of the flask. Start the stopclock and add the quicklime to the flask within 10 seconds. Close the lid and restart the stirrer after 20 seconds.

---

* See also BS 6100, Part 6, Section 6.1, "Glossary of Building and Civil Engineering Terms — Binders", 1984 (1995).

3.3. Record the temperature either at 1 min. intervals, or continuously. Note
  a) the temperature at 2 min.,
  b) the time taken to reach maximum temperature,
  c) the maximum temperature, and
  d) the "turn-down" time at which the temperature falls by 0.2 °C (with digi-
     tal recording devices, reading to 1 decimal place, this ensures that the fall
     is at least 0.1 °C).

3.4. Calculate the "temperature rise ratio" (temperature rise at 2 min. divided by
     the maximum temperature rise), which is a convenient indication of the
     shape of the curve.

4. Expression of Results

Where a high degree of consistency is required, it is recommended that all of the
above parameters are monitored, namely: 2-minute temperature, maximum temperature, time to maximum temperature, turn-down time and temperature rise
ratio.

5. Measurement of Heat Loss from the Apparatus.

Proceed as in section A.3.1, using water at 70±5 °C. Stir for 5 min. and note the
temperature. Continue stirring for a further 30 min. The temperature loss during
that 30 min. period should be between 1.5 and 3.5 °C. If it is below that range,
reduce the level of insulation in the lid. If it is above that range, increase the level
of insulation, or replace the flask.

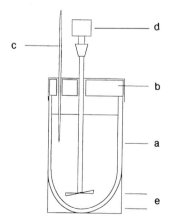

Figure A3.1. Diagram of a reactivity apparatus:
(a) Dewar flask; (b) hinged lid with 13 mm insulation;
(c) thermometer/thermocouple; (d) stirrer; (e) 20 mm gap

Appendix B
Calculation of Free Lime in Hydrated Lime

Tests required:
a) Loss on ignition at 975 °C (combined CO_2 + combined H_2O + excess H_2O)
$$= A$$
b) Combined CO_2 $= B$
c) Excess H_2O $= C$
d) Active lime $= D$

Whence:
a) Combined H_2O $= A-B-C$ $= E$
b) $Ca(OH)_2$ $= 4.113 \times E$ $= F$
c) CaO in $Ca(OH)_2$ $= F-E$ $= G$
d) Free lime $= D-G$ $= H$

For example, if
A = 24.25 %,
B = 0.80 %,
C = 0.20 %,
D = 72.75 %,
E = 24.25–0.80–0.20 = 23.25 % of combined water,
F = 4.113 × 23.25 = 95.63 % of $Ca(OH)_2$,
G = 95.63–23.25 = 72.38 % of CaO in $Ca(OH)_2$,
H = 72.75–72.38 = 0.37 % of free CaO.

All percentages are by mass.

Appendix C
Units and Conversion Factors

The units used in this book are those which are widely used in the Lime and Limestone Industries and with which the author is familiar. While this may help many readers, it may have the opposite effect on others. This appendix, therefore, includes some of the more common units, with conversions to SI and other common units.

1 Powers of Ten

Symbol	Prefix	Multiple
n	nano	10^{-9}
μ	micro	10^{-6}
m	milli	10^{-3}
c	centi	10^{-2}
k	kilo	10^{3}
M	mega	10^{6}
G	giga	10^{9}

2 Percentages

Unless stated otherwise, percentages are by mass.

3 Time

Unit	Symbol
second	s
minute	m
hour	h (hr in some diagrams)
day	d
year	a

4 Length

Unit	Symbol	Value
Angstrom	Å	10^{-10} m
micron	μm	10^{-6} m
millimetre	mm	10^{-3} m
centimeter	cm	10^{-2} m
metre	m	
kilometre	km	10^{3} m

4 Mass

Unit	Symbol	Value
milligram	mg	10^{-3} g
gram	g	
kilogram	kg	10^3 g
tonne	t	10^6 g

The short ton, used in North America $= 2,000$ lbs $= 0.907185$ tonnes.

5 Output

Expressed as t/h, t/d, or tpa (tonnes per annum).
The tpa figure allows for down-time (e.g., 90 % availability for 330 days per annum).

6 Volume

Unit	Symbol	Value
millilitre	ml	10^{-3} l
cubic centimetre	cc, or cm^3	10^{-3} l
litre	l	
cubic metre	m^3	10^3 l

7 Pressure

Unit	Symbol	Value
Pascal	Pa	10^5 N/m^2
bar	bar	100 k Pa
atmosphere	atm	1.01325 bar
atmosphere	atm	1013.25 mbar
atmosphere	atm	101.325 k Pa
atmosphere	atm	760 mm of mercury (0 °C)

8 Crushing strength

Expressed in $MN/m^2 = N/mm^2$

9 Density

Expressed in $g/cm^3 = g/ml$

10 Bulk Density

Expressed in kg/m^3

11 Specific Heat

Expressed in cal/g · °C $\equiv$ 4,189 J/kg · °C

12 Heat

1 cal (int)	≡	4.189 J
1 Btu (int)	≡	1054.9 J
	≡	251.8 cal
1 Therm	=	10^5 Btu
1 kWh	≡	3.600×10^6 J
	≡	3.600 MJ

13 Specific Heat Usage

The most widely used unit for the specific heat usage of lime kilns is kcal/kg.

1000 kcal/kg	=	1000 cal/g
	≡	39.71 Therms/t
	≡	4,189 kJ/kg

14 Heat of reaction

Expressed as kcal/kg, and as kJ/kg, or
kcal/mole and as kJ/mole (mole = 1 gram mole = 56 g for CaO).

15 Concentration

For solutions, expressed as g/l, mg/l, g/m^3, etc.
For solids, expressed as m/kg, mg/kg. and % (m/m).
For gases, expressed as mg/Nm^3 and ng/Nm^3
(i.e., with volume corrected to 1 atm and 0 °C).

16 Surface area

For powders, expressed as m^2/g, and measured either by the BET (Berkland Eyde and Teller) adsorption technique, or by the air permeability technique (BS 6463). It should be noted that the two techniques are not compatible.

Index

Abbreviations: (LS) = limestone; (QL) = quicklime; (SL) = slaked lime

Bordeaux mixture 348
Boron impurity 24
BOS 302–308
Bound aggregate 77, 406
Boundstone 15, 406
Brachiopod 10, 406
Brecciated limestone 406
Brick, sand-lime 285–288
Bricklaying mortar 73, 74
Brightness 20, 56
Brine purification 362
Brines 369
Briquettes 198
Briquetting, fuels 377
British Standard Specifications 429
Brown coal 133
BS Specifications 429
Building (LS) 68–79, (QL/SL) 258–296
Building limes 258–284, 406
– classification 278
– hydraulic lime 224, 282–284
– materials 68–79
– mortar 270–282
– plaster and stucco 280
– sands 73, 406
Bulk density 406, (LS) 19, (QL) 118, (SL) 207
Bulk fill 264
Bullheads 192
Bunker 50, 406
Burden 30, 406
– preparation (iron) 94
Burial environment 12
Burner design 176, 390
Burning zone 406
Burnt limes 406
Butane 129
Butene chlorohydrin/oxide 364
Butter 349

C

Cadmium impurity 24
– reduction 335, 342
Calcarenite 15
Calcareous materials 82
Calcareous ooze 406
Calcareous tufa 15, 406
Calcia 406
Calcilutite 15
Calcimatic kiln 161
Calcination
– calcium carbonate 139–153, 364, 406
– calorific requirements 155
– dead burning 147, 150, 188
– effect of limestone size 145, 146
– effect of steam 151
– fine limestone 106, 152
– fuels for 128–136
– influence on porosity 117, 147–150
– kinetics 142–146
– rate 126, 144–146
– reaction 139–141
– temperatures 142–146
Calcined dolomite 188
Calcining effort 145
Calcining time 145, 146
Calcining zone 156, 406
Calcirudite 15
Calcite 10, 20, 406
Calcitic limestone 16, 406
Calcitic mudstone 15, 407
Calcium 372
– aluminate cements 85, 371
– arsenate 348
– bicarbonate 317, 322
– bromide 358
– carbide 301, 355
– carbonate, porous/non-porous 110
– carbonate sampling 54
– carbonate sludge 407
– caseinate 349
– chloride 358
– citrate 360
– compounds 352–361
– dichromate 359
– hexacyanoferrate 359
– hypochlorite bleaches 355
– lactate 349, 361
– limes 407
– oleate 360
– oxide (free) 83, 246, 434
– phosphates 356
– salts 356–361
– silicate bricks 285–288
– silicate concrete 296
– silicate wall board 295
– silicon 359
– soaps 360
– stearate 360
– sulfate (LS) 104, (SL) 335
– sulfite (LS) 103, (SL) 355
– sulfonate 360
– tartrate 361
– tungstate 359
– zirconate 112
California Bearing Ratio 260–264, 407
Calorific value 133, 134, 187, 407
Calorific value meters 135
Cambrian era 9
Capping layer 74, 75
– stabilisation 263
Captive production 251
Carbide lime 82, 224, 238, 356, 366, 407
Carbide lime dough 239, 356